Concepts and Applications of Photosynthesis

Concepts and Applications of Photosynthesis

Edited by **Agatha Wilson**

R CALLISTO REFERENCE

New York

Published by Callisto Reference,
106 Park Avenue, Suite 200,
New York, NY 10016, USA
www.callistoreference.com

Concepts and Applications of Photosynthesis
Edited by Agatha Wilson

International Standard Book Number: 978-1-63239-120-9 (Hardback)

Printed in the United States of America.

Contents

Preface

This book has been an outcome of determined endeavour from a group of educationists in the field. The primary objective was to involve a broad spectrum of professionals from diverse cultural background involved in the field for developing new researches. The book not only targets students but also scholars pursuing higher research for further enhancement of the theoretical and practical applications of the subject.

Photosynthesis is described as the process by which green plants and some other organisms use sunlight to synthesize nutrients from carbon dioxide and water. Among a multitude of books dedicated to several aspects of photosynthesis, this book is unique in bringing together an update on the latest insights about this most significant biological process in the biosphere. In addition to fueling all the life supporting activities and processes of all living creatures on the planet ranging from bacteria to humans, photosynthesis has also developed and maintained our life supporting oxygenic atmosphere. This book has been methodically compiled in a manner that it comprehensively elucidates the mechanisms, methodologies, applications as well as stress effects of photosynthesis by providing valuable information to the readers.

It was an honour to edit such a profound book and also a challenging task to compile and examine all the relevant data for accuracy and originality. I wish to acknowledge the efforts of the contributors for submitting such brilliant and diverse chapters in the field and for endlessly working for the completion of the book. Last, but not the least; I thank my family for being a constant source of support in all my research endeavours.

Editor

Mechanisms

Electron Transfer Routes in Oxygenic Photosynthesis: Regulatory Mechanisms and New Perspectives

Snježana Jurić, Lea Vojta and Hrvoje Fulgosi

Additional information is available at the end of the chapter

1. Introduction

In nature, the intensity of light changes rapidly, spanning from 2000 µmol $_{PHOTONS}$ m^{-2} s^{-1} during the brightest sunlight, down to 200-500 µmol $_{PHOTONS}$ m^{-2} s^{-1}, when shaded by clouds, followed by the period of dark, to encircle with the next sunrise of barely 10 µmol $_{PHOTONS}$ m^{-2} s^{-1}. Plants, capable of performing limited movements, like leaf and chloroplast movements, respectively, have been forced to evolve different means of coping with the unpredictable nature. To better understand the acclimatization mechanisms, the intensity of light has been roughly divided into low-light and high-light conditions, shaped by the laboratory conditions of plant growth and the strength of the measuring apparatus available. Moreover, looking at the time scale, it is possible to differentiate between short-term and long-term acclimatization processes in plants. These artificial divisions are needed to simplify the most important process in photo-synthesis in vascular plants: the rate and the distribution of excitation energy between two photosystems. Photosystem units are organized into large supercomplexes with peripherally attached antenna complexes, being further assembled into megacomplexes [1]. Two types of peripheral antenna proteins associated to photosystem II (PSII) are known: the major Light-Harvesting Complex II (LHCII), which occurs as a trimeric complex containing the proteins Lhcb1, Lhcb2 and Lhcb3, and three minor monomeric complexes, namely Lhcb4 (CP29), Lhcb5 (CP26), and Lhcb6 (CP24). These peripheral complexes bind variable number of molecules of chlorophyll *a* and *b*, and of some xanthophyll molecules (for physicochemical properties of chlorophylls see the Chapter Kobayashi et al.). In Arabidopsis PSII-LHCII supercomplexes, Lhcb are organized into two rings around the dimeric PSII core complexes, with Lhcb1, Lhcb2, Lhcb4 and Lhcb5 detected in the inner ring, where Lhcb1 and Lhcb2 participate in strongly bound LHCII trimer, while the outer ring consists of Lhcb6 and of moderately bound LHCII trimer (the Lhcb1 and Lhcb3 gene products). Photosystem I (PSI) is associated with the Light-

Harvesting Complex I (LHCI) that binds 10 molecules of chlorophyll *a* or chlorophyll *b* plus a few xanthophylls per one Lhca protein. In green plants, LHCI consists of four polypeptides (Lhca1-Lhca4) from the Lhc protein superfamily. In Arabidopsis, two additional proteins have been identified (Lhca5 and Lhca6), but their contribution to LHCI is still under debate [1]. More distantly related family members are the photoprotective early light-induced stress response proteins (ELIPS), and the component of PSII, PsbS [2].

2. The way the plant protects itself: Non-photochemical quenching

The non-photochemical quenching (NPQ) is a short-term response by which plants harmlessly dissipate excess excitation energy into heat under high-light conditions. NPQ is observed in all higher plants, in lower plants, green algae and diatoms [3]. NPQ is also present in oceanic picophytoplankton species (see Chapter Kulk et al.). Basically, during the absorption of sunlight by light-harvesting complexes (LHCs) associated with reaction centres, a chlorophyll *a* molecule shifts from its ground energetic state to its singlet excited state. It can return to its ground state *via* one of several pathways: re-emission of excitation energy in the form of chlorophyll fluorescence; transfer of excitation energy to reaction centres to be utilised in photochemistry reactions; de-excitation by dissipating heat (NPQ); production of triplet excited state, which would be a highly profitable valve for excess excitation, but it indirectly produces a very reactive oxygen species (ROS), singlet oxygen, by transferring energy to the ground-state oxygen [4]. In addition to the dissipation of excitation energy, non-photochemical processes also quench or diminish chlorophyll *a* fluorescence, therefore being mainly observed at PSII. The phenomenon of quenching of chlorophyll *a* fluorescence is usually analysed in terms of three components, based on their relaxation kinetics: state-transitions (qT), ΔpH-dependent quenching (qE) and photoinhibition (qI). The majority of NPQ is believed to occur through qE in the PSII antenna pigments bound to the LHCII proteins [5, 6]. State-transitions are considered to be the component of NPQ because the fluorescence yield of PSII diminishes due to the lateral redistribution of the phosphorylated LHCII proteins and their attachment to PSI [5]. Photoinhibition exibits the slowest relaxation and it is the least defined. qI quenching is proposed to be involved in long-term down-regulation of PSII [4, 7].

2.1. ΔpH-dependent quenching

The process of qE is triggered by acidification of the thylakoid lumen under light-saturating conditions, which activates the interconversion of specific xanthophyll pigments (oxygenated carotenoids), that are mostly bound to the LHC proteins. For comparison, in cyanobacteria, strong blue-green or white light activates the orange carotenoid protein (OCP) which interacts with phycobilisome and dissipates the excess energy in the form of heat [8]. The xanthophyll cycle in plants, green and brown algae consists of the pH-dependent conversion from viola-xanthin first to antheraxanthin and then to zeaxanthin. In plants, the reactions towards zeaxanthin are catalysed by the enzyme violaxanthin de-epoxidase, while the relatively slow reactions towards violaxanthin are catalysed by the enzyme zeaxanthin epoxidase [4]. The *npq1* mutants are unable to convert violaxanthin to zeaxanthin but still exhibit qE, demonstrating that the xanthophyll cycle is not the prerequisite for qE formation [9]. Zeaxanthin was demonstrated to have an additional photoprotective function not connected with NPQ, but rather with thylakoid membrane lipids. It is hypothesised that the function of the non-protein bound zeaxanthin could be the removal of highly deleterious singlet oxygen, working together with the well-known antioxidant tocopherol [10]. The *npq4* mutant, which completely lacks the PsbS protein and qE, can survive under high-light conditions, thus implying that carotenoids may compensate to some extent the deficiency in qE formation [9].

Due to the slower kinetics of formation and relaxation of qE compared to the proton gradient, and taking into account that the light causes changes in charge distribution, which consequently alter the aggregation state of thylakoids, it was proposed that such a conformational change might accompany the qE event [6, 11]. Indeed, reports from at least three independent laboratories confirmed the structural rearrangement of the PSII-LHCII macro-organization during qE [12-14]. Time-correlated single photon counting (TCSPC) measurements revealed an additional far-red fluorescence component in the leaves of the high-light-adapted wild-type, the mutant unable to accumulate zeaxanthin at high-light (*npq1*), and in the mutant overexpressing PsbS, the protein proposed to act as a luminal pH sensor that consequently determines the level of qE [12, 15], respectively. The same component was not observed in the mutant devoid of PsbS, *npq4*, or in the dark-adapted state of abovementioned plant lines [12]. It was concluded that this fluorescence originated from the major LHCII antenna complex detached specifically from PSII, with the required presence of the PsbS protein [12]. Furthermore, it was biochemically demonstrated that the supramolecular complex B4C (Lhcb4, Lhcb6, and moderately bound LHC trimer) dissociates during light exposure, and associates back during dark period [13]. However, *npq4* mutants did not show light-induced B4C dissociation, establishing the role of PsbS as the key player in thylakoid rearrangements [13]. Finally, Ruban group used freeze-fracture electron microscopy to demonstrate that the formation of qE was indeed associated with the reorganisation of PSII and LHCII within the thylakoids [14]. Their experimental data support and update the 20-years-old hypothetical model built to explain the mechanistics of qE, the LHCII aggregation model [16]. According to the model, LHCII antenna could be found in four different states: (i) dark-adapted, unquenched; (ii) dark-adapted, partially aggregated, quenched; (iii) illuminated, partially aggregated, quenched; and (iv) illuminated, aggregated, fully quenched. Illumination causes two events necessary for LHCII aggregation to occur: conversion of violaxanthin to zeaxanthin and the protonation of LHCII, respectively [6]. Precisely, the formation of ΔpH triggers a conformational change within the LHCII antenna, which leads to the partial dissociation of LHCII trimers from the PSII-LHCII supercomplexes, and, consequently, to their aggregation (Figure 1). In parallel, de-epoxidation of violaxanthin to zeaxanthin promotes the LHCII aggregation and the formation of NPQ [14]. The question for debate would be the number and the exact position of quenched complexes. According to Holzwarth group, two quenching centres are formed: detached and aggregated LHCII antenna measured within 1 to 5 minutes, and the minor components of LHCII; still attached to the PSII core, measured within 10 to 15 minutes [12]. As PSII core does not bind any zeaxanthin molecules, it is excluded as a quenching site; however, it is well-covered by formation of the second quenching centre.

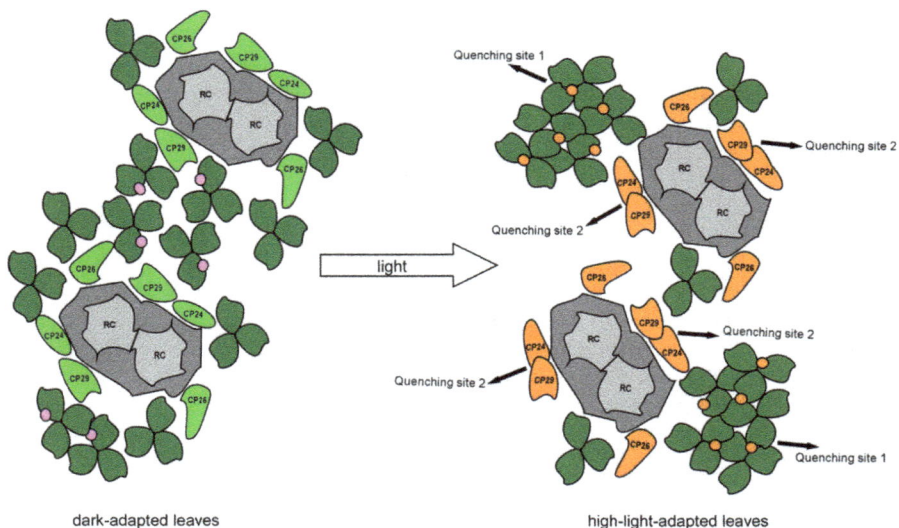

Figure 1. Structural rearrangement of the PSII-LHCII macro-organization during qE. Composition and arrangement of the LHCII trimers (dark green) and the minor antenna (light green) around the PSII RCs (light grey) in the wild-type plant during the dark period (left panel) and under the high-light conditions (right panel) were presented according to [1]. Pink circles on the left panel denote violaxanthin, while the orange circles on the right panel denote the process of conversion of violaxanthin to zeaxanthin. The change of colour of the minor antenna from light green (left panel) to orange (right panel), also denotes the conversion of violaxanthin to zeaxanthin, respectively. According to [14], in excess light, ΔpH triggers a conformational change within the LHCII antenna, which leads to the partial dissociation of LHCII trimers from the PSII LHCII supercomplexes, and, consequently, to their aggregation. In parallel, de-epoxidation of violaxanthin to zeaxanthin promotes the LHCII aggregation and the formation of NPQ (quenching site 1). Moreover, according to [12], quenching site 2 is also formed within the minor antenna still attached to the PSII.

Acidification of the thylakoid lumen also causes a conformational change in thylakoids that can be monitored at 535 nm (ΔA_{535}). This change is most likely induced by protonation of the lumen-exposed carboxylate side chains in specific PSII proteins [15]. Interestingly, *npq4* mutants, which do not express the PSII PsbS protein, also lack ΔA_{535} [17]. It was proposed that PsbS is not necessary for efficient light harvesting and photosynthesis, but it is involved in NPQ by sensing luminal pH and consequently determining the level of qE [15]. Recently, it became clear that PsbS is indispensable for the physical state of the thylakoid membranes. In dark-adapted *npq4* mutants, the formation of ordered semi-crystalline arrays of PSII was increased, while in the plants over-expressing PsbS no arrays could be found, suggesting that PsbS disrupts the ordering and promotes the protein diffusion within the membrane, leading to the NPQ formation [18].

Although the main components for qE are known, the biophysical mechanism of de-excitation of the excited molecules of chlorophyll *a* is still unidentified. It is hypothesised that either xanthophylls act indirectly as allosteric regulators of the LHCs, causing the conformational change that facilitates the de-excitation, or xanthophylls directly de-excite the excited molecules of chlorophyll *a* [4]. Models presented in Ruban and Holzwarth groups both permit the formation of internal dissipative pigment interactions, whether they occur between chlorophyll molecules or between carotenoid and chlorophyll molecules [19].

Variations in NPQ capacities and processes in different organisms suggest a strong evolutionary pressure to obtain optimized photoprotection [3]. For example, the LHCII proteins may have evolved from ancestors of contemporary stress-responsive proteins such as HLIP and ELIPS, which probably bind only carotenoids and are involved in photoprotection, while LHCII harvest light through high amounts of chlorophyll binding. In plants, PsbS binds pigments minimally and it is primarily involved in photoprotection, while the light-harvesting complex containing fucoxanthin (LHCF) in diatoms binds both chlorophyll and carotenoids in high amounts and contributes equally to the light harvesting and photoprotection. Under high-light conditions, the photosynthetic reaction centres can not accommodate all the electrons coming through electron transport chain, thus entering into the saturated, "closed" status. The already generated excitation energy becomes the burden for the photosynthetic membranes and has to be channeled safely, before destroying the reaction centres, especially the pigment core of PSII, the pair of the most potent oxidizers known to exist in nature, P680. If the vast energy is not diverted, P680 might rest in its prolonged oxidized state, P680$^+$, and oxidize the neighbour protein amino-acids and pigments, ultimately leading to the destruction of the PSII protein D1. However, if P680 can not submit the electron to the oxidized plastoquinone, due to the increased number of already reduced plastoqinones, P680 could go into triplet state, interacting with atmospheric triplet oxygen and producing deleterious singlet oxygen. Both of these processes lead to the photoinhibition, the state in which the decreasing of the electron transport is observed [6]. Photoinhibition has a specific signature that can be monitored by a lower oxygen production, analysis of the D1 protein level, formation of uncoupled chlorophyll and of triplet state of chlorophyll. However, photoinhibition provokes the formation of photoprotection processes, where most of the unwanted energy would be harmlessly dissipated as heat. This phenomenon leaves a palpable trace visible as a drop in the fluorescence intensity, measured by TCSPC. One of the approaches to study the origin of the quenching mechanism *in vivo* is the treatment of Arabidopsis with lincomycin, which blocks the synthesis of chloroplast-encoded proteins, such as the reaction centres of PSI and PSII (RCI and RCII, respectively). The thylakoids of treated plants contain diminished amounts of RCs, but are rich in antenna complexes. Although the maximum chlorophyll fluorescence lifetime in isolated PSII-LHCII supercomplex is 4 ns; when complex is an integral part of the thylakoid membranes, fluorescence lifetime decreases to 2 ns [20]. If RCIIs, when saturated, contribute to the quenching of excitation of LHCII antenna, in the system almost devoid of RCIIs, the fluorescence lifetimes should be higher than 2 ns. However, spectroscopic measurements of the long-term lincomycin-treated plants did not differ from the control measurements, *i.e.*, the fluorescence lifetime was still 2 ns, suggesting that the LHCII antenna, and not the closed RCs, are sufficient for the quenching to occur [20].

2.2. How does the plant repair the damaged D1 protein?

The D1 repair cycle is regulated *via* reversible protein phosphorylation in thylakoid membranes. When plants are exposed to the high-light stress, the PSII protein subunits become heavily phosphorylated and migrate from grana to stroma thylakoids, this process being facilitated by the actions of STN8 [21]. The migration is accompanied by sequential dephosphorylation of the CP43, D2 and D1 proteins, respectively. Turnover of the D1 protein includes degradation of the photo-damaged polypeptide and co-translational insertion of the newly synthesized protein [22]. The current model envisions that the D1 protein is proteolytically processed at both sides of the thylakoid membrane: from the N-terminal end on the stromal side by the FtsH, a member of the ATPases associated with various cellular activities-subfamily (AAA subfamily), and, on the lumen site by Deg, a member of serine proteases family that does not require ATP [23]. Recently, it was confirmed that FtsH and Deg act in a cooperative manner to efficiently cut the D1 protein under photoinhibitory conditions [24]. According to the D1 digestion model, under all light intensities D1 is processed by the FtsH complexes, to be additionally supported by the Deg proteases under photoinhibitory conditions, in the so-called escape pathway [24]. The requirement of D1 dephosphorylation prior to its proteolysis could be explained by low affinity of phosphorylated N-terminal end of D1 towards the FtsH [23].

3. Plastoquinone pool-inflicted plant responses

The most promising redox-active components are the pool of plastoquinone (PQ) and the PSI acceptor site (*e.g.* NADPH, thioredoxin, glutathione and glutaredoxin). The PQ pool regulates two temporally distant responses that acclimate the photosynthetic process to the prevailing environment: state-transitions that occur in the order of minutes (short-term response) [25, 26, 27], and photosystem stoichiometry adjustment that requires hours to days (long-term response) [28]. Both responses occur under low-light conditions, in contrast to the high-light provoked responses such as NPQ, D1 repair cycle or various stress-response programmes [29].

3.1. State-transitions

State-transitions re-distribute excitation energy between two photosystems, which are electrochemically connected in series, through the supramolecular reorganisation of the photosynthetic membranes. The molecular complexes that cause the reorganisation are light-harvesting proteins, which collect light excitation and channel it to the reaction centres. Already mentioned LHCII is the prototype of large and abundant class of chloroplast trans-membrane proteins that binds roughly half of the total chlorophyll in chloroplasts [2]. The absorption spectra of LHCI is in the far red region owing to the chlorophyll *a* enrichment, while LHCII is enriched in chlorophyll *b* and has the maximum absorption at shorter wavelengths in the red region, around 650 nm [26]. Under PQ reducing conditions, *i.e.* when PSII is predominantly excited compared to PSI, a redox-sensitive protein kinase acts to phosphorylate the apoproteins of LHCII. Upon phosphorylation, LHCII partially dissociates from PSII and associates with PSI (state I). Under PQ oxidizing conditions, *i.e.* when PSI is predominantly excited compared to PSII, the kinase is inactive, but an activated phosphatase dephosphory-lates the mobile LHCII, which moves laterally, and associates with PSII (state II) [29]. Although state-transitions occur only under low-light conditions, they share the same mechanism of LHCII aggregation with the high-light response mechanism, at least in *Chlamydomonas reinhardtii* [30]. This opens an intriguing question of plant capacity to wisely use the same mechanism for different living conditions. In 2003, Jean-David Rochaix group proposed that the thylakoid-associated serine-threonine Stt7 kinase from the alga *C. reinhardtii* is involved in phosphorylation of LHCII and in state-transitions [31]. In 2005, the same group proposed that the homolog of the Stt7 in Arabidopsis, STN7, is involved in state-transitions [32]. Since LHCII phosphorylation is reversible process, an extensive search has been conducted to identify the protein phosphatase(s) that dephosphorylates LHCII. Recently, two groups described the product of the nuclear gene At4g27800 as a long-sought plastid protein phosphatase specifi-cally involved in dephosphorylation of the mobile pool of major LHCII proteins, titled PPH1 [33], and TAP38 [34], respectively. The specificity of PPH1/TAP38 for LHCII supports the hypothesis that several phosphatases must be involved in dephosphorylation of thylakoid phosphoproteins [33, 34]. Recently, a new phosphatase PBCP, capable of *in vivo* dephospory-lation of PSII core subunits CP43, D1, D2 and PsbH, respectively, was identified [35]. It seems that the PBCP phosphatase targets at proteins phosphorylated by the protein kinase STN8 [36], thus forming a pair with opposing effects on phosphorylation of the photosynthetic core proteins [35].

3.2. Stoichiometry adjustments

Photosystem stoichiometry adjustment is a long-term response that affects the relative amounts of the two photosystems by changing the expression of photosynthetic genes both in the chloroplast and in the nucleus [29]. It was shown that the redox state of the PQ pool serves as a major signal in the regulation of LHCII photo-acclimation [37]. In order to avoid excess excitation energy and to minimize oxidative damage, the LHCII protein level decreased by approximately 50% in *Lemna persusilla* upon transfer from low-light to high-light conditions [37]. In contrast to the state-transitions, where active redox-regulated thylakoid-associated STN7 kinase phosphorylates LHCII, thus leading to LHCII detachment from PSII and migra-tion towards stroma thylakoids and PSI, under high-light STN7 is inactive, hence the rate of the phosphorylated LHCII decreases and the excess LHCII undergoes proteolytic degradation. The same stands for the PSI and PSII genes: upon reduction of the PQ pool, the expression of the PSI genes is favoured, while upon oxidation of the PQ pool, the expression of the PSII is favoured [29]. In 2005, the coupling of the long-term response of adjusting photosystem stoichiometry and the short-term response of state-transitions by LHCII kinase STN7 was proposed [36].

4. Immunophilins in photosynthesis

Immunophilins comprise a superfamily of conserved ubiquitous proteins consisting of two distinct subfamilies; cyclophilins and FKBPs (FK506/rapamycin-binding proteins), the targets of the immunosuppressive drugs cyclosporine A (CsA) and FK506/rapamycin, respectively [38]. Despite the lack of structural similarity, all cyclophilins and FKBPs share a common enzymatic, so-called PPIase or rotamase activity, catalizing *cis-trans* isomerization of prolineimidic peptide bonds [39, 40]; (Rotation around peptide bonds is energetically disfavored due to their partial double-bond character. Delocalization of amide nitrogen electrons results in aprox. 22 kcal/mol energy barrier to rotation and restrains the peptide bond in either *cis* or *trans* configurations). A more recently discovered third group of proteins (parvulins), which is insensitive to immunosuppressive drugs, also possesses PPIase activity [41]. The molecular masses of cyclophilins, FKBPs, and parvulins range normally in the order of 18-21 kDa, 12-13 kDa, and 10-13 kDa, respectively, but several complex immunophilins of higher molecular weight have been detected recently [42, 43, 41]. The biological significance of this group of enzymes is a matter of intense research. Acceleration of protein folding processes by PPIase *in vitro* [44] and *in vivo* [45] has supported a physiological role as folding catalysts facilitating the slow and generally rate-limiting uncatalysed isomerisation around Xaa-Pro peptide bonds. Immunophilins can also perform chaperone functions [46], or cooperate with other chaperone proteins [47, 48, 49].

Complex immunophilins all contain additional protein-protein interaction domains, such as leucine-zipper motifs and/or tetratricopeptide repeat domains. They may be constituents of supramolecular structures, as shown for the human or avian steroid receptor complex [48] or chaperone supercomplexes [50], and may be involved in hsp90-dependent signal transduction [49, 51, 52]. Various members of the immunophilins are involved in phosphorylation processes *via* transient interaction with kinases [53, 51, 52] and phosphatases [54, 55]. It seems likely that, at least for the components of MAP kinase signaling system (Src, Raf, and Mek), hsp90-immunophilin interactions are essential for the kinase regulation [see 52]. The immunophilin FKBP65 together with hsp90 forms a regulatory association with the serine/threonine kinase c-Raf-1 [51]. In mammals, the Ca^{2+}/calmodulin-dependent heterodimeric protein phosphatase calcineurin can bind immunophilins with complex immunosupresive drugs. This interaction inhibits the protein phosphatase activity, resulting in interruption of the signal-transduction cascades required for T-cell activation [see 54]. A unique immunophilin-related protein is the PP5 protein phosphatase. This enzyme is a major constituent of the glucocorticoid receptor-hsp90 complex, with properties of an FK506-binding immunophilin with low affinity FK506 binding activity [56].

Plants possess a substantial number of immunophilins which are localized in cytosol, chloroplasts, nucleus, mitochondria, or associated with secretory pathways. The majority of the cyclophilins are single-domain proteins, 23 isoforms in Arabidopsis [57]. AtCYP20-3, AtCYP20-2, AtCYP26-2, AtCYP28 and AtCYP37 can be found in chloroplasts, mostly in thylakoid lumen. The multidomain cyclophilin isoforms possess unique domain arrangements, as exemplified by AtCYP38. AtCYP40 contains a C-terminal tetratricopeptide repeat module

(TPR). Four additional multidomain cyclophilins contain RNA interaction domains, which implicates their involvement in nuclear RNA processing machinery.

Plant FKBPs encompass also 23 members in Arabidopsis and this family is one of the largest FKBP family identified to date. These FKBPs can also be divided into single and multidomain isoforms, consisting of 16 and 7 members, respectively [58]. Interestingly, 11 single-domain FKBPs appear to be targeted to the thylakoid lumen. Thus, this chloroplast sub-compartment appears to have very important role in immunophilin function, or the processes taking place in lumen require specific activity of this protein family.

The dephosphorylation of D1, D2 and CP43 in spinach is catalysed by a cyclophilin-regulated PP2A-like protein phosphatase, which was found to be associated with, and regulated by, a cyclophilin-like protein, TLP40 [59]. TLP40 is proposed to suppress phosphatase activity, when bound to the lumen-exposed epitope of the protein phosphatase, but to induce its activity, when released to the lumen [59]. However, it remains to be seen if the Arabidopsis PP2C phosphatase PBCP [35] is also under the control of TLP40. Vener group demonstrated that D1 was not only vulnerable to light, but also to high temperature [60]. Raising the temperature from 22 °C to 42 °C resulted in a very rapid dephosphorylation of the D1, D2 and CP43 proteins and in release of TLP40 from membrane into the thylakoid lumen. These events are proposed to trigger an accelerated repair of photodamaged PSII and to initiate other heat-shock responses in chloroplasts.

Higher plant thylakoid lumen prolineisomerases [38], or complex immunophilins, are found to be regulated by light and are responsive to various forms of environmental stress [61]. The structure of TLP40 and its association with the thylakoid membrane system implicates diverse functions and involvement in the intracellular signaling networks [62]. Binding of thylakoid membrane associated phosphatase involved in dephosphorylation of PSII core proteins might occur *via* two putative phosphatase-binding modules on the N-terminal side of TLP40 [59, 60]. In 2002, Baena-González and Aro [63] and in 2005 Aro et al. [64] further suggested that damaged D1 repair cycle includes TLP40 ortholog *At*CYP38, which lacks peptidyl–prolyl *cis/ trans* isomerase (PPIase) activity [65, 66, 67]. It was shown that *At*CYP38 is involved in the assembly of oxygen evolving complex (OEC) [68] and maintenance of PSII [69]. Most recently, the interaction of the E-loop of chlorophyll protein 47 (CP47) with *At*CYP38 was demonstrated [67]. This interaction is mediated through putative cyclophilin domain [67]. Further, *in vivo* role of *At*CYP38 has been investigated in *cyp*38 mutant Arabidopsis plants [68, 69, 70], suggesting its primary role in PSII biogenesis and repair.

5. Flow and partitioning of photosynthetic electrons

Photosynthetic apparatus has to be able to efficiently convert energy at low light and to avoid over-reduction and damage at excess light. This requires switching between different regulatory mechanisms which keep the cellular ATP pool almost at the constant level. Exposure of plants to higher light intensities than required for efficient photosynthesis results in saturation of photosynthetic electron transport (PET). The over-reduction of PET chain can lead to the formation of reactive oxygen species and may irreversibly damage the photosystems, as well as the cells and the whole organism [71, 72, 73]. To counteract and reduce the photoinhibitory damage, plants have developed several short- and long-term regulation mechanisms which include processes that modulate the structure and function of antenna complexes, including NPQ, alternative electron transport pathways and movement of chloroplasts, leaves and whole organisms away from intense light [74, 75]. Dissipation of excess light energy to heat in the antenna or in the reaction centre of PSII also counteracts the photoinhibitory damage. Retrograde signaling transduces information on the metabolic state of the organelle and induces many activities in the cytosol, nucleus and mitochondria, inducing alterations in nuclear gene expression of organelle-targeted proteins [76].

5.1. Linear, cyclic and pseudo-cyclic electron transfer routes

Various photosynthetic electron transfer routes become turned on and off, according to the need of the plant to adapt to wide-ranging quantities and qualities of light. Three major electron transfer pathways known are linear, cyclic and pseudo-cyclic electron transfer (LEF, CEF and PCEF, respectively). All three pathways are necessary for poised and sustained synthesis of ATP and NADPH and their interplay enables the flexibility of photosynthesis in meeting different metabolic demands [77].

During non-cyclic electron transport or LEF, light drives the conversion of water to oxygen at the level of oxygen-evolving complex of PSII and $NADP^+$ to NADPH on the stromal side of thylakoid membranes. The PET chain consists of PSII, the Cyt$b6f$ complex, PSI, and the free electron carriers plastoquinone (PQ) and plastocyanin (PC). Electron transport includes the two quinine binding sites and two cytochromes b6 (the 'Q cycle') that give two protons translocated for each electron transferred from PSII to PSI. Hidrogen ions are transferred across chloroplast membrane and accumulated on the luminal side of the thylakoids, where they drive ATP synthesis through a membrane ATPase. This way electron transport helps to establish a proton gradient that powers ATP production and also stores energy in the reduced coenzyme NADPH to power the Calvin-Benson cycle to produce sugar and other carbohydrates.

Arnon, who discovered photophosphorylation in isolated chloroplasts in 1954, demonstrated that there is also CEF in the thylakoids, driven solely by PSI. CEF is a light-driven flow of electrons through a photosynthetic reaction centre with the electrons being transferred from PSI to Cyt$b6f$ complex via ferredoxin (Fd), with associated formation of proton gradient. PQ is reduced by Fd or NADPH via one or more enzymes collectively called PQ reductase, rather than by PSII, as in LEF. From hidroplastoquinone (PQH2), electrons return to PSI via the Cyt$b6f$ complex. Four possible routes of CEF that may operate in parallel have been proposed so far: NAD(P)H dehydrogenase (NDH)-dependent route, Fd-dependent route, Nda2, a type 2 NAD(P)H:PQ oxidoreductase route and Cyt$b6f$ complex and FNR route [recently reviewed by 78].

CEF around PSI occurs under conditions when acceptor limitation or ATP shortage results in a highly reduced PQ pool and contributes to the formation and maintenance of a pH gradient across a membrane but does not produce NADPH. The pH gradient generated may drive the production of ATP (cyclic photophosphorylation) or may regulate photosynthesis. When more light is absorbed than can be used for assimilation, the increased ΔpH is a switch to dissipate excess of the light absorbed by chlorophyll molecules of PSII [79]. CEF is diminished when its components are completely reduced. Also, there is no CEF when its components are completely oxidized because there are no electrons to cycle [80]. In an attempt to avoid these two extreme situations, kinetics, post-translational modifications [25, 26, 81] and redox control of reaction-centre gene expression [82] are all employed in maintaining a poised PQ pool. In spite of these control mechanisms, over-reduction easily occurs when the Calvin–Benson cycle is unable to use NADPH, usually due to the lack of ATP. The major physiological significance of CET most probably lies in additional availability of ATP.

LEF in chloroplasts produces a number of reduced components associated with PSI that may subsequently participate in reactions that reduce oxygen. When Fd transfers electrons to molecular oxygen instead of $NADP^+$, PCEF linked with phosphorylation arises. O_2 is directly reduced to superoxide radical in so-called Mehler reaction. Subsequently, two superoxides dismutate to form H_2O_2 and O_2, which are by further redox processes converted to water. A reduced state of the FeS pool (Fd and PSI centres) promotes the Mehler reaction. The Mehler reaction leading to PCEF restores the redox poise when the PET chain is over-reduced, thereby allowing CEF to function and to generate ATP for the Calvin-Benson cycle, which will in turn oxidize NADPH and restore LEF [83]. Under high light PCEF could also cover an increased energy demand, as long as the antioxidant systems for H_2O_2 removal is sufficiently active (ascorbate and GSH recycling).

A great number of contemporary research topics on photosynthesis aims at elucidating novel, alternative electron transfer routes as the safest pathways for channelling of unwanted electrons. Chlororespiration is a well-known respiratory process that can maintain a trans-thylakoid proton gradient, thus acting as an effective alternative electron sink in preventing over-reduction of the PQ pool and protecting RCIIs from photodamage under photoinhibitory light conditions [84]. Two enzymes are important for chlororespiratory function: NADH dehydrogenase complex and nucleus-encoded plastid-localized terminal oxidase (PTOX), through which electrons from plastoquinol are transferred to molecular oxygen, forming water in the stroma [85]. In oat leaves incubated at high temperature and under high-light intensities, the amounts of both enzymes were increased, suggesting that, under unfavourable conditions, chlororespiration can act as a protective mechanism [85].

Apart from the Mehler reaction and the chlororespiration, respectively, photorespiration is another efficient mechanism to adjust the ATP/NADPH ratio and to consume excess energy [86]. Although the major part of energy supplied in the form of ATP and NADPH by the light reaction is consumed in the Calvin-Benson cycle, it is also needed for multiple anabolic processes in chloroplasts, such as synthesis of lipids and proteins and of many secondary metabolites. ATP/NADPH ratio in chloroplasts could be increased by indirect export of reducing equivalents in the form of malate through the malate valve [87]. Malate can be used in numerous ways in cytoplasm and mitochondria, providing NADH for nitrate reduction and/or ATP for sucrose synthesis. Also, different stages of tissue growth require different ATP/

NADPH ratios. The ATP/NADPH ratio that is available in the light can vary substantially because of the many possibilities for electron pathways and regulatory mechanisms [88]. Therefore, the cooperation of different electron transport pathways enables optimization of ATP/NADPH stoichiometry [77]. In example, sudden dark-to-light transition induces over-reduction of the PET chain in the first minute, which is relieved by electron transport to O_2 [89], and by the rapid activation of the chloroplast NADP–MDH [90]. Calculating the photosynthetic stoichiometries, it was estimated that ATP/NADPH ratio arising from LEF is about 1.28, which is not sufficient for driving the Calvin-Benson cycle [91, 77]. It is therefore obvious that the optimal operation of the Calvin-Benson cycle requires both LEF and CEF, whose tuning enables adjustment of ATP/NADPH to meet the cellular demands. Green alga *C. reinhardtii* in which intracellular ATP depletion induces a switch from LEF to CEF is a good example for such regulation [92].

5.2. Electron partitioning at the ferredoxin hub

In photoautotrophic plants, ferredoxin (Fd) accepts one electron from the stromal side of PSI involving the subunits PsaC, PsaD and PsaE [93]. Fd acts simultaneously as bottleneck and as a hub which distributes high-energy electrons to a multitude of enzymes involved in chloroplast metabolism. The hierarchy of electron distribution and subsequent regulation of channeling of photosynthetically derived electrons into different areas of chloroplast metabolism is still not defined. Electrons are preferentially directed to carbon assimilation, which requires NADPH, and so the majority of Fd is immediately oxidized by the enzyme ferredoxin:NADPH reductase (FNR), which is associated with the thylakoid membrane.

Besides its crucial metabolic role in reducing $NADP^+$ and thioredoxin (TRX) *via* FNR and ferredoxin-thioredoxin-reductase (FTR) respectively, Fd-dependent enzymatic reactions are also linked to nitrite and sulfur metabolism by ferredoxin-nitrite-reductase and sulfite reductase. Furthermore, Fd is a electron donor for fatty acid desaturase and glutamine-2-oxoglutarate amino transferase. In some green algae, upon transition from dark to the light under anaerobic conditions, Fd transfers transiently electrons to chloroplast hydrogenases, which in turn catalyse the formation of hydrogen [94, 95] dissipating excess reducing power when the Calvin–Benson cycle is not yet fully activated. Fd and NADPH can also act in CEF, returning electrons to the PQ pool *via* the PGR5 (proton gradient regulation 5)-dependent [96] and NDH complex-dependent [97, 98, 99] pathways, respectively. FNR may also act as the direct Fd:PQ reductase, establishing redox regulation and antioxidant defense point. FNR activity seems to represent a critical point in photosynthetic electron partitioning, because it is integral to most of these electron cascades and can associate with several different membrane complexes.

At least four Fd isoforms occur in plants [100]. Fd1 and Fd2 are found in leaves, while Fd3 and Fd4 appear to play roles in non-photosynthetic metabolism [101, 100]. Fd as electron distributing hub is particularly suitable to provide information on the redox state of the system to be transmitted into the regulatory network. Fd contributes to the control of chloroplast energization by feeding electrons into the CEF pathway and thereby controls both phosphorylation potential and reductive power.

5.3. TROL-FNR interaction influences the energy conductance

The last step of the photosynthetic electron transfer from Fd to $NADP^+$ is catalyzed by FNR. There are two evolutionary conserved types of FNR in the chloroplasts of higher plants: predominantly, or exclusively, thylakoid membrane-bound isoproteins and 'soluble', non-tightly bound isoproteins [101, 102]. The membrane-bound FNR is supposed to be involved in electron transport, while the soluble enzyme provides protection against oxidative stress [103]. Two chloroplast-type *FNR* genes have been found in Arabidopsis genome. In 2008, Hanke et al. investigated knock-out mutant of Arabidopsis FNR isoprotein, *fnr1*. The loss of the strong thylakoid binding was observed, which affected the channeling of photosynthetic electrons into NADPH- and Fd-dependent metabolism. Also, these mutants had complex variation in CEF, dependent on light conditions [104]. In *fnr1*, thylakoid $NADP^+$ photoreduction was greatly reduced even on addition of soluble FNR to rate saturating concentrations, which is consistent with the lack of membrane-bound FNR [104].

$NADP^+$ photoreduction activity of FNR was shown to be greater when the enzyme is associated with the thylakoid membrane and it has been proposed that binding of FNR to the thylakoid membrane regulates the enzyme activity [105, 106]. Subsequently, interactions of FNR and several photosynthetic protein complexes, such as Cyt*b6f* [107, 108], PSI [109] or NDH complex [110] have been shown. However, the factors controlling relative localization of FNR to different membrane complexes have not yet been established. It was shown that maize contains three chloroplast FNR proteins with completely different membrane association and distribution between cells, conducting predominantly CEF in bundle sheath cells and LEF in mesophyll cells [111]. Expression of maize FNRs in Arabidopsis as chimeras and truncated proteins showed that N-terminus determines recruitment of FNR to different membrane complexes, which impacts the photosynthetic electron flow [111].

It was also demonstrated that FNR interacts specifically with two chloroplast proteins, Tic62 (62 kDa component of the translocon at the inner envelope of cloroplasts) and TROL (thylakoid rhodanese-like protein), *via* a conserved Ser/Pro-rich motif [112, 113]. Both Tic62 and TROL seem to act as molecular anchors for FNR, because they form high molecular weight complexes with FNR at the thylakoid membranes. TROL possesses centrally positioned rhodanese-like domain, which is most probably involved in redox regulation of FNR binding and release [113]. We have proposed that such regulation could be important for balancing the redox status of stroma with the membrane electron transfer chain and therefore preventing the over-reduction of any of these two compartments and maintaining the redox poise [113]. TROL-FNR complex was clearly visible during the dark and it disappeared during light periods [112]. FNR-Tic62/TROL interaction is clearly pH-dependent. During high photosynthetic activity in the light, stroma becomes alkaline due to the transport of protons to the thylakoidal lumen. During the dark, stromal pH decreases again, and that is when FNR-Tic62/TROL complexes were found predominantly associated with the thylakoid membrane [112, 113]. The mechanism by which TROL influences the FNR activity could be that during the dark period, FNR is bound to the thylakoids *via* TROL and NADPH production does not occur. This stage could be sustained through the binding of small molecule, possibly oxidized PQ, to the RHO cavity. In conditions of growth-light, FNR is bound to the thylakoids *via* TROL and efficiently produces NADPH. When the light is saturating, FNR is released from TROL by a signal molecule, possibly reduced PQ that competes for the RHO binding site. Once soluble, FNR acts as NADPH consumer and released protons are passed to an unknown scavenger [78, 113, 114]. It was reported that membrane attachment of FNR is influenced by the stromal redox state (NADP⁺/NADPH ratio), which mimics variations in environmental conditions [115]. Therefore, reversible attachment of FNR to the thylakoid membrane *via* TROL and/or Tic62 provides an elegant way to store redundant molecules, not required when photosynthesis is less active or dormant.

We have already mentioned that reducing equivalents could be exported in the form of malate through the malate valve to increase the ATP/NADPH ratio in chloroplasts [87]. TROL-deficient plants grown under growth-light conditions show significant up-regulation of NADP-malic enzyme 2 that catalyses the oxidative decarboxylation of malate, producing pyruvate, carbon dioxide and NAD(P)H in cytosol [113]. Therefore, the TROL knock-out Arabidopsis mutant lines (*trol*) could act as efficient NADPH producers, fighting the possible hyper-reduction of the thylakoids by exporting the reducing energy in a form of malate to the cytosol.

The *trol* mutants show severely lowered relative electron transport rates at high-light intensities. Also, under high-light conditions, but in a short-term, NPQ amount was higher in the *trol* line, compared to the wild-type [113]. Moreover, TROL is important for NPQ, since the reversion of the plant without TROL to the plant that expresses TROL, even with certain alterations, leads to the restoration of wild-type levels of NPQ (Jurić, Fulgosi, Ruban, unpublished results). We are not dealing with extreme NPQ phenotype, but, the question remains, how is it possible that small changes in the TROL protein could modulate NPQ so effectively? Our unpublished data also suggest the possible enrolment of some LHCII subunits in the TROL-containing complexes, which, in the light of these results, should be thoroughly investigated. In addition, we observed that, after more than two weeks of exposure to the high-light conditions, *trol* plants exhibited NPQ almost at the wild-type levels, suggesting that some sort of long-term acclimation could also be involved.

6. Conclusions and perspectives

Photosynthesis is the crucial converter of sunlight into the chemical energy that is subsequently utilised to sustain life on Earth. However, without the layers of regulative pathways, it would be virtually impossible to discuss the efficient photosynthesis. Now, we are aware of the fact that more than 40-years-old "Z-scheme" has been constantly upgraded with many alternative routes allied with still not well-characterised electron sources and sinks. In addition, it seems that the thylakoid membranes are more flexible than anticipated, with PSII antenna involved directly in NPQ origin through detachment and aggregation.

One of the interesting domains in photosynthesis, but not enough explored, would be the involvement of lipids in a complex network of protein interactions. Lipids, especially phospholipids (phosphatidylglycerol; PG) and glycolipids (monogalactosyldiacylglycerol; MGDG, digalactosyldiacylglycerol; DGDG, and sulfoquinovosyldiacylglycerol; SQDG) are major building blocks of thylakoid membranes, providing the safe docking sites for proteins and protein complexes, respectively. It has been proposed that DGDG and PG are especially involved in the assembly and repair of PSII [116]. Moreover, plants lacking almost 30% of the wild-type amount of PG, caused by the point mutation in one of the genes involved in PG synthesis pathway, displayed pale green leaves and somewhat reduced capacity for photosynthesis [117], while plants that accumulated only 10% of the wild-type PG amount were incapable of surviving on the growth medium without the addition of sucrose [118].

The other attractive domain would be retrograde signalling, *i.e.* the pathways flowing from chloroplasts to the nucleus. At least five classes of chloroplast-derived signals have been studied and discussed: tetrapyrrole biosynthesis, chloroplast gene expression-dependent pathway, redox state of chloroplasts, ROS, and sugar and hormone signalling. While some molecules look more promising than others (sugars and hormones for example), ROS could be more challenging to investigate, due to its non-specificity and a power to response quickly to different types of stress in different parts of the cell.

The rich past of exploring photosynthetic processes gave us a wealth of knowledge, but, it seems that we have just scratched the surface. In years to come, it would be interesting to fill the missing blanks and to develop some new, "out-of-the-box" perspectives.

Author details

Snježana Jurić, Lea Vojta and Hrvoje Fulgosi*

Division of Molecular Biology, Ruđer Bošković Institute, Zagreb, Croatia

References

[1] Dekker JP, Boekema EJ. Supramolecular organization of thylakoid membrane proteins in green plants. Biochim Biophys Acta. 2005;1706(1-2):12-39.

[2] Barros T, Kühlbrandt W. Crystallisation, structure and function of plant light-harvesting Complex II. Biochim Biophys Acta. 2009;1787(6):753-772.

[3] Horton P, Ruban A. Molecular design of the photosystem II light-harvesting antenna: photosynthesis and photoprotection. J Exp Bot. 2005;56(411):365-373.

[4] Müller P, Li XP, Niyogi KK. Non-photochemical quenching. A response to excess light energy. Plant Physiol. 2001;125(4): 1558-1566.

[5] Szabó I, Bergantino E, Giacometti GM. Light and oxygenic photosynthesis: energy dissipation as a protection mechanism against photo-oxidation. EMBO Rep. 2005;6(7):629-634.

[6] Ruban AV, Johnson MP, Duffy CD. The photoprotective molecular switch in the photosystem II antenna. Biochim Biophys Acta. 2012;1817(1):167-181.

[7] Eberhard S, Finazzi G, Wollman FA. The dynamics of photosynthesis. Annu Rev Genet. 2008; 42:463-515.

[8] Kirilovsky D, Kerfeld CA. The orange carotenoid protein in photoprotection of photosystem II in cyanobacteria. Biochim Biophys Acta. 2012;1817(1):158-166.

[9] Xiao FG, Shen L, Ji HF. On photoprotective mechanisms of carotenoids in light harvesting complex. Biochem Biophys Res Commun. 2011;414(1):1-4.

[10] Jahns P, Holzwarth AR. The role of the xanthophyll cycle and of lutein in photoprotection of photosystem II. Biochim Biophys Acta. 2012;1817(1):182-193.

[11] Barber J. Influence of surface charges on thylakoid structure and function. Annu Rev Plant Physiol. 1982;33:261-295.

[12] Holzwarth AR, Miloslavina Y, Nilkens M, Jahns P. Identification of two quenching sites active in the regulation of photosynthetic light-harvesting studied by time-resolved fluorescence. Chem Phys Lett. 2009;483(4-6):262-267.

[13] Betterle N, Ballottari M, Zorzan S, de Bianchi S, Cazzaniga S, Dall'osto L, Morosinotto T, Bassi R. Light-induced dissociation of an antenna hetero-oligomer is needed for non-photochemical quenching induction. J Biol Chem. 2009;284(22):15255-15266.

[14] Johnson MP, Goral TK, Duffy CD, Brain AP, Mullineaux CW, Ruban AV. Photoprotective energy dissipation involves the reorganization of photosystem II light-harvesting complexes in the grana membranes of spinach chloroplasts. Plant Cell. 2011;23(4):1468-1479.

[15] Li Z, Wakao S, Fischer BB, Niyogi KK. Sensing and responding to excess light. Annu Rev Plant Biol. 2009;60:239-260.

[16] Horton P, Ruban AV, Rees D, Pascal AA, Noctor G, Young AJ. Control of the light-harvesting function of chloroplast membranes by aggregation of the LHCII chlorophyll-protein complex. FEBS Lett. 1991;292(1-2):1-4.

[17] Li XP, Björkman O, Shih C, Grossman AR, Rosenquist M, Jansson S, Niyogi KK. A pigment-binding protein essential for regulation of photosynthetic light harvesting. Nature. 2000;403(6768):391-395.

[18] Goral TK, Johnson MP, Duffy CD, Brain AP, Ruban AV, Mullineaux CW. Light-harvesting antenna composition controls the macrostructure and dynamics of thylakoid membranes in Arabidopsis. Plant J. 2012;69(2):289-301.

[19] Ahn TK, Avenson TJ, Ballottari M, Cheng YC, Niyogi KK, Bassi R, Fleming GR. Architecture of a charge-transfer state regulating light harvesting in a plant antenna protein. Science. 2008;320(5877):794-797.

[20] Belgio E, Johnson MP, Jurić S, Ruban AV. Higher plant photosystem II light-harvesting antenna, not the reaction center, determines the excited-state lifetime-both the maximum and the nonphotochemically quenched. Biophys J. 2012;102(12):2761-2771.

[21] Tikkanen M, Nurmi M, Kangasjärvi S, Aro EM. Core protein phosphorylation facilitates the repair of photodamaged photosystem II at high light. Biochim Biophys Acta. 2008;1777(11):1432-1437.

[22] Vener AV. Environmentally modulated phosphorylation and dynamics of proteins in photosynthetic membranes. Biochim Biophys Acta. 2007;1767(6):449-457.

[23] Kato Y, Sakamoto W. Protein quality control in chloroplasts: a current model of D1 protein degradation in the photosystem II repair cycle. J Biochem. 2009;146(4): 463-469.

[24] Kato Y, Sun X, Zhang L, Sakamoto W. Cooperative D1 Degradation in the Photosystem II Repair Mediated by Chloroplastic Proteases in Arabidopsis. Plant Physiol. 2012;159(4):1428-1439.

[25] Allen JF, Forsberg J. Molecular recognition in thylakoid structure and function. Trends Plant Sci. 2001;6(7):317-326.

[26] Wollman FA. State-transitions reveal the dynamics and flexibility of the photosynthetic apparatus. EMBO J. 2001;20(14):3623-3630.

[27] Mullineaux CW, Emlyn-Jones D. State-transitions: an example of acclimation to low-light stress. J Exp Bot. 2004;56(411):389-393.

[28] Allen JF, Pfannschmidt T. Balancing the two photosystems: photosynthetic electron transfer governs transcription of reaction centre genes in chloroplasts. Philos Trans R Soc Lond B Biol Sci. 2000; 355(1402):1351-1359.

[29] Pfannschmidt T, Bräutigam K, Wagner R, Dietzel L, Schröter Y, Steiner S, Nykytenko A. Potential regulation of gene expression in photosynthetic cells by redox and energy state: approaches towards better understanding. Ann Bot. 2009;103(4):599-607.

[30] Iwai M, Yokono M, Inada N, Minagawa J. Live-cell imaging of photosystem II antenna dissociation during state transitions. Proc Natl Acad Sci USA. 2010;107(5): 2337-2342.

[31] Depège N, Bellafiore S, Rochaix JD. Role of chloroplast protein kinase Stt7 in LHCII phosphorylation and state transition in Chlamydomonas. Science. 2003;299(5612): 1572-1575.

[32] Bellafiore S, Barneche F, Peltier G, Rochaix JD. State-transitions and light adaptation require chloroplast thylakoid protein kinase STN7. Nature. 2005;433(7028):892-895.

[33] Shapiguzov A, Ingelsson B, Samol I, Andres C, Kessler F, Rochaix JD, Vener AV, Goldschmidt-Clermont M. The PPH1 phosphatase is specifically involved in LHCII dephosphorylation and state-transitions in Arabidopsis. Proc Natl Acad Sci USA. 2010;107(10):4782-4787.

[34] Pribil M, Pesaresi P, Hertle A, Barbato R, Leister D. Role of plastid protein phosphatase TAP38 in LHCII dephosphorylation and thylakoid electron flow. PLoS Biol. 2010;8(1):e1000288.

[35] Samol I, Shapiguzov A, Ingelsson B, Fucile G, Crèvecoeur M, Vener AV, Rochaix JD, Goldschmidt-Clermont M. Identification of a Photosystem II Phosphatase Involved in Light Acclimation in Arabidopsis. Plant Cell. 2012;24(6):2596-2609.

[36] Bonardi V, Pesaresi P, Becker T, Schleiff E, Wagner R, Pfannschmidt T, Jahns P, Leister D. Photosystem II core phosphorylation and photosynthetic acclimation require two different protein kinases. Nature. 2005;437(7062):1179-1182.

[37] Oelze ML, Kandlbinder A, Dietz KJ. Redox regulation and overreduction control in the photosynthesizing cell: complexity in redox regulatory networks. Biochim Biophys Acta. 2008;1780(11):1261-1272.

[38] Schreiber SL. Chemistry and biology of the immunophilins and their immunosuppressive ligands. Science. 1991;251 (4991):283-287.

[39] Fischer G, Wittmann-Liebold B, Lang K, Kiefhaber T, Schmid FX. Cyclophilin and peptidyl-prolyl cis-trans isomerase are probably identical proteins. Nature. 1989;337(6206):476-478.

[40] Takahashi N, Hayano T, Suzuki M. Peptidyl-prolyl cis-trans isomerase is the cyclosporin A-binding protein cyclophilin. Nature. 1989;337(6206):473-475.

[41] Rudd KE, Sofia HJ, Koonin EV, Plunkett GI, Lazar S, Rouviere PE. A new family of peptidyl-prolyl isomerases. Trends Biochem Sci. 1995;20(1):12-14.

[42] Gething MJ, Sambrook J. Protein folding in the cell. Nature. 1992;355(6355):33-45.

[43] Kieffer LJ, Seng TW, Li W, Osterman DG, Handschumacher RE, Bayney R. Cyclophilin-40, a protein with homology to the P59 component of the steroid receptor complexx. Cloning of the cDNA and further characterization. J Biol Chem. 1993;268(17): 12303-12310.

[44] Fruman DA, Burakoff SJ, Bierer BE. Immunophilins in protein folding and immunosuppression. FASEB J. 1994;8(6):391-400.

[45] Matouschek A, Rospert S, Schmid K, Click BS, Schatz G. Cyclophilin catalyzes protein folding in yeast mitochondria. Proc Natl Acad Sci USA. 1995;92(14):6319-6323.

[46] Freskgård PO, Bergenhem N, Jonson BH, Svensson M, Carlsson U. Isomerase and chaperone activity of prolyl isomerase in folding of carbonic anhydrase. Science. 1992;258(5081):466-468.

[47] Chang HC, Lindquist S. Conservation of Hsp90 macromolecular complex in *Sacharomyces cerevisiae*. J Biol Chem. 1994;269(40):24983-24988.

[48] Johnson JL, Toft DO. A novel chaperone complex for steroid receptors involving heat shock proteins, immunophilins, and p23. J Biol Chem. 1994;269(40):24989-24994.

[49] Duina AA, Chang HC, Marsh JA, Lindquist S, Gaber RF. A cyclophilin function in Hsp90-dependent signal transduction. Science. 1996;274(5293):1713-1715.

[50] Freeman BC, Toft DO, Morimoto RI. Molecular chaperone machines: Chaperone activities of the cyclophilin Cyp-40 and the steroid aporeceptors-associated protein p23. Science. 1996;274(5293):1718-1720.

[51] Coss MC, Stephens RM, Morison DK, Winterstein D, Smith LM, Simek SL. The immunophilin FKBP65 forms an association with the serine/threonine kinase c-Raf-1. Cell Growth Diff. 1998;9(1):41-48.

[52] Pratt WB. The hsp90-based chaperone system-involvement in signal transduction from variety of hormone and growth factor receptors. Proc Soc Exp Biol Med. 1998;217(4):420-434.

[53] Owens-Grillo JK, Hoffmann K, Hutchison KA, Yem AW, Deibel MR Jr, Handschumacher RE, Pratt WB. The cyclosporin A-binding immunophilin CyP-40 and the FK506-binding immunophilin hsp56 bind to a common site on hsp90 and exist in independent cytosolic heterocomplexes with the untransformed glucocorticoid receptor. J Biol Chem. 1995;270(35):20479-20484.

[54] Schreiber SL. Immunophilin-sensitive protein phosphatase action in cell signaling pathways. Cell. 1992;70(3):365-368.

[55] Wilson LK, Benton BM, Zhou S, Thorner J, Martin GS. The yeast immunophilin Fpr3 is a physiological substrate of the tyrosine-specific phosphoprotein phosphatas Ptp1. J Biol Chem. 1995;270(42):25185-25193.

[56] Silverstein AM, Galigniana MD, Mei-Shya C, Owens-Grillo JK, Chinkers M, Pratt WB. Protein phosphatase 5 is a major component of glucocorticoid receptor-hsp90 complex with properties of an FK606-binding immunophilin. J Biol Chem. 1997;271(26):16224-16230.

[57] Romano P, Gray J., Horton P, Luan S. Plant immunophilins: functional versatility beyond protein maturation. New Phytologist. 2005;166(3):753-769.

[58] He Z, Li L, Luan S. Immunophilins and parvulins. Superfamily of peptidyl prolyl isomerases in Arabidopsis. Plant Physiol. 2004;134(4):1248-1267.

[59] Vener AV, Rokka A, Fulgosi H, Andersson B, Herrmann RG. A cyclophilin-regulated PP2A-like protein phosphatase in thylakoid membranes of plant chloroplasts. Biochemistry. 1999;38(45):14955-14965.

[60] Rokka A, Aro EM, Herrmann RG, Andersson B, Vener AV. Dephosphorylation of photosystem II reaction center proteins in plant photosynthetic membranes as an im-

mediate response to abrupt elevation of temperature. Plant Physiol. 2000;123(4): 1525-1536.

[61] Luan S, Albers MW, Schreiber SL. Light-regulated, tissue-specific immunophilins in a higher plant. Proc Natl Acad Sci USA. 1994;91(3):984-988.

[62] Fulgosi H, Vener AV, Altschmied L, Hermann RG, Andersson B. A novel multi-functional chloroplast protein: identification of 1 40 kDaimmunophilin-like protein located in the thylakoid lumen. EMBO J. 1998;17(6):1577-1587.

[63] Baena-Gonzàlez E, Aro EM. Biogenesis, assembly and turnover of photosystem II units. Philos Trans R Soc Lond B Biol Sci. 2002;357(1426):1451–1460.

[64] Aro EM, Suorsa M, Rokka A, Allahverdiyeva Y, Paakkarinen V, Saleem A, Battchikova N, Rintamäki E. Dynamics of photosystem II: A proteomic approach to thylakoid protein complexes. J Exp Bot. 2005;56(411):347-356.

[65] Shapiguzov A, Edvardsson A, Vener AV. Profound redox sensitivity of peptidyl-prolylisomerase activity in arabidopsis thylakoid lumen. FEBS Lett. 2006;580(15): 3671-3676.

[66] Edvardsson A, Shapiguzov A, Petersson UA, Schroder WP, Vener AV. Immunophilin AtFKBP13 sustains all peptidyl–prolylisomerase activity in the thylakoid lumen from Arabidopsis thaliana deficient in AtCYP20-2. Biochemistry. 2007;46(33): 9432-9442.

[67] Sirpiö S, Khrouchtchova A, Allaverdiyeva Y, Hansson M, Fristedt R, Vener AV, Scheller HV, Jensen PE, Haldrup A, Aro EM. AtCYP38 ensures early biogenesis, correct assembly and sustenance of photosystem II. Plant J. 2008;55(4):639-651.

[68] Fu A, He Z, Sun Cho H, Lima A, Buchanan BB, Luan S. A chloroplast cyclophilin functions in the assembly and maintenance of photosystem II in Arabidopsis thaliana. Proc Natl Acad Sci USA. 2007;104(40):15947-15952.

[69] Lepeduš H, Tomašić A, Jurić S, Katanić Z, Cesar V, Fulgosi H. Photochemistry of PSII in CYP38 Arabidopsis thaliana Mutant. Food Technol Biotechnol. 2009;47(3):275-280.

[70] Schreiber U, Bilger W, Neubauer C Schreiber U, Bilger W, Neubauer C. Chlorophyll fluorescence as a nonintrusive indicator for rapid assessment of in vivo photosynthesis. In: Schulze E-D, Caldwell MM. (eds.) Ecophysiology of Photosynthesis. Vol 100. Springer: Berlin Heidelberg New York; 1994. p49-70.

[71] Asada K. Production and scavenging of reactive oxygen species in chloroplasts and their functions. Plant Physiol. 2006;141(2):391–396.

[72] Sonoike K. Photoinhibition of photosystem I. Physiol Plant. 2011;142(1):56–64.

[73] Vass I. Molecular mechanisms of photodamage in the Photosystem II complex. Biochim Biophys Acta. 2012;1817(1):209–217.

[74] Baker NR, Harbinson J, Kramer DM. Determining the limitations and regulation of photosynthetic energy transduction in leaves. Plant Cell Environ. 2007;30(9):1107–1125.

[75] Rochaix JD. Reprint of: Regulation of photosynthetic electron transport. Biochim Biophys Acta. 2011; 1807(8):375–383.

[76] Barajas-López JD, Blanco NE, Strand A. Plastid-to-nucleus communication, signals controlling the running of the plant cell. Biochim Biophys Acta. 2013;1833(2):425-437.

[77] Allen JF. Cyclic, pseudocyclic and noncyclic photophosphorylation: new links in the chain. Trends Plant Sci. 2003;8(1):15–19.

[78] Vojta L, Fulgosi H. Energy Conductance from Thylakoid Complexes to Stromal Reducing Equivalents, Advances in Photosynthesis - Fundamental Aspects, Dr Mohammad Najafpour (Ed.), ISBN: 978-953-307-928-8, InTech, 2012, Available from: http://www.intechopen.com/books/advances-in-photosynthesisfundamental-aspects/energy-conductance-from-thylakoid-complexes-to-stromal-reducing-equivalents

[79] Heber U, Walker D. Concerning a dual function of coupled cyclic electron transport in leaves. Plant Physiol. 1992;100(4):1621-1626.

[80] Whatley FR. Photosynthesis by isolated chloroplasts: the early work in Berkeley. Photosynth Res. 1995;46(1-2):17–26.

[81] Finazzi G, Rappaport F, Furia A, Fleischmann M, Rochaix JD, Zito F, Forti G. Involvement of state transitions in the switch between linear and cyclic electron flow in Chlamydomonasreinhardtii. EMBO Rep. 2002;3(3):280–285.

[82] Pfannschmidt T, Nilsson A, Allen JF. Photosynthetic control of chloroplast gene expression. Nature. 1999;39:625-628.

[83] Ort DR, Baker NR. A photoprotective role for $O(2)$ as an alternative electron sink in photosynthesis? Curr Opin Plant Biol. 2002;5(3):193-198.

[84] Ivanov AG, Sane PV, Hurry V, Oquist G, Huner NP. Photosystem II reaction centre quenching: mechanisms and physiological role. Photosynth Res. 2008;98(1-3):565-574.

[85] Quiles MJ. Stimulation of chlororespiration by heat and high light intensity in oat plants. Plant Cell Environ. 2006;29(8):1463-1470.

[86] Scheibe R, Dietz KJ. Reduction-oxidation network for flexible adjustment of cellular metabolism in photoautotrophic cells. Plant Cell Environ. 2012;35(2):202-216.

[87] Scheibe R. Malate valves to balance cellular energy supply. Physiol Plant. 2004;120(1):21-26.

[88] Kramer DM, Evans JR. The importance of energy balance in improving photosynthetic productivity. Plant Physiol. 2011;155(1):70-78.

[89] Steiger HM, Beck E. Formation of hydrogen peroxide and oxygen dependence of photosynthetic CO2 assimilation by intact chloroplasts. Plant Cell Physiol. 1981;22: 561-576.

[90] Scheibe R, Stitt M. Comparison of NADP-malate dehydrogenase activation, QA reduction and O_2 evolution in spinach leaves. Plant Physiol Biochem. 1988;26:473–481.

[91] Joliot P, Joliot A. Cyclic electron transfer in plant leaf. Proc Natl Acad Sci USA. 2002;99(15):10209-10214.

[92] Hemschemeier A, Happe T. Alternative photosynthetic electron transport pathways during anaerobiosis in the green alga *Chlamydomonas reinhardtii*. Biochim Biophys Acta. 2011;1807(8):919-926.

[93] Sétif P, Fischer N, Lagoutte B, Bottin H, Rochaix JD. The ferredoxin docking site of photosystem I. Biochim Biophys Acta. 2002;1555(1-3):204-209.

[94] Zhang L, Happe T, Melis A. Biochemical and morphological characterization of sulfur-deprived and H2-producing Chlamydomonas reinhardtii (green alga). Planta. 2002;214(4):552-561.

[95] Tolleter D, Ghysels B, Alric J, Petroutsos D, Tolstygina I, Krawietz D, Happe T, Auroy P, Adriano JM, Beyly A, Cuiné S, Plet J, Reiter IM, Genty B, Cournac L, Hippler M, Peltier G. Control of hydrogen photoproduction by the proton gradient generated by cyclic electronflow in *Chlamydomonas reinhardtii*. Plant Cell. 2011;23(7):2619-2630.

[96] Munekage Y, Hojo M, Meurer J, Endo T, Tasaka M, Shikanai T. PGR5 is involved in cyclic electron flow around photosystem I and is essential for photoprotection in Arabidopsis. Cell. 2002;110(3):361–371.

[97] Burrows PA, Sazanov LA, Svab Z, Maliga P, Nixon PJ. Identification of a functional respiratory complex in chloroplasts through analysis of tobacco mutants containing disrupted plastid ndh genes. EMBO J. 1998;17(4):868–876.

[98] Sazanov LA, Burrows PA, Nixon PJ. The chloroplast Ndh complex mediates the dark reduction of the plastoquinone pool in response to heat stress in tobacco leaves. FEBS Lett. 1998;429(1):115–118.

[99] Shikanai T, Endo T, Hashimoto T, Yamada Y, Asada K, Yokota A. Directed disruption of the tobacco ndhB gene impairs cyclic electron flow around photosystem I. Proc Natl Acad Sci USA. 1998;95(16):9705–9709.

[100] Hanke GT, Hase T. Variable photosynthetic roles of two leaf-type ferredoxins in arabidopsis, as revealed by RNA interference. Photochem Photobiol. 2008;84(6): 1302-1309.

[101] Hanke GT, Okutani S, Satomi Y, Takao T, Suzuki A, Hase T. Multiple iso-proteins of FNR in *Arabidopsis*: evidence for different contributions to chloroplast function and nitrogen assimilation. Plant Cell Environ. 2005;28(9):1146–1157.

[102] Okutani S, Hanke GT, Satomi Y, Takao T, Kurisu G, Suzuki A, Hase T. Three maize leaf ferredoxin:NADPH oxidoreductases vary in subchloroplast location, expression, and interaction with ferredoxin. Plant Physiol. 2005;139(3):1451-1459.

[103] Rodriguez RE, Lodeyro A, Poli HO, Zurbriggen M, Peisker M, Palatnik JF, Tognetti VB, Tschiersch H, Hajirezaei MR, Valle EM, Carrillo N. Transgenic tobacco plants overexpressing chloroplastic ferredoxin-NADP(H) reductase display normal rates of photosynthesis and increased tolerance to oxidative stress. Plant Physiol. 2007;143(2): 639-649.

[104] Hanke GT, Endo T, Satoh F, Hase T. Altered photosynthetic electron channelling into cyclic electron flow and nitrite assimilation in a mutant of ferredoxin:NADP(H) reductase. Plant Cell Environ. 2008;31(7):1017-1028.

[105] Forti G, Bracale M. Ferredoxin-ferredoxin NADP reductase interaction. FEBS Lett. 1984;166(1):81–84.

[106] Nakatani S, Shin M. The reconstituted NADP photoreducing system by rebinding of the large form of ferredoxin-NADP reductase to depleted thylakoid membranes. Arch Biochem Biophys. 1991;291(2):390–394.

[107] Clark RD, Hawkesford MJ, Coughlan SJ, Bennett J, Hind J. Association of ferredoxin-NADP+ oxidoreductase with the chloroplast cytochrome b-f complex. FEBS Lett. 1984;174(1):137–142.

[108] Zhang H, Whitelegge JP, Cramer WA. Ferredoxin: NADP+ oxidoreductase is a subunit of the chloroplast cytochrome b6f complex. J Biol Chem. 2001;276(41):38159–38165.

[109] Andersen B, Scheller HV, Moller BL. The PSI E subunit of photosystem I binds ferredoxin:NADP+ oxidoreductase. FEBS Lett. 1992;311(2):169–173.

[110] Quiles MJ, Cuello J. Association of ferredoxin-NADP oxidoreductase with the chloroplastic pyridine nucleotide dehydrogenase complex in barley leaves. Plant Physiol. 1998;117(1):235–244.

[111] Twachtmann M, Altmann B, Muraki N, Voss I, Okutani S, Kurisu G, Hase T, Hanke GT. N-Terminal Structure of Maize Ferredoxin:NADP+ Reductase Determines Recruitment into Different Thylakoid Membrane Complexes. Plant Cell. 2012;24(7): 2979-2991.

[112] Benz JP, Stengel A, Lintala M, Lee YH, Weber A, Philippar K, Gugel IL, Kaieda S, Ikegami T, Mulo P, Soll J, Bölter B. ArabidopsisTic62 and ferredoxin-NADP(H) oxidoreductase form light-regulated complexes that are integrated into the chloroplast redox poise. Plant Cell. 2009;21(12):3965–3983.

[113] Jurić S, Hazler-Pilepić K, Tomašić A, Lepeduš H, Jeličić B, Puthiyaveetil S, Bionda T, Vojta L, Allen JF, Schleiff E, Fulgosi H. Tethering of ferredoxin:NADP+ oxidoreduc-

tase to thylakoid membranes is mediated by novel chloroplast protein TROL. Plant J. 2009;60(5):783–794.

[114] Vojta L, Horvat L, Fulgosi H. Balancing chloroplast redox status – regulation of FNR binding and release. Periodicum biologorum. 2012;114(1):25–31.

[115] Stengel A, Benz P, Balsera M, Soll J, Bölter B. TIC62 redox-regulated translocon composition and dynamics. J Biol Chem. 2008;283(11):6656–6667.

[116] Mizusawa N, Wada H. The role of lipids in photosystem II. Biochim Biophys Acta. 2012;1817(1):194-208.

[117] Xu C, Härtel H, Wada H, Hagio M, Yu B, Eakin C, Benning C. The pgp1 mutant locus of Arabidopsis encodes a phosphatidylglycerophosphate synthase with impaired activity. Plant Physiol. 2002;129:594–604.

[118] Hagio M, Sakurai I, Sato S, Kato T, Tabata S, Wada H. Phosphatidylglycerol is essential for the development of thylakoid membranes in Arabidopsis thaliana. Plant Cell Physiol. 2002;43(12):1456-1464.

Chapter 2

The Path of Carbon in Photosynthesis - XXX - α-Mannosides

Arthur M. Nonomura and Andrew A. Benson

Additional information is available at the end of the chapter

1. Introduction

For decades, discoveries have been reported in the series, *The Path of Carbon in Photosynthesis*, including a chapter elucidating a competitive mechanism for binding and releasing sugars from lectins [1], and we present current research that further supports this mode of activity. From the inception of the series [2] and onward, the program has been based on interdisciplinary discourse resulting in achievements of the first order [3], legendary advances of the Path including publications describing the initial products of carbon fixation and with diagrammatic summarization [4]. As a result of the search for carbon fixation intermediates by feeding single carbon fragments (C_1) from ^{14}C-methanol to *Scenedesmus* and *Chlorella*, methanol was later applied to improve the growth rate of "Showa" [5]. Colonies of "Showa" were proven to accumulate *in vitro* concentrations of 30% to 40% botryococcenes, the highest in the field of hydrocarbon sources for renewable automobile and aviation fuels [6,7] and, as an adjunct to C_1-cultivation, foliar applications of 15 Molar C_1 formulated with fertilizers were developed [8] and independently verified [9,10]. Consistent with field observations, foliar C_1 inhibited glycolate formation [11]; and thereafter, the application of nuclear magnetic resonance to follow *in vivo* metabolism of methanol identified methyl-β-D-glucoside (MeG) [12]. As a consequence of our survey of substituted glycosides, it was shown that not only do glycosides improve productivity, but they also are transported in plants from root to shoot and from shoot to root [13-15]. Furthermore, formulations of polyalkylglycoside and mixed polyacylglycopyranose (MPG) were far more potent than MeG [1].

Having established a history of consistent responses to these substrates, we had often taken note of significant differences that were clearly distinguishable to the naked eye; and so we sought methods to photodocument the events by developing systems for cultivation in glass microbeads (μBeads). Thus, we present images of plants treated with indoxyl-β-D-glucoside

(IG) as compared to controls. Previously, significant responses of plants to IG had been reported [15,16]; therefore, in our applications of µBeads, the purpose of this section is to exhibit responses of representative plants without further statistical treatment. In addition to serving as solid support media, µBeads refract light to the improvement of photosynthetic efficiency. Not only does the boost to solar intensity from µBeads have the potential to improve productivity, when increased to saturation, it can have the opposite effect of inhibiting growth by photorespiration. Therefore, in consideration of the critical balance that must be achieved, we cultivated plants in µBeads with safeners, selecting appropriately structured substituted sugars.

α-Glycosides have higher binding affinities to lectins over β-glycosides, therefore, we undertook experiments comparing mixed α-and β-anomers to α-mannosides. Mannose polyacetates and methyl-α-D-mannoside were applied to plants because they are closely related to compounds for which we had established dosing. Additionally, responses to low concentrations of arylglucosides, such as IG, provided a starting range of dose requirements for an arylmannoside; and consistent with our hypothesis for specific affinities of lectins, we discovered the highest potencies with µM α-mannosides.

2. Materials and methods

Plants were cultured in research facilities according to previously described methods [1]. Consistency of response to treatments was achieved by supplementation with chelated Ca and Mn. Solutions for foliar applications included phytobland surfactants, but formulas for roots did not. Controls were placed in the same location and all plants were given identical irrigation, fertigation, and handling. Plants were matched to control populations, treated within a week of emergence of cotyledon and true leaves. After treatment, individual plants were scheduled for harvest and analysis. For biomass, plants were dried overnight and weighed. The performance of compounds was evaluated by comparing statistical means of individual dry weights of shoots and roots. All plants were regularly given modified Hoagland water-culture nutrients [17]. Foliar spray applications of identical volumes, either 100 or 186 liters/hectare (L/ha), were mechanically applied. Manual sprays were spray-to-drip volumes of approximately 800 L/ha. For all populations, means of different treatment groups were compared using two-tailed Student's t-test with p-values significant within 95% confidence intervals. Counts of populations are "n" values and standard error is denoted "±SE." Specialty chemicals from Sigma (St. Louis, MO, USA), included the following: tetramethyl-β-D-glucoside (TMG); tetraacetyl-D-glucopyranose (TAG); pentaacetyl-α-D-mannopyranose (MP); p-amino-phenyl-α-D-mannoside (APM); methyl-α-D-mannoside (MeM) and methylglucoside (MeG). MPG was synthesized with modification [1,18]; and 2,3,4,6-tetra-O-acetyl-D-mannopyranose mixed α- & β-anomers was from Toronto Research Chemicals, North York, ON, Canada. As required, MP and APM were dissolved in a lower aliphatic alcohol prior to dilution in aqueous media. Vascular plants included Canola Nexera 500, *Brassica napus* L., a shoot crop; radish 'Cherry Belle' *Raphanus sativus* L., a root crop; rice, *Oryza sativa* L., a cereal crop; corn TMF 114, *Zea mays* L. ssp. *Mays*; ryegrass, *Lolium multiflorum* Lam., cv Gulf; paperwhite, *Narcissus papyra-*

ceus Ker Gawl; ornamental coleus, *Solenostemon scutellaroides* (L.) Codd; and these species were maintained as previously described [13,16].

Rapid Radish Assay –Radiolabeled methylglucoside is transported into leaves within minutes [14], therefore, we developed a bioassay that could be run within a few days. Furthermore, in the course of surveying different species of plants, we observed germination of radish within 24 h. Therefore, we treated radish with substituted α-mannosides in water culture nutrients after emergence of hypocotyls as a means of testing our lectin cycle. Radish was sown on 25 - 30 cm Pyrex® dishes with lids or 150 X 15 mm polystyrene Petri dishes filled to depths of seeds with ½X modified Hoagland water-culture nutrients until emergence of hypocotyls. The nutrient solution served as the stock diluent and nutrient control. Overnight, approximately a fifth of the population germinated and those that had shed seed coats were selected for trials. Sprouts were matched for size of cotyledons and hypocotyl and were transferred to Nutrient Control solution or experimental α-mannosides. In glass dishes, experiments with MP were undertaken completely in water-culture solution. Experiments with APM in water-culture solution were undertaken by sowing sprouted radish seeds on filter paper moistened with treatment or Nutrient Control solutions. Assays were maintained under environmental conditions as follow: Photosynthetically active radiation 100 μEin m^{-2} s^{-1}, diel cycle of 16:8 h light:dark, 26:26° C.

Glass microbeads - Materials and methods for utilization of clear glass microbeads (μBeads) were previously described [16]. Briefly, μBeads were obtained with the following specifications: Nominal modal diameters 500 - 700 μm; density 2.5 g/cc; pH 9; and sodalime glass. Reference to the size of a μBead refers to its μm diameter. Reflected light intensity (I) was measured out-of-doors directly over bare sandy loam as compared to 1 cm layer of μBeads where solar I was in the range of 1700 to 1800 μEin m^{-2} sec^{-1}. For drainage, containers were perforated with holes smaller than the μBeads. Prior to sowing seeds, μBeads were saturated with pH 6 "nutribead" (modified Hoagland) solution [17] for drip fertigation (<1 L/h) or hourly misting. After >8 h uptake of treatments, fertigation resumed in a manner consistent with pH-control and cultures were regularly given equal volumes of nutribead solution. Controls were placed side-by-side and cultivated likewise. Basal plates of bulbs were immersed into moistened 700 μm μBeads to initiate rooting, after which they were treated. For photography, μBeads were saturated with water and individual plants were manually lifted out. When roots were dipped in a beaker of water, most of the μBeads dropped off. Representative plants were selected visually from among experimental populations for macrophotography. To avoid injury from dehydration, plants were photographed within a minute and returned to water.

3. Results

The investigations include summaries of previously described experiments with poly-substituted glycopyranoses formulated with nutrients [1]. Manual spray-to-drip foliar treatments were applied to even stands of 5 cm tall radish, as follow: Nutrient Control with 1 g/L surfactants; and 0.3 mM TMG and 1 mM TMG with 1 g/L surfactants. Foliar applications of 1 mM TMG to radish shoots resulted in a significant (n=36; ±SE 0.07; p=0.05) 27% enhancement of mean weights of roots over those of the Nutrient Control. The low concentration of 0.3 mM TMG showed no significant difference (n=36; ±SE 0.05; p=0.8) from Control in either growth of radish shoots or roots. The test was calibrated to deliver a volume typical for row crops, 186 L/ha, as follows: Nutrient Control as compared to 3 mM TMG, identically supplemented. The application of foliar 3 mM TMG resulted in a significant (n=18; p=0.03) increase over the Nutrient Control. TAG is similar to TMG except that it is substituted around the pyranose-ring with four acyls instead of alkyls. Foliar 10 mM TAG and Nutrient Control solutions were applied to shoots of radish and harvested a week later and results showed a significant (n=11, p=0.004) 27% increase of root mean dry weight as compared to Nutrient Control. The growth response of the roots of radish to foliar TAG, therefore, was similar to that of TMG. Tests were extended to various species of plants and, on Canola, responses to foliar applications of 3 mM MPG, 4 mM TAG and 309 mM MeG were compared. Results are graphically depicted in Figure 1. Three treated populations each showed significant (p=0.000) shoot wet weight increases over Nutrient Control, as follow: 3 mM MPG n=37, 18% increase; 4 mM TAG, n=35, 20% increase; and 309 mM MeG, n=36, 14% increase. Foliar 3 - 4 mM polyacetylglucopyranoses showed activity comparable to the higher dose of 309 mM MeG.

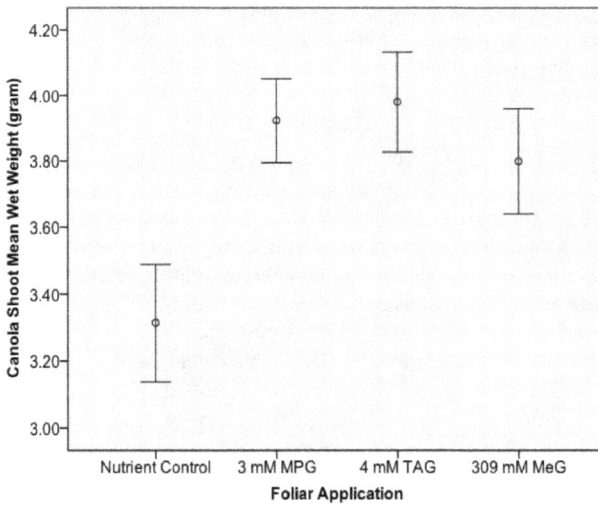

Figure 1. Foliar applications with low concentrations of polyacetylglucopyranoses, 3 mM MPG and 4 mM TAG, were comparable to treatments with high methylglucosides, 309 mM MeG, resulting in significant shoot enhancements over Nutrient Control. Error bars indicate ±SE.

Treatment of rice with the application of MPG to roots was compared to MeG. Roots exposed to formulations of 500 µM MPG and 50 mM MeG showed significant (n=27; p=0.000) increases in shoot yields of approximately 15% over controls. Roots of corn immersed in 1 mM MPG were compared to Nutrient Control, individual shoots were harvested after two weeks, and results showed significant (n=21; mean dry weight p=0.00; mean wet weight p=0.000) increases of 12% in vegetative yields of shoots over the population of Nutrient Control.

Rapid Radish Assay – Methods with radish enabled repetitious runs in the laboratory to determine the range of effective doses and to yield robust data. As mixed α- and β-anomers, TAM was compared to the α-anomer, MP. Within one day of exposure to 1 mM TAM or 100 μM MP, early greening of the cotyledon leaves was visually discernible from leaves of the Nutrient Control. After 48 h, sprouts treated to 1 mM TAM or to 100 μM MP showed advanced growth responses as compared to the nutrient Control. Application of 1 mM TAM in water-culture to radish sprouts resulted in statistically significant enhancement of mean dry weight (n=41; 8.8 mg) of whole plants over mean dry weight of the nutrient Control (n=41; 7.4 mg; p=0.002). At a lower dose, treatment with 100 μM TAM resulted in no significant difference of mean dry weight (n=41; 8.1 mg; p=0.11) from the nutrient Control. Application of MP to radish sprouts resulted in significant enhancement of mean dry weight (n=41; 8.2 mg) of whole plants over the mean dry weight of the nutrient Control (n=41; 7.4 mg; p=0.05). Therefore, the α-anomer showed higher potency than the mixed anomers and these effective doses of 100 μM MP and 1 mM TAM are compared to the nutrient Control in Figure 2.

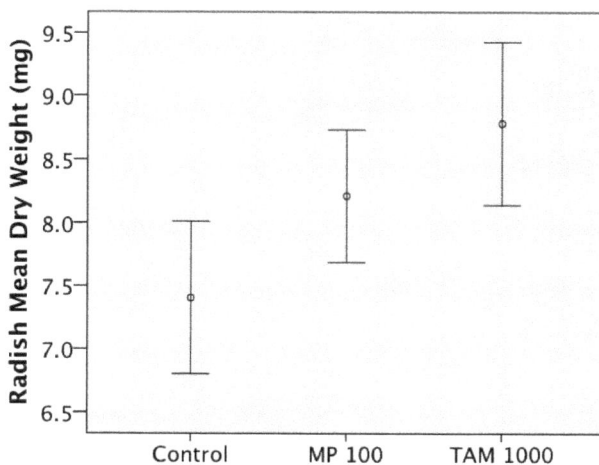

Figure 2. Treatment of radish by 100 μM pentaacetyl-α-D-mannpyranose (MP 100) and 1000 μM tetraacetyl-D-mannopyranose, mixed α/β-anomers (TAM 1000) with nutrients resulted in enhanced whole plant mean dry weight over that of the nutrient Control after 2 days. The α-mannose was more potent than the mixed α/β-anomers. Error bars indicate ±SE.

Owing to enhanced growth and deeper pigmentation in response to treatments with manno-sides, we sought a higher potency response, therefore, undertaking rapid radish assays with methyl-α-D-mannoside. Soon after first morning light, exposure of radish sprouts to 500 μM MeM resulted in notable greening of the cotyledon leaves within ~24 h. After 48 h, sprouted germlings treated to 25 μM to 500 μM MeM showed advanced growth responses as compared to Nutrient Control, roots and shoots showing robust enhancement of growth over the nutrient Control, as follow: Application of 500 μM MeM to radish sprouts resulted in statistically significant 11% enhancement of mean dry weight (n=10; 10.3 mg) of whole plants over nutrient Control mean dry weight (n=10; 7.9 mg; p=0.000). Treatment with 100 μM MeM resulted in a highly significant 17% enhancement of mean dry weight (n=15; 10.9 mg) over the mean dry weight of the Nutrient Control (n=35; 8.7 mg; p=0.003); 50 μM MeM resulted in significant 11% enhancement of mean dry weight (n=10; 11 mg) over mean dry weight of the Nutrient Control (n=10; 9.9 mg; p=0.03); and 25 μM MeM resulted in a significant 12% enhancement of mean dry weight (n=20; 10 mg) over mean dry weight of the Nutrient Control (n=35; 8.7 mg; p=0.03). Results of dosing radish roots with 25 μM and 100 μM MeM are graphically summarized in Figure 3.

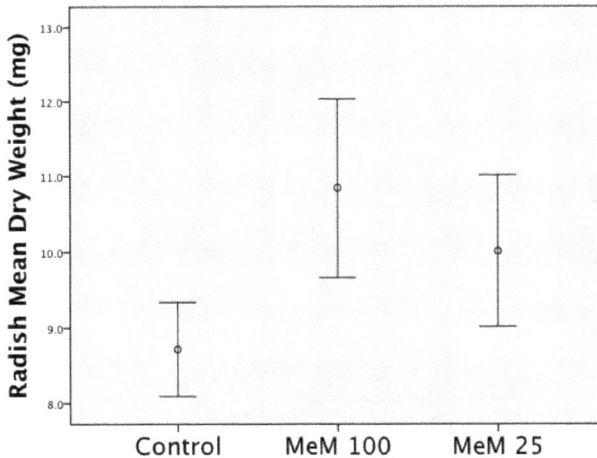

Figure 3. Immersion of radish sprouts in methyl-α-D-mannoside (MeM) resulted in significantly increased whole plant mean dry weights of approximately 17% and 12% over the population of the Nutrient Control in 48 h. Error bars indicate ±SE.

Figure 4. Treatment of radish with methyl-α-D-mannoside, right, showed enhanced pigmentation and general growth as compared to Nutrient Control, left. By harvest time on the 2nd day, expansion of treated cotyledon leaves and roots was clearly advanced over the Control. Scale bar = 1 cm.

Early on at 24 h, rapid responses were exemplified by visual comparisons of treated and control radish, shown in Figure 4. In one day, treatments with 500 µM MeM, right, showed deeper pigmentation, longer roots and larger expansion of cotyledon leaves as compared to the Nutrient Control, left.

Experience with IG now guided the next experiments to test another arylmannoside, p-amino-phenyl-α-D-mannoside (APM), for higher potency than the aforementioned 25 µM MeM. Immersion of radish sprouts in 100 µM APM resulted in a statistically significant 10% increase of mean dry weight (n=10; 11 mg) over the Nutrient Control (n=10; 9.9 mg; p=0.01). Hydroponic culture of radish sprouts on filter paper moistened with 10 µM APM in nutrient solution resulted in a significant 13% increase of mean dry weight (n=20; 10.3 mg) over Nutrient Control (n=40; 8.7 mg; p=0.01); but, the mean dry weight of 5 µM APM (n=20; 9.4 mg) was not significantly difference to that of the Nutrient Control (n=40; 8.7 mg; p=0.06). Results of the high potency response of radish roots to 10 µM APM are summarized in Figure 5.

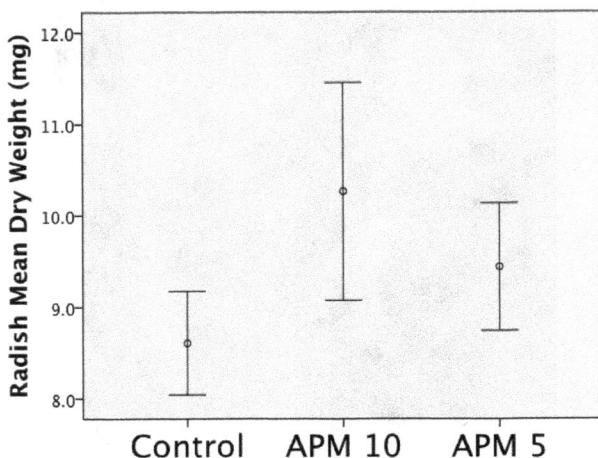

Figure 5. Treatment of radish sprouts in 10 µM *p*-amino-phenyl-α-D-mannoside (APM 10) enhanced whole plant mean dry weight over the nutrient Control after 2 days. The population treated with 5 µM APM showed no significant difference of mean dry weight as compared to the nutrient Control, but APM 10 showed the highest potency of the series, thus far. Error bars indicate ±SE.

Representative selections from the population treated with an arylmannoside are compared to a nutrient Control, exhibited in Figure 6. A radish germling treated with 10 µM APM, right, showed longer roots and larger expansion of cotyledon leaves as compared to the nutrient Control, left. Also, healthy root hairs are evident. In this experiment, we established the highest potency of the currently tested series of compounds and the growth responses that resulted may be attributable to the specific binding affinities of α-mannosides to lectins.

Glass Microbeads The various µBeads that we tested provided support for hydroponic culture of plants with erect plants anchored by their roots in µBeads and detailed results were previously reported [16]. Aeration appeared to be adequate in our container cultures and we found that the larger the µBeads, the longer the durations of pH-stability. For example, 700 µm µBeads maintained neutrality for a full day or longer, but 100 µm µBeads rose above pH 8 within a few hours. When starting seeds in 700 µm µBeads, maintenance of moisture at the surface is critical to germination because the top layer tends to drain completely of water, possibly leaving the seeds to desiccate. At harvest, roots were immersed in full beakers of water, whereupon, µBeads rolled off of the roots and dropped to the bottom of the beaker. Cuttings of coleus propagated in 500 µm µBeads with daily exchanges of nutribead solution showed roots, intact hairs and caps within two weeks, as displayed in Figure 7.

Figure 6. Within 2 days, treatment of radish sprouts by 10 μM *p*-amino-phenyl-α-D-mannoside, right, showed advanced growth as compared to Nutrient Control, left. Scale bar = 1 cm.

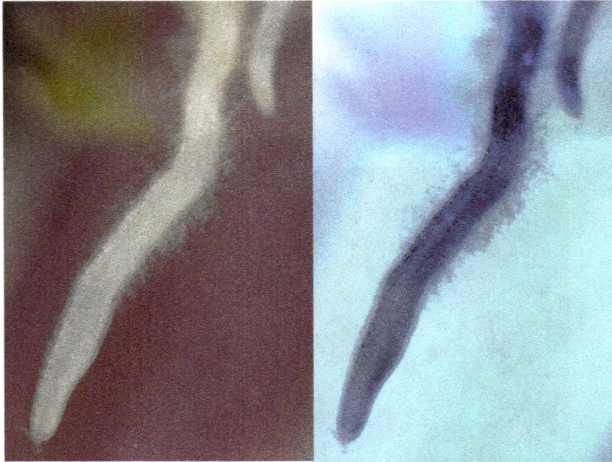

Figure 7. Root hairs of coleus after propagation in 500 µm µBeads are shown. A dip in water rolled µBeads off of roots, leaving the plant intact. The true color image, left, shows root hairs covering the top two-thirds of the white root; and the inverted color image, right, displays the root hairs in silhouette.

Paperwhite narcissus, was cultured in 700 µm µBeads in clear plastic 11 cm tall cylinders with <700 µm diameter perforations for drainage. Roots and shoots are exhibited in Figure 8.

Safety Handling µBeads must be performed according to protocols that include reviews of Material Safety Data Sheets prior to experimentation. If spilled, these glass spheres are slippery underfoot. Therefore, spills must be picked up immediately with a vacuum cleaner. Bearing in mind that glass is over twice as dense as water, when lifting a full sack or a 20 L bucket of µBeads, take precautions to preserve healthy backs by requesting assistance. For laboratory utilization, sterilize µBeads separately from liquids, preferably by heating the dry glass in 200° C ovens overnight. Allow several hours for both µBeads and sterile aqueous solutions to cool to room temperature. Moisten µBeads only after cooling to <40° C to prevent bumping. Eruptions of wet µBeads in an autoclave may damage valves, controls, glassware, and instrumentation. Avoid touching µBeads to mucous membranes and eyes. Wear eye protection. Don a dust mask to prevent inhalation of µBeads and glass dust.

Refractive Index In kilns, glass beads are melted to form clear glass spheres with highly polished surfaces. Each µBead is a micro-lens that refracts light. Moreover, diffuse reflection of light across the surface of a µBead may send a fraction of the light in all directions. Light may be directed according to the index of refraction of the glass from which µBeads are manufactured. For example, a µBead with a high index of refraction exhibits reflex reflectivity, sending light back toward its source. In contrast, a µBead with a lower index of refraction may send a beam at a right angle to the incoming ray. In Figure 9, theoretical paths of light through a µBead of high index of refraction, ~1.9, are compared to a µBead with a lower index of refraction. Coincidentally, familiarity with reflex reflectivity at night from µBead-coated road markers and signs had been the primary barrier to consideration of utilization of μBeads in the light of the day, but the application improves solar illumination of plants. To be true, the diagram of Figure 9 is portrayed in two dimensions, but the refractive illuminatory effects of dispersed layers of μBeads are three-dimensional (3D). Solar illumination is diffuse and, therefore, a contiguous layer of μBeads refracts spherically in all directions. This 3D charac-teristic may be observed by viewing the phenomenon through a polarizing filter by which millions of refractions from μBeads spread over a 1-m² concrete pad may be seen as an aura

Figure 8. Cultivation of paperwhite narcissus in μBeads show a representative nutrient control, left, with a crown of roots up to ~5 cm in length around the basal plate; and, by comparison, when treated with IG, right, with roots elongated ~7 cm.

of a halo. The refraction of sunlight is exhibited in Figure 10 in which an aura surrounds the 16 mm wide-angle lens of the handheld camera at the center, 15-30 cm above the dome of light. Out of doors, measurements of intensities directly over substrates at 2.5 cm distance were as follow: Above sandy loam, 270 to 300 μEin m^{-2} s^{-1} and over μBeads, 360 to 380 Ein m^{-2} s^{-1}; sunlight refracted upward from the ground at approximately 20% higher light intensity than sandy loam. The additional light intensity from surface refraction may induce midday wilting for plants placed under direct sunlight and cultivated in μBeads, but may be corrected by preparing plants with applications of glycosides.

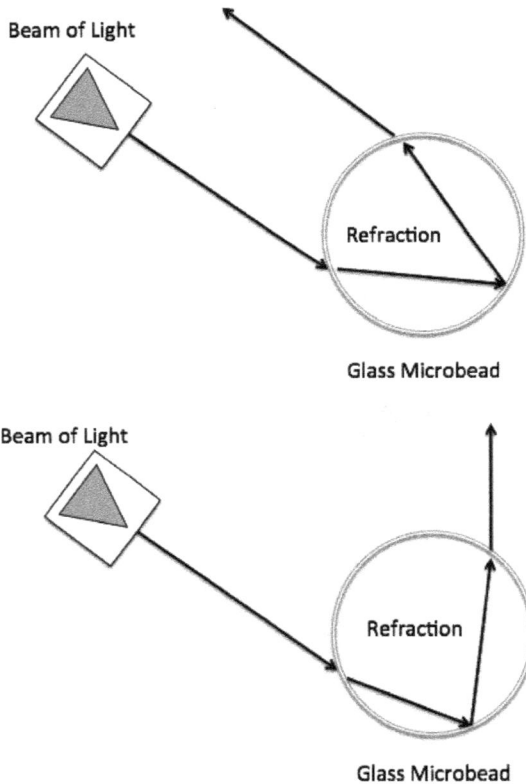

Figure 9. The index of refraction of μBeads determines the paths of beams of light. A μBead with a high index of refraction, approximately 1.9, sends light back in the general direction of its source, top, a phenomenon known as reflex reflectivity. A μBead with a lower index of refraction, approximately 1.5, may send light out at approximately a right angle to its approach, bottom. In each diagram, the symbol for a point source of light is a triangle in a box, labeled, "Beam of Light;" The circle labeled "Glass Microbead" represents a single μBead; and "Refraction" of a beam of light through the μBead follows the direction of the linear black arrows. Under environments with diffuse lighting, a μBead with a lower index of refraction may be a practical consideration.

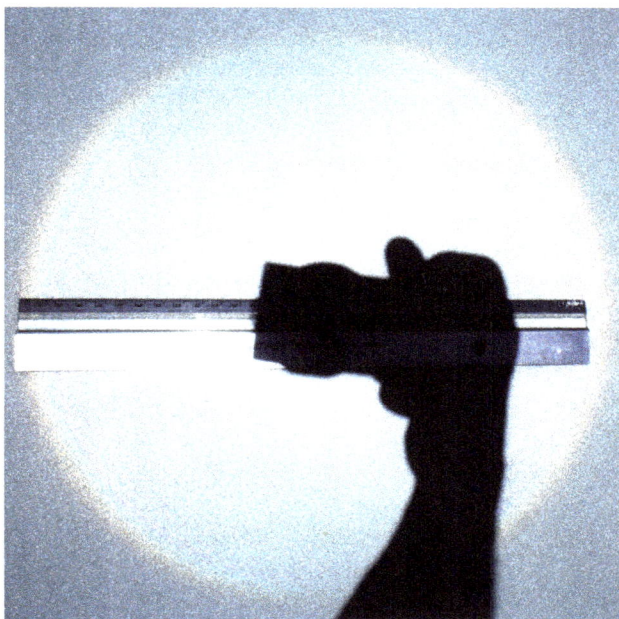

Figure 10. An aura above a layer of μBeads is shown through polarizing filters. The spectral halo, best described as a three-dimensional rainbow of colors, was the result of upward projections of light by refraction through millions of μBeads spread in a 2 – 3 mm layer over a flat 1 m² level concrete area. A 30 cm ruler spans the diameter of the circle of light and the black silhouette is of the author's camera and forearm. The hemisphere is brightest toward its center; moreover, all points of the 1 m² covered with μBeads were approximately 20% higher in PAR intensity than adjacent surfaces.

4. Discussion

Innovative glassware has been a hallmark of research in photosynthesis and the application of μBeads to refract light to the foliage emphasizes an integral role. At the start, we were faced with several problems; for example, the raw material source of μBeads, recycled sodalime glass, is alkaline. In practice, we found that the smaller the μBead, the larger the relative surface area from which to extract native alkalinity; and their value in daylight had not been considered previously. As pH-stability became an important consideration, it became evident that the largest μBeads would be the preferred media for green plants. Treatment of μBeads with nutribead solution overcame the alkalinity problem while providing a buffered environment for cultivation. Continuous fertigation is a means of stabilizing the medium; and, ideally, automated pH controllers may be implemented to efficiently meter flow rates in a manner that permits high density planting. Such is the case exhibited in Figure 11, showing five paperwhite narcissus plants in a small container with their bulbs nearly apressed. As well, dense cultivation is applicable to protistans as previously demonstrated on "Showa" [1] where frequent flow through of a pH-adjusted nutribead solution is matched by even drainage. Features of daylight enhancement are demonstrated in Figure 10, and because I was enhanced, application of µBeads to crops may entail broadcasting a thin 1-10 mm layer over the ground. As the index of refraction may be specified to direct light at different angles, µBeads of a lower index of refraction may be useful to start crops at subpolar latitudes during seasons for which the angle of solar illumination is low and bending light to a wider angle may distribute illumination advantageously. The application of µBeads in conjunction with glycoside formulations may be requisite to the continued growth of plants exposed to saturated-I, whether or not the overexposure is intentional. It is also important for this system of dual treatments with µBeads and glycoside formulations to maintain a soil at a pH that is amenable for growth. Clearly, for the cultivation of plants, µBeads may be of benefit significant enhancements of ambient light may be achieved by refraction through a multitude of glass spheres.

Figure 11. Five bulbs were planted in close proximity and the paperwhites blossomed while cultivated in 700 µm µBeads. Roots showed through µBeads in the bottom half of the container. Colors from fluorescent illumination contributed to the blue and red hues of the moist µBeads that filled the container.

Applications of polyalkylglucopyranose to shoots of radish resulted in significant root enhancements over controls and, conversely, applications of polyacylglucopyranoses to roots of corn resulted in significant increases of shoots as compared to controls. Similar to findings of our previous experiments with C_1 fragments and various glycosides [8,13,15], polyalkyl- and polyacylglycopyranoses required supplementation with nitrogen for significant improvements of growth. The production of ninhydrin-stained products may be from incorporation of nitrogen into protein, drawing attention to lectin as a protein complex from which stores of glucose could be displaced repeatedly by chemical competition with a glycoside. Lectin must be abundant and ubiquitous because, as we have found that Canola and corn respond to treatments of substituted glycosides, lectins occur in C_3 and C_4 plants. Moreover, not only do plant lectins bind β-glycoside, they bind preferentially to the α-anomer. As much as a quarter of the protein content of seeds and up to ten percent of the protein content of leaves may be attributable to lectins; however, even with such abundance, the provenance of vacuolar lectins was that they served no endogenous role in plants [19,20]. Notably, plant lectins have structural requirements for specific divalent cations to bind sugars [21] and we are currently confirming these requirements with subtractive formulations of corresponding plant nutrient in conjunction with applications of glycosides to plants. The results of our current investigations are consistent with the highly specific binding affinities of mannosides to lectins, the corresponding potencies indicative of their tendencies toward proportionally higher orders of binding to lectins than for glucosides.

A case in point, the lectin from *Canavalia ensiformis*, concanavalin A (con A), specifies α-trimannoside [23] and we have this core of complex glycans currently under examination. As the sequence of amino acids in this protein complex for recognition of mannosides is conserved in plants, the structural basis for specific recognition of mannosides in correspondence to the results of our experimental biology add compelling support of The Lectin Cycle. In conclusion, under conditions in which the cellular sugar concentration of a plant is diminished, chemical competition with substituted sugars may act to release other sugars from lectins—and this is an essential process to sustain viability. Consequently, binding affinities of lectins may service the natural displacements of sugars in periodic competitions, allowing energy to be reapportioned for growth resulting from metabolism of the freed sugars. For example, the concentration of methyl-β-D-glucoside (MeG) remains nearly constant in the plant [22] and as a result of photorespiratory depletion of the concentration of glucose competition for binding to lectin by MeG arises and glucose is released. Then, under conditions more conducive to photosynthesis, critical concentrations of glucose are re-established to sufficiently high levels that a high concentration of glucose outcompetes the substituted glucoside. MeG is released and glucose wins a storage site on the lectin. To an extent, the timely and direct provision of free glucose may mitigate the effects of any impoverishment, whether by photorespiration, heat, drought, darkness, or other forms of stress that consume glucose faster than it is replenished by photosynthesis. The Lectin Cycle, schematically represented in Figure 12, may repeat many times in a day, releasing sugar from lectins at each lengthy photorespiratory event, followed by sugar refresh from the Benson-Calvin Cycle upon resuming photosynthesis. Indeed, Nature's response to major environmental stimuli by means of chemical competition is well known. For example, photosynthesis turns to photorespiration as a result of oxygen outcompeting carbon dioxide for Rubisco. In a plant, the higher the quantity of lectins, the more capable it may be of storing and releasing sugars to endure prolonged depletion of glucose. This understanding led us to development of our rapid bioassay because we intended to exploit the high content of lectins in cotyledon leaves for the release of sugar. Additionally, when exogenous chemical competitors for binding sites on lectins are applied to plants, especially by the input of substrates that do not naturally occur in plants, such as p-amino-phenyl-α-D-mannoside, the duration of the effect may be substantially extended. Therefore, responses to treatments with glycosides must be carefully measured against the conformation of binding sites, biochemical structure, and their orders of preferences for prospective sugars. From another perspective, empirically formulated dosages of crops may possibly reflect the content and binding determinations of major lectins in a cultivar and our search in the future will be focused on the details of descriptions of the functions of substituted sugars in relation to defining suitability to lectins.

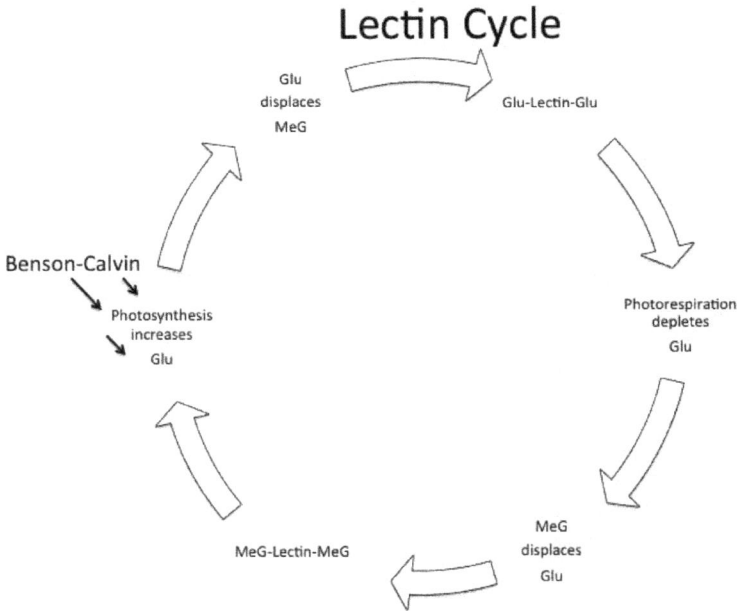

Figure 12. In The Lectin Cycle, various substrates displace glucose. The **Benson-Calvin** Cycle, left, contributes **Glu** (Glucose), of which, some is bound to lectins for storage, **Glu-Lectin-Glu**. Stress such as **Photorespiration depletes Glu**, right. Reduced concentrations of Glu create competition for lectin sites. For example, Methyl Glucoside (MeG) outcompetes Glu for binding sites as it is reduced to critically low concentrations. Thus, **MeG displaces Glu**, bottom right; and MeG binds to lectin, **MeG-Lectin-MeG**, bottom left. On return to photosynthesis, the **Benson-Calvin Cycle** once again contributes a sufficiency of glucose that raises the concentration of Glu to a competitive level and **Glu displaces MeG**, top left; thus, completing the Lectin Cycle.

5. Historical note

Steps toward management of the photosynthetic ecosystem were taken when coauthor Benson applied the first available ^{14}C to plants [24, 25] and, most certainly, one of the great joys of life is to have made such extraordinary contributions early in the atomic era. For a time, Benson held the entire concentrated supply of [14]CO_2 because these were the most rare of all materials. Only the eminently prepared and bravest knew how to handle manmade atomic particles and this required the creation of equipment that had never been known before. For example, when Benson designed the "lollipop" to feed algae [14]CO_2 with even illumination [1] he developed a method for the "atomic culture" to quickly drop into methanol to stop the reactions at every step of The Path. The keys to the success of this apparatus were attributable to (1) flattening the glass vessel, thus creating an efficient photobioreactor; and (2) enlarging the bore of stopcock, permitting drainage of the entire volume in a second. An original apparatus is exhibited in Figure 13, showing the flat round face and the thin side view resembling a "lollipop" from which the glassware was so appropriately named. For this demonstration, the historically significant flat panel was filled with *Haematococcus thermalis* Lemmerman. Hence, we celebrate this 70th year of Benson's originating concept and, felicitously, it is with best wishes that we also recognize his 50 years as a Member of the United States National Academy of Sciences [26] as well as his 78 years with the University of California; for, not only has the Path stood the test of time, Professor Emeritus Benson has, too!

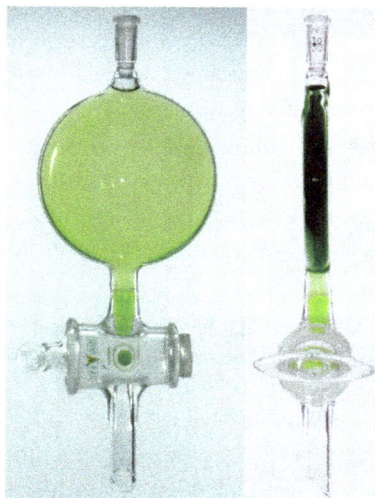

Figure 13. The "lollipop" is a laboratory apparatus for the purpose of cultivating algae to track the path of [14]CO_2. This first glass photobioreactor was designed by Andrew A. Benson and is exhibited to the left in face view, filled with *Haematococcus thermalis*. The large 4 mm bore of the stopcock is featured by fill with the green alga. Viewed from the side, right, the characteristic thin layer of the algal culture is revealed.

Author details

Arthur M. Nonomura and Andrew A. Benson

Scripps Institution of Oceanography, University of California San Diego, La Jolla, California, USA

References

[1] Nonomura, A., Cullen, B. A., Benson, A. A. 2012. The Path of Carbon in Photosynthesis. XXVIII. Responses of Plants to Polyalkylglycopyranose and Polyacylglycopyranose. In Advances in Photosynthesis - Fundamental Aspects, M. M. Najafpour (Ed.), ISBN: 978-953-307-928-8, Part 3, The Path of Carbon in Photosynthesis, Chapter 13, pages 259-271; InTech, Janeza Trdine 9, 51000 Rijeka, Croatia; URL http://www.intechopen.com/articles/show/title/the-path-of-carbon-in-photosynthesis-xxviii-responses-of-plants-to-polyalkylglycopyranose-and-poly

[2] Calvin, M. and Benson, A. A. 1948. The Path of Carbon in Photosynthesis, Science 107(2784): 476-480.

[3] NobelPrize. 2012. The Nobel Prize in Chemistry 1961. URL: http://nobelprize.org/nobel_prizes/chemistry/laureates/1961

[4] Bassham, J. A. 2003. Mapping the carbon reduction cycle: a personal retrospective. Photosynthesis Research 76: 35–52.

[5] Nonomura, A. M. 1988. Botryococcus braunii var. Showa (Chlorophyta), a new isolate from Berkeley, California, U.S.A. Japanese Journal of Phycology 36(4):285-291.

[6] Wolf, F. R., Nonomura, A. M., Bassham, J. A. 1985. Growth and branched hydrocarbon pro- duction in a strain of Botryococcus braunii (Chlorophyta). Journal of Phycology 21(3): 388-396.

[7] Eroglu, E., Okada, S., Melis, A. 2011. Hydrocarbon productivities in different Botryococcus strains: comparative methods in product quantification. Journal of Applied Phycology 23:763-775. DOI 10.1007/s10811-010-9577-8 URL: http://www.springerlink.com/content/41551k4246p77627/fulltext.pdf

[8] Nonomura, A. M., Benson, A. A. 1992. The Path of Carbon in Photosynthesis. XXIV. Improved crop yields with methanol. Proceedings of the National Academy of Sciences, USA 89(20): 9794-9798. URL: http://www.pnas.org/content/89/20/9794.long

[9] Devlin, R. M., Karczmarczyk, S. J., Lopes, P. R. 1995. The Effect of Methanol on the Growth and Pigment Content of Bachelor's-Button (Centaurea cyanus) and Geranium (Pelargonium hortorum). PGRSA Quarterly 23(2):127-136.

[10] Ligocka, A., Zbiec, I, Karczmarczyk, S., Podsialdo, C. 2003. Response of some culti-vated plants to methanol as compared to supplemental irrigation. Electronic Journal of Polish Agricultural Universities, Agronomy 6(1). URL: http://www.ejpau.media.pl

[11] Benson, A. A., Nonomura, A. M. 1992. The Path of Carbon in Photosynthesis: Metha-nol inhibition of glycolic acid accumulation, in Murata, N. (ed.), Research in Photo-synthesis 1, Proceedings of the IX International Congress on Photosynthesis, Kluver, Nagoya, Japan, P-522.

[12] Gout, E., Aubert, S., Bligny, R., Rebeille, F., Nonomura, A., Benson, A. A., Douce, R. 2000. Metabolism of methanol in plant cells. Carbon-13 nuclear magnetic resonance studies. Plant Physiology 123(1): 287-296.

[13] Benson, A. A., Nonomura, A. M., Gerard, V. A. 2009. The Path of Carbon in Photo-synthesis. XXV. Plant and Algal Growth Responses to Glycopyranosides, Journal of Plant Nutrition 32(7): 1185-1200. URL: http://www.tandfonline.com/doi/pdf/10.1080/01904160902943205

[14] Biel, K. Y., Nonomura, A. M., Benson, A. A., Nishio, J. N. 2010. The Path of Carbon in Photosynthesis. XXVI. Uptake and transport of methylglucopyranoside throughout plants. Journal of Plant Nutrition 33(6): 902–913. URL: http://www.tandfonline.com/doi/pdf/10.1080/01904161003669897

[15] Nonomura, A. M., Benson, A. A., Biel, K. Y. 2011. The Path of Carbon in Photosyn-thesis. XXVII. Sugar-conjugated plant growth regulators enhance general productivi-ty. Journal of Plant Nutrition 34:(5): 653—664. URL: http://www.tandfonline.com/doi/abs/10.1080/01904167.2011.540622

[16] Nonomura, A. M., Benson, A. A. 2012. The Path of Carbon in Photosynthesis. XXIX. Glass Microbeads. Journal of Plant Nutrition 35(12): 1896-1909. URL: http://www.tandfonline.com/doi/abs/10.1080/01904167.2012.706685

[17] Hoagland, D. R. and D. I. Arnon. 1950. The Water-Culture Method for Growing Plants without Soil. California Agricultural Experiment Station Circular 347, The Col-lege of Agriculture, University of California, Berkeley. 32 pp. URL: http://plant-bio.berkeley.edu/newpmb/faculty/arnon/Hoagland_Arnon_Solution.pdf

[18] Hyatt, J. A. & Tindall, G. W. 1993. The Intermediacy of Sulfate Esters in Sulfuric Acid Catalyzed Acetylation of Carbohydrates. Heterocycles 35(1): 227-234.

[19] Lannoo, N., Van Damme, Els J. M. 2010. Review: Nucelocytoplasmic plant lectins. Bi-ochimica et Biophysica Acta 1800: 190-201.

[20] Lannoo, N., Vandenborre, G., Miersch, O., Smagghe, G., Wasternack, C., Peumans, W. J., Van Damme, Els J. M. 2007. The Jasmonate-Induced Expression of the Nicoti-ana tabacum Leaf Lectin. Plant Cell Physiology 1–12. URL: http://www.pcp.oxford-journals.org

[21] Brewer, C. F., Sternlicht, H., Marcus, D. M., Grollman, A. P. 1973. Binding of 13C-En-riched α-Methyl-D-Glucopyranoside to Concanavalin A as Studied by Carbon Mag-

netic Resonance. Proceedings of the National Academy of Sciences, USA 70 (4): 1007-1011. URL: http://www.pnas.org/content/70/4/1007.full.pdf

[22] Aubert , S., Choler, P., Pratt, J., Douzet, R.; Gout, E., Bligny, R. 2004. Methyl-β-D-glu-copyranoside in higher plants: accumulation and intracellular localization in Geum montanum L. leaves and in model systems studied by 13C nuclear magnetic reso-nance. Journal of Experimental Botany 55(406): 2179–2189.

[23] Naismith, J. H., Field, R. A. 1996. Structural Basis of Trimannoside Recognition by Concanavalin A. The Journal of Biological Chemistry 271(2): 972–976.

[24] Andrew A. Benson 2002a. Paving the Path. Annual Review of Plant Biology 53: 1-25.

[25] Andrew A. Benson 2002b. Following the Path of Carbon in Photosynthesis: A person-al story. Photosynthesis Research 73: 29-49.

[26] Buchanan, B. B., Douce, R., Lichtenthaler, H. K. 2007. Andrew A. Benson. Photosyn-thesis Research 92:143-144.

Chapter 3

Physicochemical Properties of Chlorophylls in Oxygenic Photosynthesis — Succession of Co-Factors from Anoxygenic to Oxygenic Photosynthesis

Masami Kobayashi, Shinya Akutsu, Daiki Fujinuma,
Hayato Furukawa, Hirohisa Komatsu, Yuichi Hotota,
Yuki Kato, Yoshinori Kuroiwa, Tadashi Watanabe,
Mayumi Ohnishi-Kameyama, Hiroshi Ono,
Satoshi Ohkubo and Hideaki Miyashita

Additional information is available at the end of the chapter

1. Introduction

Chlorophylls are essential components in oxygenic photosynthesis, and chlorophyll (Chl) a (Fig. 1) is the major chlorophyll in cyanobacteria, algae and terrestrial plants. However, primary charge separation is initiated by a few specially-tailored chlorophylls in the reaction centers (RCs). Excitation of the primary donor reduces the primary and secondary electron acceptors, which again are often specially-tailored chlorophylls; $e.g.$, the 13^2-epimer of Chl a, Chl a' (Fig. 1), constitutes the primary electron donor of P700 in PS I as a heterodimer of Chl a/a' (Kobayashi et al. 1988; Jordan et al. 2001), and a metal-free Chl a, pheophytin (Phe) a (Fig. 1), functions as the primary electron acceptor in PS II (Klimov et al. 1977a,b; Zouni et al. 2001) (Fig. 2).

In 1996, a unique cyanobacterium, $Acaryochloris\ marina$, with Chl d as the dominant chlorophyll was discovered in colonial ascidians (Miyashita et al. 1996). Though P740 in PS I of $A.\ marina$ is composed of Chl d', pheophytin in PS II is not d-type but a-type (Akiyama et al. 2001). One of the difficulties in finding Chl d in nature was its overlap with Chl b on a reversed-phase HPLC trace. To make matters worse, Chl d had been thought to have a lower oxidation potential than Chl a even though no experimental evidence was available, mainly because a midpoint potential, E_m, for P740 in $A.\ marina$ was shown to be +335 mV (Hu et al. 1998), marvelously more negative than that for P700 (ca. +470 mV) in the Chl a-type cyanobacteria. The fact that the Q_Y-band of Chl d is at the longest wavelength compared with Chls a and b seems to have led to some misapprehensions concerning the oxidation potential of Chl d; one estimated that Chl d had the lowest oxidation potential among Chls a, b and d. In 2007, however, the E_{ox} value of Chl d *in vitro* was first determined and found to be higher than that of Chl a (Kobayashi et al. 2007), and hence the E_m of P740 was re-examined; the value was ca. + 435 mV (+430 mV: Benjamin et al. 2007, Telfer et al. 2007, and +439 mV: Tomo et al. 2008), being far positive of the initial report (+335 mV) and almost equal to the Chl a-type P700 values, around +470 mV (Brettel 1997, Krabben et al. 2000, Ke 2001, Itoh et al. 2001, Nakamura et al. 2004) (see Fig. 7 in Ohashi et al. 2008).

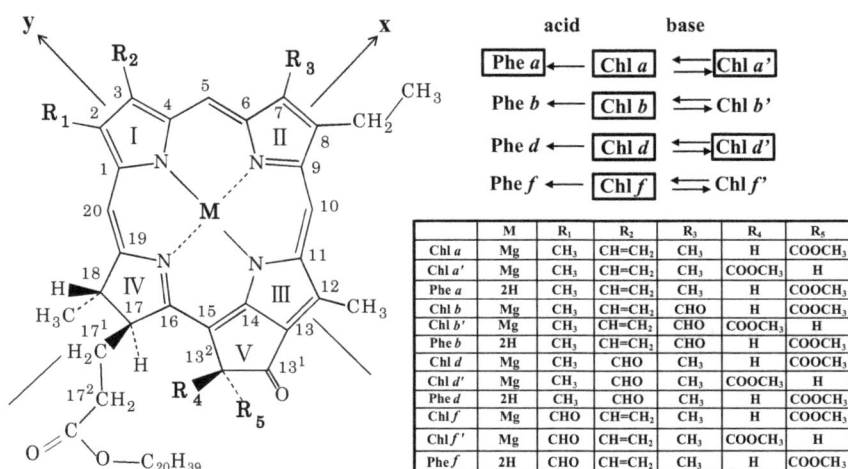

	M	R_1	R_2	R_3	R_4	R_5
Chl a	Mg	CH₃	CH=CH₂	CH₃	H	COOCH₃
Chl a'	Mg	CH₃	CH=CH₂	CH₃	COOCH₃	H
Phe a	2H	CH₃	CH=CH₂	CH₃	H	COOCH₃
Chl b	Mg	CH₃	CH=CH₂	CHO	H	COOCH₃
Chl b'	Mg	CH₃	CH=CH₂	CHO	COOCH₃	H
Phe b	2H	CH₃	CH=CH₂	CHO	H	COOCH₃
Chl d	Mg	CH₃	CHO	CH₃	H	COOCH₃
Chl d'	Mg	CH₃	CHO	CH₃	COOCH₃	H
Phe d	2H	CH₃	CHO	CH₃	H	COOCH₃
Chl f	Mg	CHO	CH=CH₂	CH₃	H	COOCH₃
Chl f'	Mg	CHO	CH=CH₂	CH₃	COOCH₃	H
Phe f	2H	CHO	CH=CH₂	CH₃	H	COOCH₃

Figure 1. Molecular structure and carbon numbering of chlorophylls, according to the IUPAC numbering system. Naturally occurring chlorophylls are designated by squares.

In 2010, a red-shifted chlorophyll was discovered in a methanolic extract of Shark Bay stromatolites, and was named Chl f (Chen et al. 2010); a Chl f-containing filamentous cyanobacterium was purified and named as *Halomicronema hongdechloris* (Chen et al. 2012). In 2011, a Chl f-like pigment was discovered in a unicellular cyanobacterium, strain KC1, isolated from Lake Biwa (Ohkubo et al. 2011), but there were also difficulties in its identification because there were not systematic physicochemical data as to Chls a, b, d and f acquired under the same conditions. For example, the optical peak wavelengths of the red-shifted chlorophyll purified from the strain KC1 in methanol are almost the same as the reported values of Chl f (λ = 406 nm and 706 nm), while the reported ratio of Soret/Q_Y-bands exhibited a large difference; the pigment purified from KC1 showed 0.9, but the reported ratio for Chl f was surprisingly high, 1.9. What is worse, the ^{1}H-NMR chemical shifts of formyl-H of Chl b and f in acetone d_6 at 263 K being presented in this chapter are 11.31 and 11.22, respectively, but the corresponding

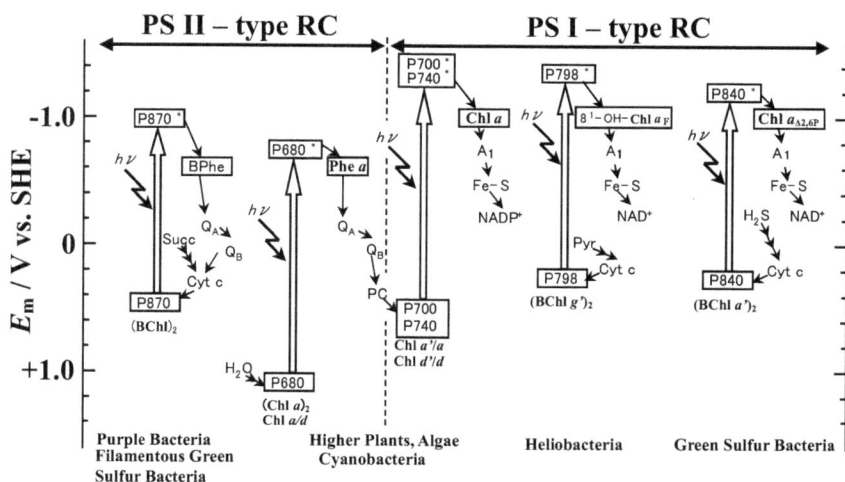

Figure 2. Schematic comparison of photosynthetic electron transport in PS I-type RC and PS II-type RC. Components are placed according to their estimated or approximate midpoint potentials. The arrows indicate the direction of electron flow. In order to simplify the figure, some primary electron donors, P970, P850 and P865 are omitted here: P970 and P850 are the primary electron donors of BChl b and Zn-BChl a containing purple bacteria, respectively; P865 is the primary electron donor of green filamentous bacteria. Figure adapted from Akutsu et al. (2011).

values reported were 11.22 for Chl b in CDCl$_3$ at unknown temperature and 11.35 for Chl f in CD$_2$Cl$_2$/d$_5$-pyridine(97/3, v/v) at 293 K, respectively, where our data are directly counter to the reported pair. These facts exhibit that chemical shift is highly sensitive to both solvents and temperature.

In this chapter, we present systematic and essential physicochemical properties of chlorophylls *in vitro* obtained under as common conditions as possible, introducing the detailed experimental procedures; HPLC, absorption, circular dichroism, mass, NMR and redox potential. We also introduce the succession of co-factors in PS I RCs from the viewpoints of minor but key chlorophylls (Fig. 3). In the future, our basic data will help researchers to identify the molecular structures of newly discovered chlorophylls and determine and/or predict their characteristics.

2. Chlorophylls in oxygenic photosynthesis

2.1. Popular chlorophylls: Chlorophylls *a* and *b*

In 1818, the term chlorophyll (Chl), the green (Greek *chloros*) of leaf (Greek *phyllon*), was introduced for the pigments extracted from leaves with organic solvents (Pelletier and Caventou 1818). In 1903, a Russian botanist Tsvet(Tswett) (in Russian meaning "colour") separated leaf pigments by chromatography (from Greek *chroma* and *graphein* meaning "color"

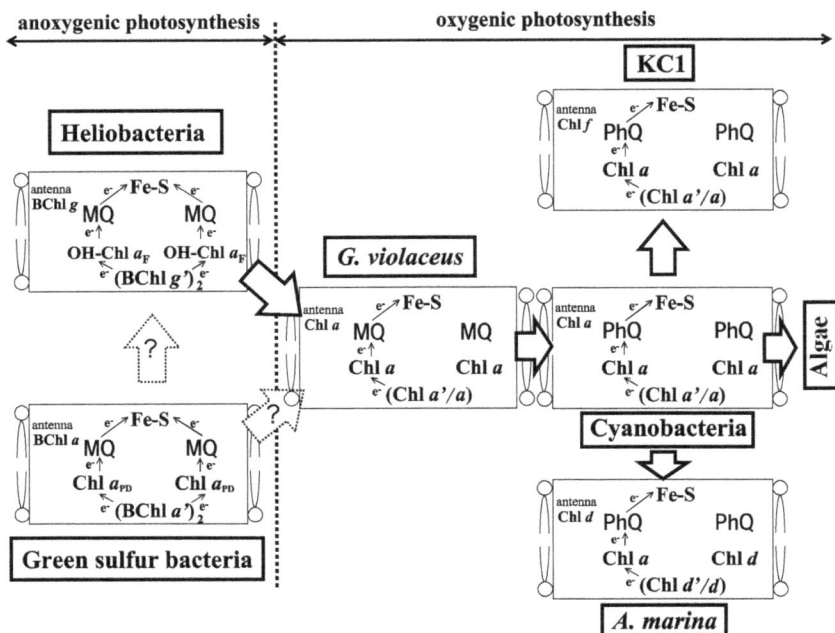

Figure 3. Schematic arrangement of co-factors in the PS I-type RCs. Our hypothesis about the evolution of the PS I-type RCs from the viewpoint of the molecular modifications of chlorophylls and quinones are designated by solid arrows.

and "to write", respectively) into the blue Chl a, the green Chl b and several yellow to orange carotenoids.

2.2. Recently discovered chlorophylls in algae

2.2.1. Chlorophyll d in a cyanobacterium, Acaryochloris marina

In 1943, Chl d (Fig. 1) was first reported as a minor pigment in several red macroalgae (Manning and Strain 1943). In 1996, a unique cyanobacterium, *Acaryochloris marina*, was isolated from colonial ascidians containing Chl d as the dominant chlorophyll (Miyashita et al. 1996; see Introduction in a review by Ohashi et al. 2008 for the short history of Chl d discovery).

2.2.2. Chlorophyll f in a cyanobacterium, strain KC1, isolated from Lake Biwa

A Chl f-producing cyanobacterium, strain KC1, was discovered by Miyashita, one of the authors of this chapter. The discovery story was a kind of fortunate accident, similar to that of Chl d-producing cyanobacterium, *A. marina* (Miyashita et al. 1996, Ohashi et al. 2008). Strain KC1 was a by-product during the hunting of chlorophyll d-producing cyanobacteria in freshwater environment. Chl d had been only produced by cyanobacteria in the genus *Acaryochloris*, and the strains in *Acaryochloris* had been isolated only from saline environments such as the marine or salty lake but not from freshwater environments at all (Murakami et al. 2004, Miller et al. 2005, Mohr et al. 2010, Behrendt et al. 2011). Moreover, the strain *A. marina* MBIC11017 dose not grow in freshwater media and requires sodium chloride for its growth at more than 1.5% (w/v) in the medium (Miyashita et al. 1997). Chlorophyll d was, however, detected in the sediment at the bottom of Lake Biwa, the largest freshwater lake in Japan (Kashiyama et al. 2008), bringing us the idea that Chl d-containing algae exist in this lake. We collected algal mats and lake water from a shore zone of Lake Biwa at 24th, Apr. 2008. The samples were suspended in several media for freshwater algae, diluted and dispensed into cell culture plate or on agar plates. Those culture/agar plates were kept in an incubator with near infrared (NIR) light as the sole light source, because we expected that a Chl d-containing alga should grow faster than other algae under such light condition. Colony of *Acaryochloris* cells grown under NIR light looked yellow-green rather than blue-green of common cyanobacteria, and we also saw several yellow-green colonies (Oct. 2008). Our attempt to isolate a Chl d-containing freshwater *Acaryochloris* sp. from Lake Biwa turned out to be a success (Feb. 2010) (details will be reported elsewhere). Miyashita checked the culture/agar plates, which were left for a long time in the incubator under NIR light, and some unusual cyanobacterial colonies were found (Oct. 2009). The colonies were different from that of *Acaryochloris* sp. in color; being dark-blue-green rather than yellow-green. The cells grew under NIR light as the sole light source when we put them in a freshwater medium. Morphological features of the cells were similar to those of *Acaryochloris* sp. in that the cell was unicellular, spherical to subspherical and aggregated. We expected that the organism was a new Chl d-containing cyanobacterium which was closely related to the genus *Acaryochloris*, phylogenetically. The results of HPLC analysis disappointed us, since the cyanobacterium possessed no Chl d at all, and Chl a as the major chlorophyll like common cyanobacteria. Immediately thereafter, however, came an excitement when an unusual chlorophyll was detected (22nd. Jan. 2010). The pigment showed typical two absorption peaks in the Soret (406 nm) and Q_Y (707 nm) regions in MeOH; they were clearly different from those of known chlorophylls. We concluded that the pigment was a new chlorophyll that should be named "Chl f". We started mass culture of the cells for chemical characterization such as detailed spectral properties, molecular mass and chemical structure. At around the same time, Chen et al. (2010) reported the discovery of Chl f in a methanolic extract of Shark Bay stromatolites incubated under NIR light for the initial purpose of the isolation of new Chl d-containing phototrophs. Discoveries are mostly accidental.

Note that in the strain KC1 Chl f was not induced under white fluorescent light even if NIR LED was also used as additional light. Further, neither Chl f' nor Phe f was detected, suggesting that Chl a' and Phe a function as P700 and the primary electron acceptor in PS II, respectively, as in common cyanobacteria (Akutsu et al. 2011). The results indicate that Chl f may function as not an electron transfer component but an antenna part. Chl f is not the major photopigment, and may function as an accessory chlorophyll, although the function of Chl f in energy storage is under debate, because uphill energy transfer is needed to deliver the excitation energy to Chl a molecules in the RC (Chen and Blankenship 2011).

2.3. Specially-tailored chlorophylls associated with reaction centers

2.3.1. Prime-type chlorophylls as the primary electron donors in PS I

2.3.1.1. Chlorophyll a' and P700

The 13^2-epimer of Chl a, Chl a' ("a-prime") (Fig. 1), was first reported in 1942 (Strain and Manning 1942). In 1988, it was proposed that Chl a' constitutes P700 as a heterodimer of Chl a/a' (Fig. 2) (Kobayashi et al. 1988), and the idea has been confirmed in 2001 (Jordan et al. 2001). As seen in Fig. 2, it has also been shown that P798 consists of BChl g' in the RC of heliobacteria in 1991 (Kobayashi et al. 1991), and that P840 consists of BChl a' in green sulfur bacteria in 1992 (Kobayashi et al. 1992, 2000), suggesting that prime-type chlorophylls are essential as the primary electron donors in the PS I-type RCs (see Figs. 2 and 3). For more details, see Chapter 4 in Kobayashi et al. (2006).

2.3.1.2. Chlorophyll d' and P740 in Acaryochloris marina

Chl d', the 13^2-epimer of Chl d (Fig. 1), was always detected in A. $marina$ as a minor component, while Chl a' was absent (see Fig. 6(C)) (Akiyama et al. 2001). P740, the primary electron donor of PS I in A. $marina$, was initially proposed to be a homodimer of Chl d (Hu et al. 1998), then a homodimer of Chl d' (Akiyama et al. 2001), and finally a Chl d/d' heterodimer (Fig. 2) (Akiyama et al. 2002, 2004; Kobayashi et al. 2005, 2007: Ohashi et al. 2008), just like the Chl a/a' for P700 in other cyanobacteria and higher plants (Figs. 2 and 3): a dimer model for P740 was supported by FTIR spectroscopy (Sivakumar et al. 2003). The finding of Chl d' in A. $marina$ appears to ensure our hypothesis that prime-type chlorophyll is a general feature of the primary electron donor in the PS I-type RCs (see Figs. 2 and 3). The homology of PsaA and PsaB between A. $marina$ and other cyanobacteria is low (Swingley et al. 2008), which may reflect the replacement of Chl a by Chl d, also Chl a' by Chl d', in the PS I RC of A. $marina$ (see Fig. 3).

It is interesting to note that the primary electron acceptor, A_0, in PS I of A. $marina$ is not Chl d but Chl a (Figs. 2 and 3), which was first shown by laser photolysis experiment (Kumazaki et al. 2002), and then supported by flash-induced spectral analysis (Itoh et al. 2007). The results support our hypothesis that Chl a-derivative is a general feature of A_0 in the PS I-type RCs (see Figs. 2 and 3).

2.3.1.3. Evolutionary relationship between chlorophylls and PS-I type reaction centers

Here we introduce our hypothesis about the evolution of the PS I-type RCs based on the structures of chlorophylls and quinones (Fig. 3). The prime-type chlorophylls, bacteriochlorophyll (BChl) a' in green sulfur bacteria, BChl g' in heliobacteria, Chl a' in Chl a-type PS I, and Chl d' in Chl d-type PS I, function as the special pairs, either as homodimers, (BChl a')$_2$ and (BChl g')$_2$ in anoxygenic organisms, or heterodimers, Chl a/a' and Chl d/d' in oxygenic photosynthesis. BChl g/g' may be a convincing ancestor of Chl a/a', because the BChl g/g' ⊠ Chl a/a' conversion takes place spontaneously under mild conditions *in vitro* (Kobayashi et al. 1998). Further, a Chl a/a' ⊠ Chl d/d' conversion also occurs with ease *in vitro* (Kobayashi et al. 2005: Koizumi et al. 2005), supporting the succession from the Chl a-type cyanobacteria to *A. marina* (Fig. 3). Chl f is produced in very small amounts in a Chl a-type special cyanobacterium, only when cultivated under NIR light, suggesting that Chl f appeared after the birth of Chl a. As mentioned above, the primary electron acceptors, A_0, are Chl a-derivatives even in anoxygenic PS I-type RCs. The secondary electron acceptors are naphthoquinones, and the side chains appear to have been modified after the birth of cyanobacteria, leading to succession from menaquinone to phylloquinone in PS I of oxygenic photosynthetic species (Ohashi et al. 2010).

In Fig. 4, BChl g/g' and their derivatives functioning in natural photosynthesis are illustrated. Chlorophyll a/a' are produced from BChl g/g' by isomerization, and Chl d/d' are then produced from Chl a/a' by oxidation. These three primed chlorophylls function as the primary electron donors of PS I-type RCs. Note that Chls b' and f' are not found in natural photosynthesis. Chlorophylls b and f are produced from Chl a by oxidation, but function as antenna pigments.

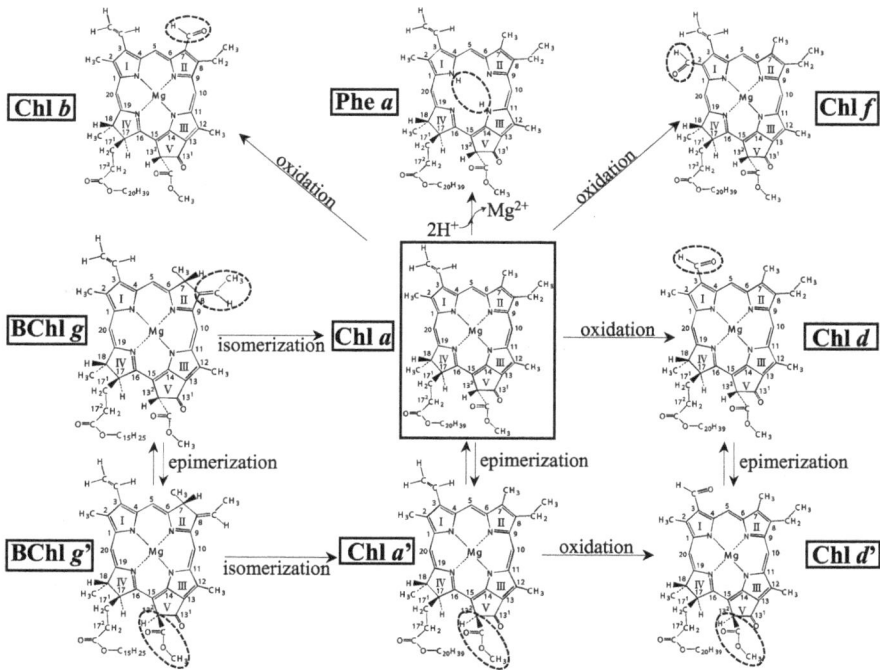

Figure 4. Bacterlochlorophyll g/g' and their derivatives functioning in natural photosynthesis.

2.3.2. Pheophytin a as the primary electron acceptor in PS II

In 1974, pheophytin (Phe) *a* (Fig. 1), a demetallated Chl *a*, was first postulated to be the primary electron acceptor in PS II (Fig. 2) (van Gorkom 1974), and the idea was experimentally confirmed in 1977 (Klimov et al. 1977b). In 1975, bacteriopheophytin (BPhe) *a* was found to function as a primary electron acceptor in the RC of purple bacteria (Fig. 2) (Parson et al. 1975; Rockley et al. 1975; Fajer et al. 1975; Kaufmann et al. 1975), and shortly thereafter BPhe *b* was also found to perform the same function (Fig. 2) (Klimov et al. 1977c). In 1986, BPhe *a* was also found to function in green filamentous bacteria (Fig. 2) (Kirmaier et al. 1986; Shuvalov et al. 1986). In 2001, the primary electron acceptor of PS II in *A. marina* was first found to be Phe *a* (Fig. 2) (Akiyama et al. 2001) and later supported by Tomo et al. (2007). For more details, see Chapter 4 in Kobayashi et al. (2006) and a review by Ohashi et al. (2008).

It is of interest to note that Phe *a* as well as Chl *a'*, *d'* and Chl *d* are artifacts easily produced *in vitro* (see Fig. 4): Phe *a* can be readily produced from Chl *a* under acidic conditions, primed chlorophylls from non-primed ones by epimerization under basic conditions, and Chl *d* from Chl *a* under oxidative conditions. These artifacts, however, function as key components in natural photosynthesis, while Phes *b*, *d*, *f* and Chls *b'*, *f'* are not found in nature.

3. Physicochemical properties of chlorophylls *in vitro*

3.1. HPLC

In the late 1970s, the high performance liquid chromatography (HPLC) technique was applied to the separation of plant pigments. In many cases the reversed-phase HPLC was preferred (Eskins et al. 1977; Shoaf et al. 1978; Schoch et al. 1978), and is still the main option to date. In that system, however, an eluent gradient is usually required for simultaneous separation of Chls and Phes and the gradient system is unfavorable for quantitative analysis, since the molar absorptivities of pigments strongly depend on solvents. In this context, an isocratic eluent system is favorable. In 1978, a simultaneous separation of Chls and Phes by normal-phase HPLC was attained by an isocratic procedure (Iriyama et al. 1978). In 1984, the isocratic normal-phase HPLC was established as a powerful tool for chlorophyll analysis (Watanabe et al. 1984).

3.1.1. Mixture of chlorophylls and pheophytins

Chls *a* and *b* were extracted with acetone/methanol (7/3, *v/v*) mixture at 277 K from parsley (*Petroselinum crispum* Nym.), Chl *d* from *A. marina*, and Chl *f* from a cyanobacterium strain KC1 grown under near-infrared LED light. The extract was applied to a preparative-scale HPLC (Senshupak 5251-N, 250 mm x 20 mm i.d.) and eluted with hexane/2-propanol/methanol (100/2/0.4, *v/v*) at a flow rate of 7 mL min^{-1} at 277 K, as described elsewhere (Kobayashi et al. 1991). Other authentic pigments, Chl *a'*, Chl *f'*, Phe *a* and Phe *f*, were prepared by epimerization and pheophytinization of Chl *a* and Chl *f* as described elsewhere (Watanabe et al. 1984).

A mixture of Chls and Phes was injected into a silica HPLC column (YMC-pak SIL, 250 x 4.6 mm i.d.) cooled to 277 K in an ice-water bath. The pigments were eluted isocratically with degassed hexane/2-propanol/methanol (100/0.7/0.2, v/v) at a flow rate of 0.9 mL min^{-1}, and were monitored with a JASCO UV-970 detector (λ = 670 nm) and a JASCO Multiwavelength MD-915 detector (λ = 300 - 800 nm) in series.

As illustrated in Fig. 6(F), eight Chls and four Phes are clearly separated. One can easily see that *Synechocystis* sp. PCC6803 possesses Phe *a* and Chl *a'* as well as Chl *a*, and that *Chlorella vulgaris* has also Chl *b*.

3.1.2. Pigment extract from A. marina

Pigments were extracted from cell suspension (ca. 10 µL) by sonication in a ca. 300-fold volume of acetone/methanol (7/3, v/v) mixture for 2 min in the dark at room temperature. The extract was filtered and dried *in vacuo*. The whole procedure was completed within 5 min. The solid material thus obtained was immediately dissolved in 10 µL of chloroform, and injected into a silica HPLC column.

As seen in Fig. 6(C), *A. marina* has three minor chlorophylls, Phe *a*, Chl *a* and Chl *d'*, in addition to the major Chl *d* (Akiyama et al. 2001). Pheophytin *a* functions as the primary electron acceptor in PS II, Chl *a* as the primary electron acceptor in PS I, and Chl *d'* as the primary electron donor P740 as a heterodimer of Chl *d/d'*, like the Chl *a/a'* in P700 (Fig. 2).

3.1.3. Pigment extract from strain KC 1

Cells of the cyanobacterium strain KC1 were grown in BG-11 medium in a glass cell culture flask (1 L) at 297 K with continuous air-bubbling. Cells were incubated under continuous white fluorescent light (50 µmol photons/m^2/s) or near-infrared LED light (see Fig. 5A, Tokyorika, Tokyo). Cells at the early stationary phase were harvested by centrifugation. See Akutsu et al. (2011) for more details.

Typical HPLC traces for acetone/methanol extracts from cells of the cyanobacterium strain KC1 cultivated under white fluorescent light or NIR LED light are shown in Figs. 6D and E, respectively. A large amount of Chl *a*, as well as small amounts of Chl *a'* and Phe *a*, were detected in both cells. We should note that only the strain KC1 grown under NIR LED light showed the presence of Chl *f* as a minor pigment and that Chl *f'* and Phe *f* were not detected at all (Fig. 6E).

3.2. Absorption spectra in four solvent varieties

The absorption spectrum is the simplest, most useful and extensively used analytical property to characterize chlorophylls. Absorption spectra of Chls show the electronic transitions along the x axis of the Chl running through the two nitrogen (N) atoms of rings II and IV, and along the y-axis through the N atoms of rings I and III (see Fig. 1). The two main absorption bands in the blue and red regions are called Soret and Q bands, respectively, and arise from $\pi \rightarrow \pi^*$ transitions of four frontier orbitals (Weiss 1978; Petke et al. 1979; Hanson 1991).

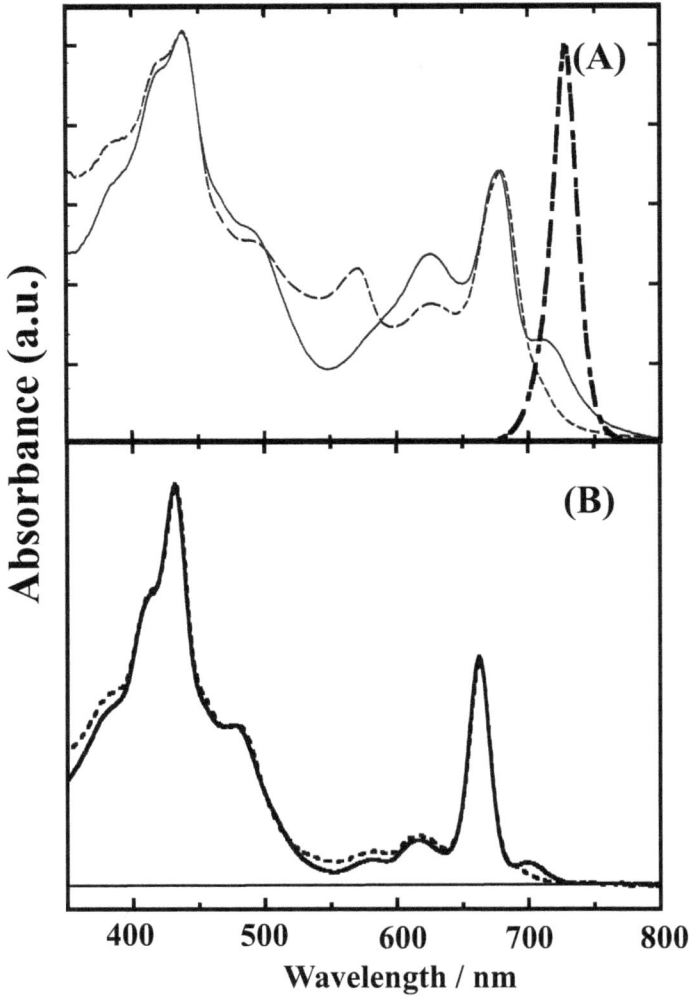

Figure 5. Absorption spectra of (A) the cells of strain KC1 grown under white fluorescent light (- - -), near infrared (NIR) LED light (—) and (B) acetone solution of acetone/methanol extracts from the corresponding KC1 cells. Emission spectrum of NIR LED (-•-•-) is inserted in (A).

Physicochemical Properties of Chlorophylls in Oxygenic Photosynthesis — Succession of Co-Factors
from Anoxygenic to Oxygenic Photosynthesis

57

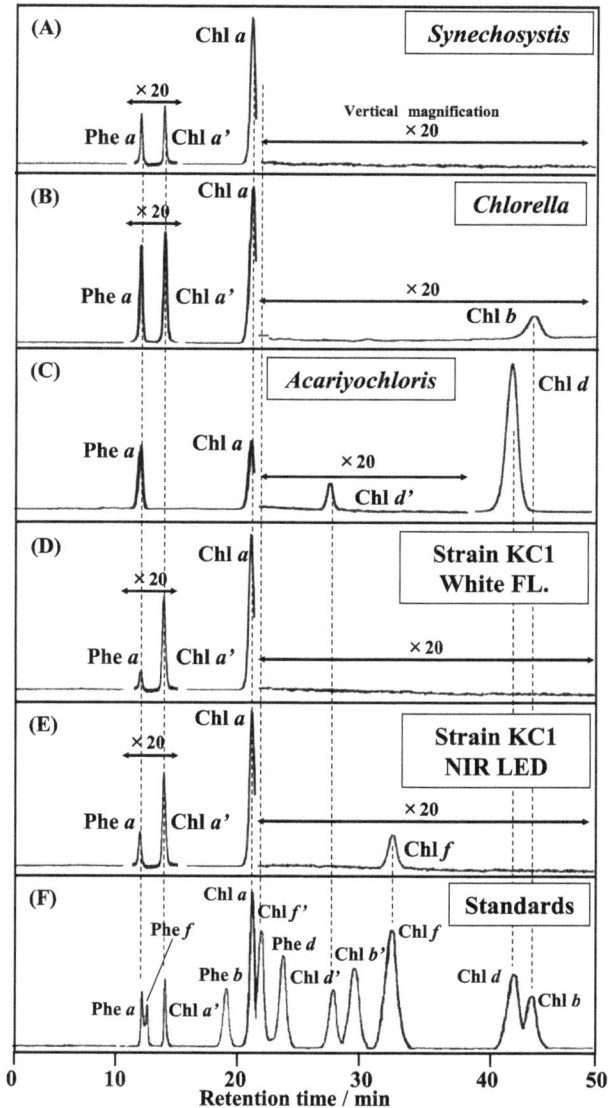

Figure 6. Normal-phase HPLC profiles for acetone/methanol extracts of (A) *Synechocystis* sp. PCC6803, (B) *Chlorella vulgaris*, (C) *A. marina*, (D) the cyanobacterium strain KC1 grown under white fluorescent light, (E) the cyanobacterium strain KC1 grown under near-infrared LED light, and (F) a mixture of Chls and Phes. Detection wavelength is 670 nm.

3.2.1. Chlorophylls a, b, d and f

Absorption spectra of Chls a, b, d and f in four kinds of solvent measured at room temperature are shown in Fig. 7 (top). As compared to Chl a, Chl b shows red-shifted Soret bands and blue-shifted weak Q_Y bands, while the Q_Y bands of Chls d and f are intensified and shifted to longer wavelengths. The Qx exhibits practically no intensity. The ratios of Soret/Q_Y band intensities show remarkable differences, $e.g.$, in diethyl ether ca. 1.3 in Chl a, ca. 2.8 in Chl b, ca. 0.85 in Chl d, and ca. 0.65 in Chl f (Table 1). Note that the Soret/Q_Y-band ratios in Chl b is more than 2 and those in Chls d and f are below 1 in all solvents in Fig. 7, while the ratios of Chl a in three solvents excluding methanol are around 1.3, but slightly below 1 in methanol. So one can easily distinguish Chl b from Chls a, d and f; also Chl a from Chls d and f, by their absorption spectra in any solvents used here.

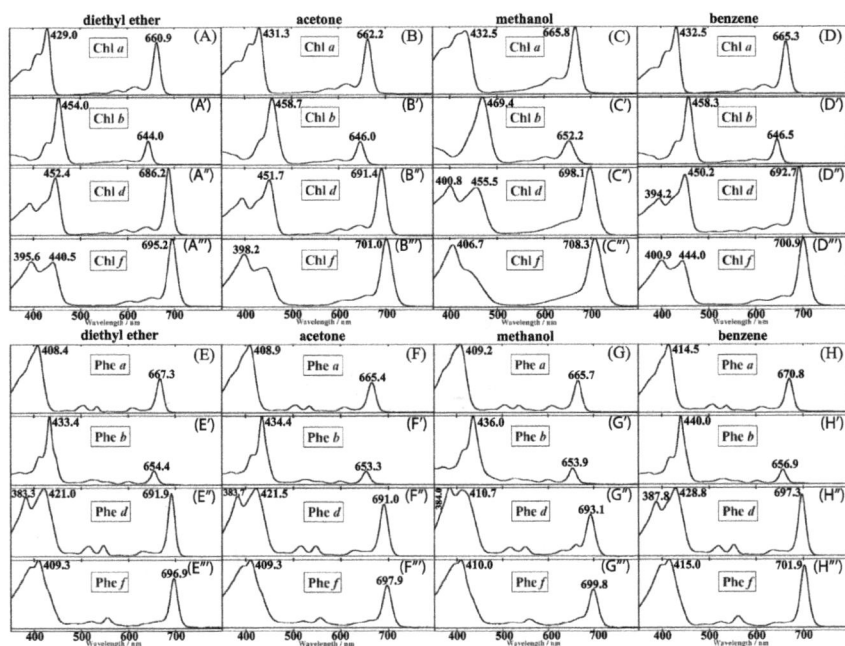

Figure 7. Comparison of the absorption spectra of Chls a, b, d, f (top), and Phes a, b, d, f (bottom) in diethyl ether, acetone, methanol and benzene at room temperature. Spectra were scaled to the Soret- or Q_Y-band maximum.

Compound	$\lambda_{max,blue}$ [nm]	$\lambda_{max,red}$ [nm]	
	$(\varepsilon[10^3\ M^{-1}cm^{-1}])$	$(\varepsilon[10^3\ M^{-1}cm^{-1}])$	Ref
Chl a	428.4	660.3	Watanabe et al. (1984)
	(115)	(89.8)	ibid.
	429.1[b]	661.6[b]	ibid.
	(100)[b]	(81.3)[b]	ibid.
	432.5[d]	665.4[d]	ibid.
	(101)[d]	(79.7)[d]	ibid.
	429.0	660.9	Kobayashi et al.(2006)
	(1.00)[a]	(0.775)[a]	ibid.
	431.3[b]	662.2[b]	This work
	(1.00)[a,b]	(0.828)[a,b]	ibid.
	432.5[c]	665.8[c]	ibid.
	(0.944)[a,c]	(1.00)[a,c]	ibid.
	432.5[d]	665.3[d]	ibid.
	(1.00)[a,d]	(0.785)[a,d]	ibid.
Phe a	408.4	667.9	Watanabe et al. (1984)
	(107)	(52.6)	ibid.
	409.2[b]	665.9[b]	ibid.
	(104)[b]	(46.0)[b]	ibid.
	414.8[d]	671.6[d]	ibid.
	(108)[d]	(53.1)[d]	ibid.
	408.4	667.3	Kobayashi et al.(2006)
	(1.00)[a]	(0.497)[a]	ibid.
	408.9[b]	665.4[b]	This work
	(1.00)[a,b]	(0.440)[a,b]	ibid.
	409.2[c]	665.7[c]	ibid.
	(1.00)[a,c]	(0.464)[a,c]	ibid.
	414.5[d]	670.8[d]	ibid.
	(1.00)[a,d]	(0.798)[a,d]	ibid.
Chl b	451.9	641.9	Watanabe et al. (1984)
	(159)	(56.7)	ibid.
	455.8[b]	644.6[b]	ibid.
	(136)[b]	(47.6)[b]	ibid.

Compound	$\lambda_{max,blue}$ [nm]	$\lambda_{max,red}$ [nm]	
	$(\varepsilon[10^3 \text{ M}^{-1}\text{cm}^{-1}])$	$(\varepsilon[10^3 \text{ M}^{-1}\text{cm}^{-1}])$	Ref
	457.9[d]	646.2[d]	ibid.
	(152)[d]	(56.2)[d]	ibid.
	452.4	642.5	Kobayashi et al.(2006)
	(1.00)[a]	(0.355)[a]	ibid.
	458.7[b]	646.0[b]	This work
	(1.00)[a,b]	(0.355)[a,b]	ibid.
	469.4[c]	652.2[c]	ibid.
	(1.00)[a,c]	(0.355)[a,c]	ibid.
	458.3[d]	646.5[d]	ibid.
	(1.00)[a,d]	(0.364)[a,d]	ibid.
Phe b	432.7	654.6	Watanabe et al. (1984)
	(172)	(34.8)	ibid.
	433.8[b]	653.3[b]	ibid.
	(153)[b]	(29.3)[b]	ibid.
	439.5[d]	656.7[d]	ibid.
	(152)[d]	(4.718)[d]	ibid.
	433.2	654.5	Kobayashi et al.(2006)
	(1.00)[a]	(0.202)[a]	ibid.
	434.4[b]	653.3[b]	This work
	(1.00)[a,b]	(0.195)[a,b]	ibid.
	436.0[c]	653.9[c]	ibid.
	(1.00)[a,c]	(0.245)[a,c]	ibid.
	440.0[d]	656.9[d]	ibid.
	(1.00)[a,d]	(0.216)[a,d]	ibid.
Chl d	447	688	Smith and Benitez (1955)
	(87.6)	(98.9)	ibid.
	447	688	French (1960)
	(87.6)	(98.5)	ibid.
	390, 445	686	Miyashita et al. (1997)
	392[b], 447[b]	688[b]	ibid.
	400[c], 455[c]	697[c]	ibid.
	445.6	686.2	Kobayashi et al.(2006)

Compound	$\lambda_{max,blue}$ [nm]	$\lambda_{max,red}$ [nm]	
	$(\varepsilon[10^3\ M^{-1}cm^{-1}])$	$(\varepsilon[10^3\ M^{-1}cm^{-1}])$	Ref
	$(0.853)^a$	$(1.00)^a$	ibid.
	$394.3^b, 451.7^b$	691.4^b	This work
	$(0.559)^{a,b}(0.826)^{a,b}$	$(1.00)^{a,b}$	ibid.
	$400.8^c, 455.5^c$	698.1^c	ibid.
	$(0.735)^{a,c}(0.706)^{a,c}$	$(1.00)^{a,c}$	ibid.
	$394.2^d, 450.2^d$	693.7^d	ibid.
	$(0.532)^{a,d}, (0.885)^{a,d}$	$(1.00)^{a,d}$	ibid.
Phe d	421	692	Smith and Benitez (1955)
	(84.9)	(72.2)	ibid.
	421	692	French (1960)
	(84.9)	(72.2)	ibid.
	382.7, 421.3	692.0	Kobayashi et al.(2006)
	$(0.881)^a,(1.00)^a$	$(0.911)^a$	ibid.
	$383.7^b, 421.5^b$	691.0^b	This work
	$(0.888)^{a,b}, (1.00)^{a,b}$	$(0.761)^{a,b}$	ibid.
	$384.0^c, 410.7^c$	693.1^c	ibid.
	$(1.00)^{a,c}, (0.964)^{a,c}$	$(0.637)^{a,c}$	ibid.
	$387.8^d, 428.8^d$	697.3^d	ibid.
	$(0.802)^{a,d}, (1.00)^{a,d}$	$(0.915)^{a,d}$	ibid.
Chl f	395.6, 440.5	695.2	This work
	$(0.657)^a (0.648)^a$	$(1.00)^a$	ibid.
	$398.2^b, 442.0^b$	701.0^b	ibid.
	$(0.780)^{a,c}, (0.576)^{a,b}$	$(1.00)^{a,b}$	ibid.
	$400.9^d, 444.0^d$	700.9^d	ibid.
	$(0.668)^{a,d}, (0.658)^{a,d}$	$(1.00)^{a,d}$	ibid.
	406.7^c	708.3^c	Akutsu et al. (2011)
	$(0.904)^{a,c}$	$(1.00)^{a,c}$	ibid.
	406	706	Chen et al. (2010)
	$(1.00)^{a,d}$	$(0.527)^{a,d}$	ibid.
Phe f	409.3	696.9	This work
	$(1.00)^a$	$(0.727)^a$	ibid.
	409.3^b	697.9^b	ibid.

Compound	$\lambda_{max,blue}$ [nm] ($\varepsilon[10^3\ M^{-1}cm^{-1}]$)	$\lambda_{max,red}$ [nm] ($\varepsilon[10^3\ M^{-1}cm^{-1}]$)	Ref
	(1.00)[a,b]	(0.610)[a,b]	ibid.
	410.0[c]	699.8[c]	ibid.
	(1.00)[a,c]	(0.561)[a,c]	ibid.
	415.0[d]	701.9[d]	ibid.
	(1.00)[a,d]	(0.776)[a,d]	ibid.

a: relative values

b: in acetone

c: in methanol

d: in benzene

Table 1. Absorption properties of chlorophylls in diethylether at room temperature

It is somewhat difficult to distinguish Chl f from Chl d, when one roughly compares the absorption spectrum of Chl f in diethyl ether (Fig. 7A''') with that of Chl d in methanol (Fig. 7C''), because their spectral shapes are very similar. In contrast, in diethyl ether one can easily distinguish Chl f from Chl d without spectrophotometer, because Chl f looks blue-green as Chl a, while Chl d light-green as Chl b, indicating that the naked eye is often powerful for colour judgement.

The Soret bands include several intense bands. In diethyl ether and benzene, the Soret band of Chl f is clearly split into two bands, most probably the so-called B-bands (longer wavelength) and η-bands (shorter wavelength), while Chl d shows such a split not in those solvents but in methanol, and hence one can easily distinguish them by comparing their optical spectra in the same solvents, *e.g.*, diethyl ether (Figs. 7A'' and A''').

In Fig. 5 are shown the absorption spectra of the strain KC1 grown under white fluorescent light and NIR LED light. The cells grown under NIR LED light show a clear shoulder over 700 nm, extending up to almost 800 nm (Fig. 5A). Absorption spectra in acetone solution of acetone/methanol extracts from the KC1 cells cultivated under NIR LED also exhibit a longer wavelength peak in the range of about 690 to 720 nm (Fig. 5B), due to the presence of Chl f. We had better pay attention that the NIR LED emission spectrum seen in Fig. 5A overlaps the absorption spectrum of the strain KC1 cells grown under white fluorescent light, indicating that the cells without Chl f can absorb NIR LED light, where some Chl a molecules possessing longer wavelength absorption may act as a trigger for Chl f biosynthesis under NIR LED light. If this hypothesis holds, a much longer wavelength LED could not induce Chl f biosynthesis. In such a study, one should give a lot of care to the shorter wavelength foot of emission spectrum of NIR LED not to overlap the absorption of cells at all.

We should note that inductive effects on the absorption wavelengths and intensities of Q_Y-bands of chlorophylls strongly depend on the nature and position of substituent(s) on the macrocycle, due to the presence of two different electronic transitions polarized in the x and y directions (the axes of transition moments are depicted in Fig.1) (Gouterman 1961, Gouterman et al. 1963; Weiss 1978; Petke et al. 1979; Hanson 1991, Kobayashi et al. 2006b). Replacement of the electron-donating group, $-CH_3$, on ring II of Chl a by the electron-withdrawing group, $-CHO$, yielding Chl b, causes the blue-shift and significant intensity reduction of the Q_Y-band (Fig. 7). In contrast, replacement of $-CH_3$ on ring I of Chl a by $-CHO$, yielding Chl f, causes the red-shift and intensity increase of the Q_Y-band (Fig. 7). A similar phenomenon is clearly seen in Chl d, where $-CH=CH_2$ on ring I of Chl a is replaced with $-CHO$. These results indicate it is a general feature that substitution by the electron-withdrawing group on ring II causes the blue-shift and intensity reduction of the Q_Y-band and that the same substitution on ring I leads the opposite, namely, the red-shift and intensity increase of the Q_Y-band. Moreover, it looks that substitution on ring I by the electron-withdrawing group generates the well-split Soret band, while showing heavy dependence on solvent as described above.

3.2.2. Pheophytins a, b, d and f

The free base related to Chl is called Phe. First of all, we emphasize that in natural photosynthesis only Phe a functions, and Phes b, d and f are not functional. In general, the more structured shape and red shifted Soret band of Chls distinguishes them from the corresponding Phes. In contrast to Chls, the η bands in the Soret band was poorly resolved in any Phes except Phe d (Fig. 7). Removal of the central Mg increases deviation from planarity and reduces the molecular symmetry, thus increasing Soret and Q_X transition. The Soret/Q_Y-band ratios noticeably increases by pheophytinization; in diethyl ether Phe b shows the highest value of around 5, Phe a the secondary highest about 2, and Phe d the lowest near 1 (compare bottom with top in Fig. 7, see also Table 1). Therefore, contamination of pheophytins in a Chl sample is often noticed from the optical spectra.

As seen in Fig. 7, Phe b can be easily distinguished from Phe a by its blue shifted Q_Y-band, red shifted Soret band, and by the marginally higher Soret/Q_Y band ratio. Phes d and f can be distinguished from Phes a and b by their red shifted Q_Y band and intense Q_Y bands, i.e., the Soret/Q_Y-band ratios in Phes d and f are not high and almost the same as those seen in Chl a. We should also pay attention to Phe f, because in methanol its optical shape is somewhat similar to Phe a (compare Figs. 7G with 7G'''), although they can be distinguished by the Q_Y wavelength difference. We must emphasize again that Phes possess relatively strong and characteristic Q_X-bands in the region of 490-570 nm; the Q_X bands in Phes a and d are better resolved to the $Q_X(0,0)$ and $Q_X(1,0)$ transitions. Phe d also shows significant splitting of the Soret and Q_X bands in all solvents as illustrated in Fig. 7. In contrast, the Q_X band corresponding to the $Q_X(1,0)$ transition (shorter wavelength) of Phe f looks unclear. It is of interest to note that in diethyl ether Phe d assumes a pale pink color, while both Phes b and f show a dull color. These characteristics will help us to discern Phes from Chls, and among Phes.

3.3. Circular dichroism spectra

Circular dichroism (CD) spectra are very useful for distinction between the primed chloro-phyll, e.g., Chl a', and the corresponding non-primed one, Chl a, although the absorption characteristics of the primed derivatives (Chls a', b', d', f', Phes a', b', d' and f') are identical with those of the non-primed ones (Wolf and Scheer 1973; Weiss 1978; Watanabe et al. 1984; Kobayashi et al. 2006b).

A spectropolarimeter Model FDCD-309 (JASCO) was used for CD measurements. Benzene was chosen as the solvent, in view of the sufficiently slow interconversion between epimeric species in this medium (Watanabe et al. 1984). The spectra were recorded from 800 nm to 300 nm at a scan rate of 200 nm/min with 20 scans at room temperature; time for measurement was ca. one hour.

The CD spectra of Chl a/a', b/b', d/d' and f/f' in benzene are illustrated in Fig. 8. It is immediately seen that, for a given pair of epimers, the CD spectra are considerably different, although the absorption spectra are practically identical with each other. For each of Chls a', b', d' and f', an intense negative CD is associated with $Q_Y(0,0)$ and a well-defined weakly negative satellite with $Q_Y(1,0)$. On the other hand, the non-primed species, Chls a, b, d and f, show complicated, very weak negative and/or positive activities at these transitions. In Fig. 9, all pheophytins show negative activities, and primed ones reveal stronger and red-shifted signals compared to the non-primed ones, although the absorption spectra of the primed derivatives are also identical with those of the non-primed ones. The findings suggest that the Q_Y maximum transition consists of at least two bands, and shorter wavelength band shows stronger activity in primed Phe and longer wavelength band stronger in non-primed Phe.

A series of Q_X transitions occur in the "valley" of the absorption spectrum as described in section 2. The positive CD activities derived from the $Q_X(0,0)$ absorption (called bands III, see Fig. 2 in Petke et al. 1979) appear at 579, 535, 594, 557, 548, 604, 567 and 559 nm for Chl a', Phe a', Chl d', Phes d, d', Chl f', Phes f and f', respectively, while Chl f and Phe a exhibit weakly negative activities at 609 nm and 535 nm, respectively. Such definite activities are not observed in Chls a, b/b', d and Phe b/b'; optical activities of b-type pigments, Chl b/b' and Phe b/b', are extremely weak, which may be related to the very diffuse feature of their absorption spectra. The CD activity associated with the $Q_X(0,1)$ absorption satellites (band IV) at the shorter wavelength also is very weak and vague in all the pigments examined.

The Chl and Phe Sorets contain many π-π^* transitions characterized by a complex mixture of configurations. According to the results of molecular orbital calculations (Weiss 1978; Petke et al. 1979; Hanson 1991), band B in the Soret absorption consists of two nearly degenerate electronic transitions, $B_X(0,0)$ and $B_Y(0,0)$. All the primed derivatives gave single and strongly positive CD spectra at this absorption peak, suggesting that the two transitions contribute to CD spectra in a similar manner (Watanabe et al. 1984). In contrast, the CD spectra of non-primed species, except Chl f and Phes, apparently reflect the existence of the two transitions: they show a maximum and a minimum with the center wavelength roughly coinciding with the Soret absorption maximum. Different feature of CD spectrum for Chl f among Chls may come from its characteristically splitted Soret absorption arising from B-bands and so-called

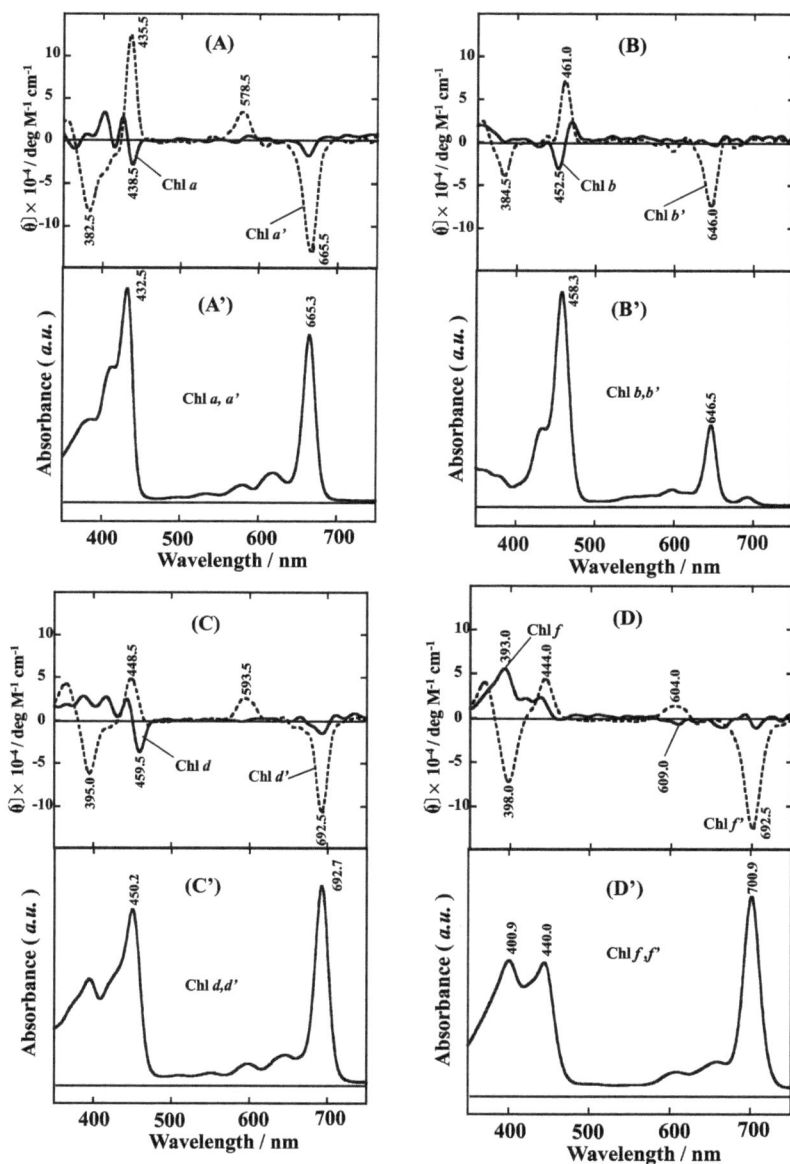

Figure 8. CD spectra for (A) Chls a/a', (B) Chls b/b', (C) Chls d/d', (D) Chls f/f', and (A')-(D') the corresponding absorption spectra in benzene at room temperature. [θ] denotes the molar ellipticity.

η-bands in benzene. Primed Chls exhibit relatively intense negative CD spectra in the near ultraviolet at η-bands of the Soret region (Welss 1978; Petke et al. 1979), whereas non-primed ones exhibit positive activities. Phes a and f also show positive CD spectra at η-bands in the near ultraviolet region, but such a tendency is not clear in Phe b, and Phe d exhibits negative activity (Fig. 9), although all the primed Phes show negative activity and Phe d' shows the most intense activity.

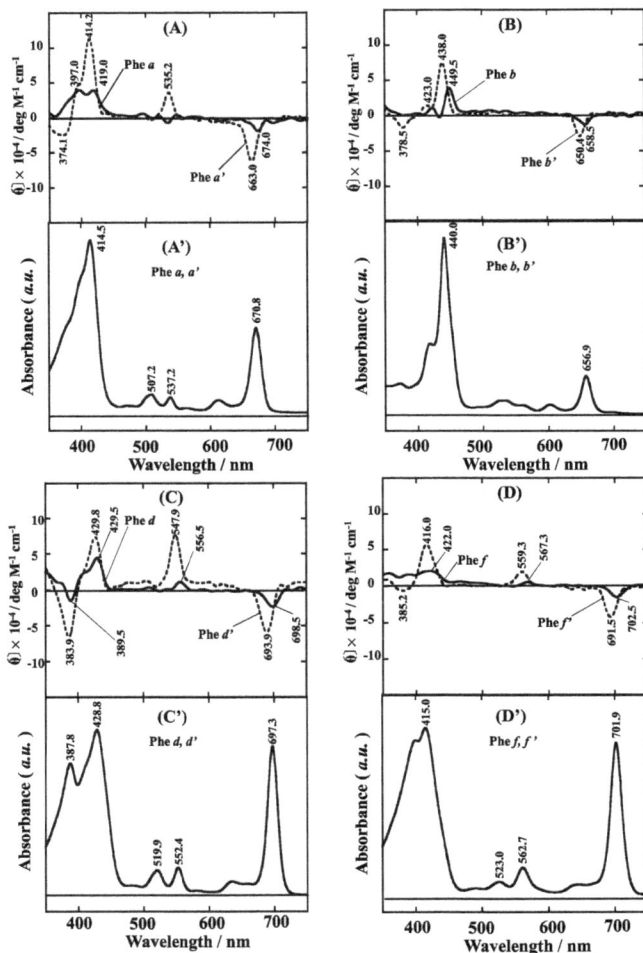

Figure 9. CD spectra for (A) Phes a/a', (B) Phes b/b', (C) Phes d/d', (D) Phes f/f', and (A')-(D') the corresponding absorption spectra in benzene at room temperature. [θ] denotes the molar ellipticity.

3.4. Mass spectra

Chlorophylls in natural photosynthesis are sometimes present in very small amounts, and hence the use of mass spectrometry (MS) can be advantageous since only minute samples are required. MS can provide accurate and useful information not only on molecular weights and elemental compositions but also on the nature of functional groups attached to the macrocycle (*e.g.* phytol) and of the central metal (see reviews by Smith 1975; Hunt and Michalski 1991; Porra and Scheer 2000; Kobayashi et al. 2006).

LC/MS experiments were performed on an LCQ mass spectrometer (Thermo Fisher Scientific Inc., MA, U.S.A.) equipped with an HPLC system (HP1100, Agilent, CA, U.S.A.) connected with a diode array detector. Each sample dissolved in dichloromethane before analysis was applied on a JASCO Finepak SIL C18S column (150 mm x 4.6 mm i.d.) cooled to 277 K in an ice-water bath, and separated using a mixture of ethanol/methanol/2-propanol/water (86/13/1/3, *v/v*) at a flow rate of 300 μL min^{-1}. The eluate was monitored by the UV-Vis absorption in a range of 220-800 nm, and was introduced into the mass spectrometer from 5 to 55 min after sample injection. Atmospheric pressure chemical ionization (APCI) mass and MS/MS spectra were recorded in the positive-ion mode in the mass range of m/z 150-2,000. Helium was used as collision gas for MS/MS experiments, followed by the isolation of ions over a selected mass window of 2 Da. The mass spectrometer was initially tuned using a standard Chl *a* solution as follows: APCI vaporizer temp., 723 K; spray voltage, 4 kV; capillary temperature, 423 K; capillary voltage, 8 V; sheath gas (nitrogen); flow rate, 56 (arbitrary unit); auxiliary gas flow rate, 9 (arbitrary unit).

As illustrated in Fig. 10 (left), Chl *a* ($C_{55}H_{72}MgN_4O_5$, monoisotopic mass; 892.535. Hereafter the value in the bracket shows the monoisotopic mass of the molecule or the ion) gives the protonated molecule ([M+H]$^+$) at m/z 893.2 producing the dominant fragment ion at m/z 615.1. The mass difference 278 between [M+H]$^+$ and the product ion corresponds to $C_{20}H_{38}$. This shows the presence of a phytyl chain in Chl *a*. The other product ions at m/z 583.0 and m/z 555.2 corresponding to [M+H-278-32]$^+$ and [M+H-278-60]$^+$, respectively, are supposed to be the results of the loss of carboxymethyl group followed by the cleavage of phytol. The losses of 278, 310 and 338 from the precursor ion in MS/MS spectra are seen in all the pigments examined here reveals the presence of a phytyl chain.

It is interesting to note that a typical fast atom bombardment (FAB)-mass spectrum of Chl *a* shows two intense molecular ion peaks, [M]$^+$ at m/z 892.6 and [M + H]$^+$ at 893.6, as well as the dominant fragment ion [M-$C_{20}H_{38}$+H]$^+$ at m/z 614.3 (see Fig. 4 in Kobayashi et al. 2000). Both spectra, however, reveal that Chl *a* has a phytyl chain ($C_{20}H_{39}$). We should make sure that the mass spectra of chlorophyll molecular ion peak(s) and the fragment ion peak(s) vary by m/z 1.0 according to the ionization methods.

Chlorophyll *d* ($C_{54}H_{70}MgN_4O_6$, 894.515) gives the protonated molecule ([M+H]$^+$) at m/z 895.2 and the prominent fragment ion at m/z 617.1. Though chlorophylls *b* and *f* (both $C_{55}H_{70}MgN_4O_6$, 906.515) are eluted at different LC retention times, they show the same mass and MS/MS spectral patterns, [M+H]$^+$ at m/z 907 and the dominant product ion at m/z 629, which correspond to [$C_{55}H_{71}MgN_4O_6$]$^+$ (907.522) and [M-$C_{20}H_{38}$+H]$^+$ (629.225). These results

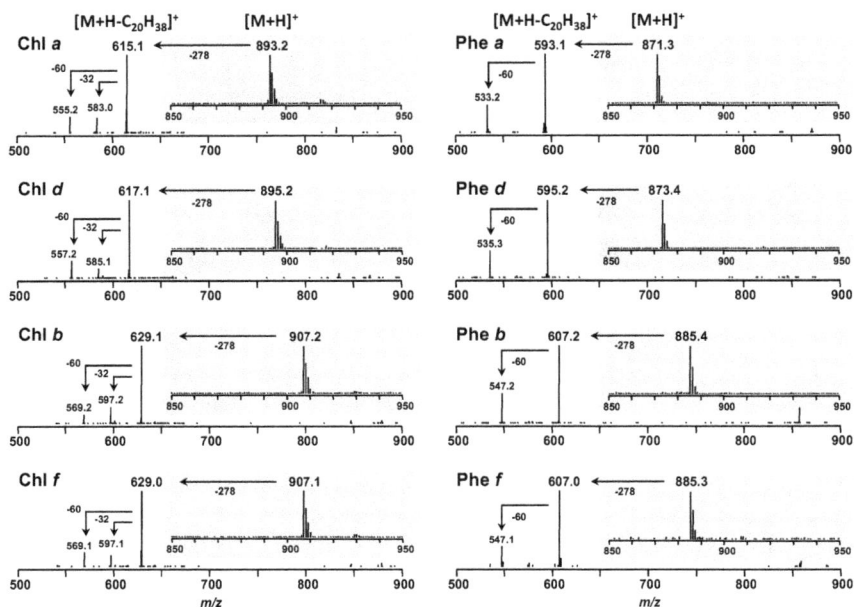

Figure 10. MS/MS spectra of Chls a, b, d, f (left column) and Phes a, b, d and f (right column). Each mass spectrum of the chlorophyll fraction is shown in the shaded square. MS/MS spectra of the protonated molecules ([M+H]⁺) of Chls a, b, d and f give product ions of [M+H-278]⁺, [M+H-278-32]⁺ and [M+H-278-60]⁺. Pheophytins a, b, d and f give product ions of [M+H-278]⁺ and [M+H-278-60]⁺.

suugest that Chl f also possesses a phythyl long chain in such molecules as Chls a, b and d, and that most probably one -CH₃ moiety of Chl a is substituted for -CHO group in Chl f like Chl b, yielding [2-formyl]-Chl a, [12-formyl]-Chl a or [18-formyl]-Chl a.

As seen in Fig. 10 (right), the corresponding pheophytins prepared by acid treatment clearly showed the absence of magnesium (Fig. 1). For example, [M+H]⁺ of Phe a is observed at m/z 871.3 which is 22 Da smaller than that of Chl a, showing the substitution of Mg with two H atoms by pheophytinization (see Fig. 1). Pheophytins b and f ($C_{55}H_{74}N_4O_6$, 884.545) and Phe d ($C_{54}H_{72}N_4O_6$ 872.545) show the similar pattern, supporting that all of them do not possess Mg as central metal.

3.5. Nuclear magnetic resonance spectra

Nuclear magnetic resonance (NMR) spectroscopy can offer ample information about the molecular structure. Coupled use of NMR with HPLC, absorption-, CD- and mass-spectro-metries has not only definitively identified the structures of several major naturally-occurring Chls but has also assisted recent studies of minor Chl pigments, present in minute quantities, such as electron donors and acceptors in the RC.

Physicochemical Properties of Chlorophylls in Oxygenic Photosynthesis — Succession of Co-Factors
from Anoxygenic to Oxygenic Photosynthesis

69

The NMR spectra were recorded on a Bruker Avance 800 spectrometer (Bruker Biospin, Karlsruhe, Germany), with a frequency of ^{1}H at 800 MHz and ^{13}C at 201 MHz, equipped with TCI CryoProbe using a microtube (Shigemi Inc., Tokyo) and about 0.5 mg of sample in 0.3 mL of acetone-d_6 with tetramethylsilane (TMS) as an internal standard. The chemical shifts are given in δ-scale [ppm] downfield from TMS. The measurements were performed at 273 K. The typical experimental conditions for the ^{1}H NMR spectra were 256 scans, a spectral width of 17 ppm, 128k data points. The ^{13}C spectra were acquired using a power gated decoupling with 48k scans. The spectral width of 220 ppm was acquired in 64k data points. The 2D-homonuclear (Nuclear Overhauser and Exchange Spectroscopy (NOESY)) and 2D-heteronuclear (H,C-Heteronuclear Single Quantum Coherence (H,C-HSQC) and H,C-Heteronuclear Multiple Bond Correlation (H,C-HMBC)) experiments were performed for the structural assignments of the ^{1}H and ^{13}C signals using standard 2D-NMR pulse sequences of Bruker software.

3.5.1. ^{1}H-NMR

As observed in one-dimensional ^{1}H-NMR spectra (Fig. 11, Table 2), marked differences are seen in the signals arising from the formyl group. Each low-field singlet signal characteristic of the formyl moiety observed around 11 ppm in the spectra of Chls b (7^1), d (3^1) and f (2^1) is absent from the spectrum of Chl a. Similarly, double doublet signal of 3^1-H vinylic proton at 8-8.5 ppm in the spectra of Chls a, b and f is not seen in the spectrum of Chl d. These results reconfirm that Chl d is 3-desvinyl-3-formyl Chl a ([3-formyl]-Chl a).

Here we note that the 3^1-H vinylic proton shows a large downfield shift in Chl f (8.534 ppm) and that is slightly upfield shifted in Chl b (8.043 ppm), as compared to Chl a (8.162 ppm), suggesting that Chl f should be formylated along the y-axis and that interaction between -CH=CH$_2$ and -CHO in Chl f is rather strong than that in Chl b. The -CHO substitution position in Chl f is hence most probably at C2, next to the 3-vinyl group.

The pair signals of 3^2- and 3^2-H vinylic protons are well resolved in the spectra of Chls a (6.242 ppm and 6.028 ppm) and b (6.302 ppm and 6.055 ppm), while the corresponding pair signals in Chl f show low resolution (6.365 ppm and 6.324 ppm), suggesting that the environment of 3-vinyl moiety in Chls a and d is very similar to each other, but is profoundly different from that in Chl f, most probably due to the presence of formyl moiety at the neighboring C2 in Chl f. The ^{1}H singlet signals for 7^1-CH$_3$ in Chls a, d and f are at 3.3 ppm and absent from the spectrum of Chl b, while the ^{1}H signal for 7^1-CHO in Chl b appears on a much lower field at 11.3 ppm. Another marked difference seen in Chl f spectrum is the disappearance of the singlet signal of 2^1-CH$_3$ proton; the corresponding signals are observed at 3.3-3.72 ppm in the spectra of Chls a, b and d, implying that 2^1-CH$_3$ of Chl a is substituted in Chl f for some other moiety, most probably -CHO. No other change is clear in the one-dimensional ^{1}H-NMR spectra.

3.5.2. ^{13}C-NMR

In the ^{13}C-NMR spectra (Fig. 12, Table 3), marked differences are noted in the range of 0 ppm to 20 ppm, 120 ppm to 140 ppm, and 180 ppm to 200 ppm, relating to the -CH$_3$, -CH=CH$_2$ and -CHO moieties. Compared to Chl a, in the spectrum of Chl f, the ^{13}C signal of 2^1-CH$_3$ is absent, the 7^1-CH_3 and $3^1,2$-CH=CH_3 carbon signals appear at 11 ppm, 130 ppm and 126 ppm, respectively, similar to Chl a (11 ppm, 131 ppm and 120 ppm). Further, a new carbon signal appears at 189 ppm, close to the signals of -CHO in Chls b (188 ppm) and d (190 ppm), supporting the presence of -CHO at C2 in Chl f.

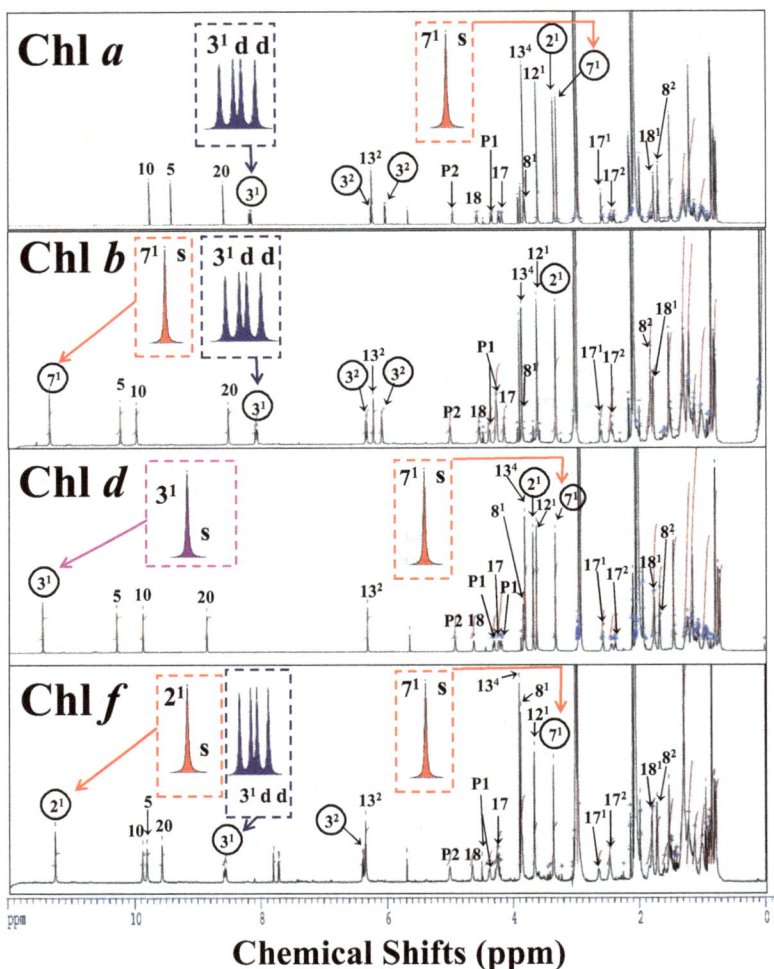

Figure 11. ^{1}H-NMR spectra of Chls a, b, d and f measured in acetone-d_6 at 273 K. Signals corresponding to ^{1}H atoms of the macrocycle are labeled with the numbers of the corresponding carbons. The peak at 2.06-2.11 ppm is acetone and 3.07-3.10 ppm is H_2O, respectively.

IUPAC no.of carbon atom	Chl a	Chl b	Chl d	Chl f
2^1	3.343 (3.36)[1] (s)	3.316 (3.40)[2*](s)	3.724 (3.68)[3](s)	11.215 (11.35)[4**](s)
3	-	-	-	-
3^1	8.162 (8.18)[1](dd)	8.043(7.95)[2*](dd)	11.460 (11.40)[3](s)	8.534(dd)
3^2	6.242 (6.24)[1](dd), 6.028 (6.03)[1](dd)	6.302(6.25)[2*](dd), 6.055 (6.04)[2*](dd)	-	6.324(dd), 6.365(dd)
4	-	-	-	-
5	9.410 (9.40)[1](s)	10.192 (10.04)[2*](s)	10.294(10.20)[3](s)	9.770 (9.79)[4**](s)
7^1	3.300 (3.30)[1](s)	11.305 (11.22)[2*](s)	3.365 (3.33)[3](s)	3.351(s)
8	-	-	-	-
8^1	3.817 (3.82)[1](q)	4.243	3.876 (3.86)[3](q)	3.754(q)
8^2	1.696 (1.69)[1](t)	1.815	1.723 (1.73)[3](t)	1.705(t)
10	9.749 (9.75)[1](s)	9.934 (9.64)[2*](s)	9.873 (9.8)[3](s)	9.838 (9.86)[4**](s)
11	-	-	-	-
12	-	-	-	-
12^1	3.619 (3.61)[1](s)	3.606 (3.65)[2*](s)	3.668 (3.65)[3](s)	3.637(s)
13	-	-	-	-
13^2	6.234 (6.24)[1](s)	6.189 (6.19)[2*](s)	6.335 (6.28)[3](s)	6.318(s)
13^3	-	-	-	-
13^4	3.829 (3.83)[1](s)	3.842 (4.02)[2*](s)	3.851 (3.83)[3](s)	3.887(s)
17	4.175 (4.16)[1]	4.128	4.242 (4.25)[3]	4.230
17^1	2.589 (2.60)[1], 2.461 (2.45)[1]	2.43,2.593	2.484,2.622	2.467,2.632
17^2	2.431 (2.35)[1], 2.159 (2.05)[1]	2.08,2.44	1.98,2.418	2.08,2.47
18	4.572 (4.57)[1](q)	4.524(q)	4.660 (4.63)[3](q)	4.634(q)
18^1	1.772 (1.77)[1], 1.762(1.76)[1](d)	1.768,1.759(1.78)[2*](d)	1.812, 1.802(1.82)[3](d)	1.800, 1.791(d)
20	8.582 (8.58)[1](s)	8.480(8.20)[2*](s)	8.867 (8.81)[3](s)	9.533 (9.77)[4**](s)
P1	4.342 (4.33)[1],4.224 (4.21)[1]	4.364, 4.247	4.343 (4.26)[3], 4.227 (4.36)[3]	4.361, 4.263
P2	4.955 (4.95)[1]	4.980	4.944 (5.04)[3]	4.987
P3	-	-	-	-
$P3^1$	1.509 (1.51)[1]	1.519	1.505 (1.54)[3]	1.525
P4	1.822 (1.82)[1]	1.845	1.832 (1.85)[3]	1.845
P5	1.31	1.31	1.30	1.195
P6	0.97,1.17	0.98,1.18	1.97,1.16	0.97,1.17
P7	1.31	1.33	1.31	1.324
$P7^1$	0.811,(0.81)[1], 0.803(0.80)[1]	0.785, 0.777	0.778, 0.770 (0.79)[3]	0.785, 0.777
P8	1.01,1.23	1.02,1.22	1.01,1.22	1.01,1.22
P9	1.15,1.28	1.15,1.28	1.14,1.28	1.15,1.27
P10	1.01,1.23	1.02,1.22	1.01,1.22	1.01,1.22
P11	1.31	1.32	1.32	1.32
$P11^1$	0.783(0.79)[1],0.774 (0.78)[1]	0.809,0.801	0.806, 0.797 (0.81)[3]	0.807, 0.798
P12	1.01,1.23	1.02,1.22	1.01,1.22	1.01,1.22
P13	1.23,1.28	1.23,1.28	1.23,1.28	1.23,1.28
P14	1.12	1.12	1.12	1.12
P15	1.500 (1.50)[1]	1.489	1.497 (1.51)[3]	1.495
$P15^1$	0.854(0.86)[1],0.845 (0.84)[1]	0.851,0.842	0.850 (0.85)[3], 0.842 (0.85)[3]	0.849, 0.841
P16	0.854(0.86)[1],0.845 (0.84)[1]	0.851,0.842	0.850 (0.85)[3], 0.842 (0.85)[3]	0.849, 0.841

[1]Kobayashi et al. (2000), [2]Wu et al. (1985), [3]Miyashita et al. (1997), [4]Chen et al. (2010)

* in $CDCl_3$, ** in CD_2Cl_2/d_5-pyridine(97/3, v/v)

Table 2. [1]H-chemcal shifts of Chls a, b, d and f in acetone-d_6 at 273K

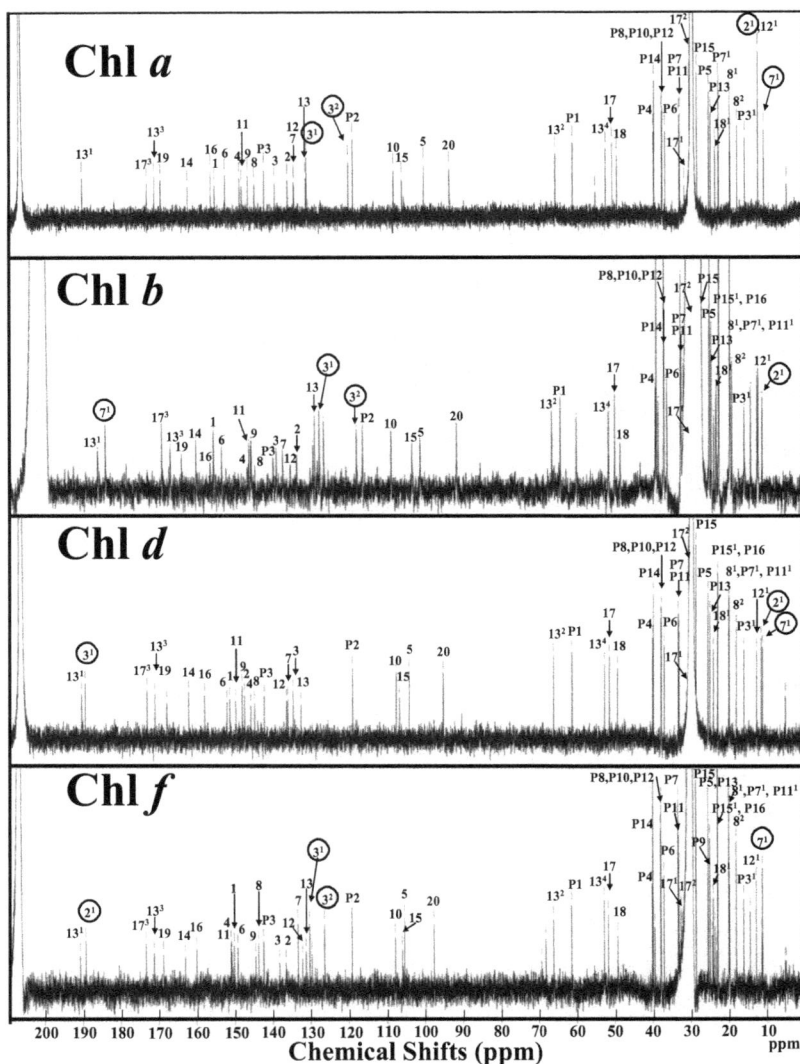

Figure 12. NMR spectra of Chls a, b, d and f measured in acetone-d_6 at 273 K. Signals corresponding to ^{13}C atoms of the molecules are labeled. The peak at 30-32 ppm is acetone.

Physicochemical Properties of Chlorophylls in Oxygenic Photosynthesis — Succession of Co-Factors
from Anoxygenic to Oxygenic Photosynthesis

73

IUPAC no. of carbon atom	Chl a	Chl b	Chl d	Chl f
1	155.47(155.46)[1]	159.43	151.47(151.81)[2]	151.02
2	136.28(136.24)[1]	136.98	147.66(147.33)[2]	136.49
2[1]	12.65(12.70)[1]	12.33	12.58(11.75)[2]	189.27
3	139.76(139.68)[1]	142.53	134.77(135.12)[2]	138.17
3[1]	131.30(131.29)[1]	130.83	189.59(189.54)[2]	130.22
3[2]	120.33(120.33)[1]	120.80	-	126.36
4	148.96(148.99)[1]	150.15	146.00(146.36)[2]	150.20
5	100.58(100.51)[1]	103.63	106.24(104.41)[2]	105.39
6	152.78(152.80)[1]	157.25	152.27(152.55)[2]	149.24
7	134.68(134.65)[1]	140.76	136.19(136.31)[2]	133.35
7[1]	11.15(11.16)[1]	188.62	11.25(11.39)[2]	11.17
8	145.08(145.02)[1]	148.32	145.02(145.13)[2]	143.83
8[1]	19.91(19.88)[1]	19.51	19.94(20.12)[2]	19.87
8[2]	18.12(18.18)[1]	19.91	18.12(18.12)[2]	18.07
9	146.74(146.74)[1]	149.29	148.28(148.64)[2]	144.59
10	108.50(108.53)[1]	111.53	107.68(107.8)[2]	107.82
11	148.34(148.34)[1]	149.94	149.97(150.35)[2]	159.99
12	134.57(134.46)[1]	138.66	136.50(136.63)[2]	132.12
12[1]	12.65(12.66)[1]	12.47	12.79(12.89)[2]	12.82
13	131.53(131.41)[1]	132.13	132.76(133.04)[2]	131.19
13[1]	190.30(190.37)[1]	190.66	190.49(190.50)[2]	190.68
13[2]	66.00(65.95)[1]	65.77	66.19(66.47)[2]	66.13
13[3]	171.32(171.36)[1]	171.04	171.20(171.33)[2]	173.30
13[4]	52.67(52.71)[1]	52.79	52.79(52.84)[2]	52.83
14	162.55(162.58)[1]	164.11	162.30(162.68)[2]	168.67
15	106.35(106.27)[1]	105.94	106.82(107.04)[2]	106.03
16	156.50(156.54)[1]	160.38	158.12(158.60)[2]	163.04
17	50.98(50.92)[1]	51.14	51.54(52.05)[2]	51.74
17[1]	30.28(30.03)[1]	30.28	30.66(30.64)[2]	30.90
17[2]	30.66(30.09)[1]	30.66	31.27(31.37)[2]	30.35
17[3]	173.29(173.39)[1]	173.35	173.26(173.45)[2]	189.25
18	49.73(49.69)[1]	49.65	49.44(49.80)[2]	49.26
18[1]	23.01(23.88)[1]	23.68	24.13(24.25)[2]	24.20
19	169.69(167.74)[1]	168.03	168.09(168.34)[2]	171.15
20	93.78(93.79)[1]	93.86	95.26(95.36)[2]	97.60
P1	61.31(61.32)[1]	61.42	61.33(61.67)[2]	61.35
P2	120.17(119.12)[1]	119.19	119.13(119.53)[2]	119.18
P3	142.46(142.48)[1]	143.42	142.49(142.70)[2]	142.49
P3[1]	16.11(16.10)[1]	16.16	16.11(16.40)[2]	16.13
P4	40.19(40.19)[1]	40.23	40.19(40.46)[2]	40.20
P5	25.54(25.52)[1]	25.57	25.53(25.86)[2]	25.55
P6	37.11	37.14	38.11(37.40)[2]	37.11
P7	33.43(33.44)[1]	33.44	33.22(33.47)[2]	33.50
P7[1]	20.03(20.02)[1]	20.03	19.98(20.19)[2]	20.02
P8	37.94	37.95	37.94(38.21)[2]	37.94
P9	25.04(25.06)[1]	25.06	25.04(25.25)[2]	25.04
P10	38.01	38.02	38.01(38.26)[2]	38.01
P11	33.22(33.23)[1]	33.25	33.42(33.65)[2]	33.42
P11[1]	19.98(19.98)[1]	19.99	20.02(20.23)[2]	19.98
P12	37.87	37.89	37.88(38.14)[2]	37.88
P13	25.49(25.52)[1]	25.49	25.49(25.64)[2]	25.48
P14	40.00(39.98)[1]	40.00	40.00(40.27)[2]	40.00
P15	28.65(28.66)[1]	28.65	28.65(28.85)[2]	28.65
P15[1]	23.01[3] (28.66)[1]	23.01[4]	23.01[5](23.15)[2]	23.00[6]
P16	22.90[3] (22.91)[1]	22.91[4]	22.90[5] (23.07)[2]	22.90[6]

[1]Kobayashi et al. (2000), [2]Miyashita et al. (1997)

[3-6] assignment interchangeable

Table 3. ^{13}C-chemcal shifts of Chls a, b, d and f in acetone-d_6 at 273K

The signals of 7^1-CH_3 of Chls a, d and f show almost the same chemical shifts, 11.15 ppm, 11.24 ppm and 11.17 ppm, respectively, indicating the interaction between the -CHO and 7^1-CH_3 moieties in Chls d and f are negligibly small, and hence the -CHO substituent is not so close to the 7^1-CH_3 moiety in Chls d and f. The chemical shifts of 3^2-vinyl carbons in Chls a and b are almost identical (120.3-120.8 ppm), but Chl f shows a slight but significant downfield shift (126.3 ppm), suggesting that the formyl substituent in Chl f is positioned very close to the 3-vinyl group, most probably on C2. The signals of -CHO moiety of Chls d and f exhibits almost the same chemical shift (189.59 ppm and 189.27 ppm), but a slight upfield C-formyl signal is observed at 188.62 ppm in Chl b, indicating that the environment of -CHO in Chls d and f is very similar, supporting that the -CHO moieties of Chls d and f are positioned at the same ring I, while that of Chl b is at ring II.

3.5.3. NOESY

Two-dimensional NMR spectra provide further information about a molecule than one-dimesional NMR spectra. NOESY is one of several types of two-dimensional NMR, where the nuclear Overhauser effect (NOE) between nuclear spins is used to establish the correlations. The cross-peaks in the two-dimensional spectrum connect resonances from spins that are spatially close to each other.

To obtain further evidence for the structural identity of Chl f, the signals were investigated using NOESY spectra. First of all, we had better see well-defined coherent correlations on the NOESY spectrum of Chl a. Here, we will trace the coherent correlations from meso-20-H, because Chl f posseses -CHO most probably at C2 near to C20. Coherent correlations can be easlily traced from 20-H on the NOESY spectrum of Chl a (Fig. 13), where the signal of 20-H at 8.582 ppm shows three cross peaks with the signals of 18^1-H at 1.771 ppm, 2^1-H at 3.343 ppm, and 18-H at 4.572 ppm. Good coherent correlations can also be traced from meso-5-H and meso-10-H; (1) the signal of 5-H at 9.410 ppm shows three cross peaks with the signal of 7^1-H at 3.300 ppm, 3^2-H at 6.242 ppm, and 3^1-H at 8.162 ppm, and (2) the signal of 10-H at 9.749 ppm shows two cross peaks with the signal of 8^2-H at 1.696 ppm and 12^1-H at 3.619 ppm. In Chl b and Chl d, similar nice correlations are seen (Fig. 13).

To obtain further evidence for Chl f, a NOESY spectrum is illustrated in Fig. 13. As expected, nice coherent correlations can be traced from meso-20-H, the signal of 20-H at 9.553 ppm shows three cross peaks with the signal of 18-H at 4.634 ppm, 18^1-H at 1.816 ppm, and 2^1-H, most probably assigned to -CHO moiety, at 11.215 ppm. The signal of meso-5-H at 9.770 ppm shows three cross peaks with the signal of 7^1-H at 3.351 ppm, 3^2-H at 6.347 ppm, and 3^1-H at 8.534 ppm. Similarly, coherent correlations of meso-10-H with 8^1-H (3.754 ppm) and 12^1-H (3.637 ppm) are clearly seen. These results indicate that the C-20 methine, among the three methines, is nearest to the -CHO moiety, supporting the substitution position by -CHO is most likely at C2 in Chl f.

Physicochemical Properties of Chlorophylls in Oxygenic Photosynthesis — Succession of Co-Factors
from Anoxygenic to Oxygenic Photosynthesis

75

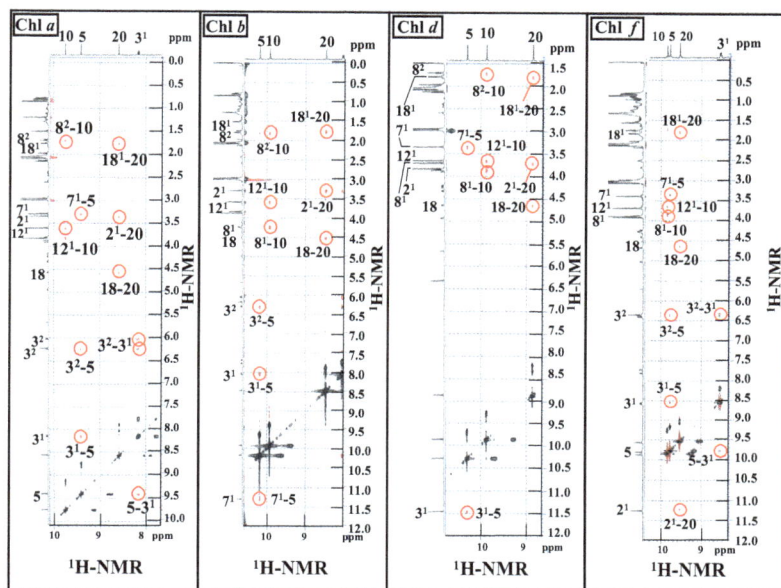

Figure 13. ^{1}H-^{1}H-NOESY spectra of Chls a, b, d and f measured in acetone-d_6 at 273 K.

3.5.4. HSQC

HSQC is a two-dimensional inverse correlation technique that allows for the determination of connectivity between two different nuclear species, and HSQC is selective for direct coupling. As illustrated in the ^{1}H-^{13}C HSQC spectra of Chls a, b, d and f (Fig. 14), all substituents on the macrocycle show the corresponding cross peaks. The results support that one methyl group of Chl a is replaced with a formyl one in Chl f.

3.5.5. HMBC

HMBC is also a two-dimensional inverse correlation method that allows for the determination of connectivity between two different nuclear species like HSQC, but HMBC gives longer range coupling (2-4 bond coupling) than HSQC.

Three meso-Hs in Chl a exhibit one to three cross peaks as seen in Fig. 15A. The formyl-H in Chls b, d and f shows two to four cross peaks. For example, in the ^{1}H-^{13}C HMBC spectrum of Chl f, the ^{1}H-signal for 2^1-CHO has four cross peaks, one is with the signal for 3^1-vinylic carbons, one is with the signal for 1C carbon, and the other two are with the signal for 2^1-CHO carbon, where the cross peak split is due to the direct coupling. Conclusively, the Chl f-like pigment isolated from the strain KC1 has been identified as Chl f, namely, 2-desmethyl-2-formyl-Chl a ([2-CHO]-Chl a).

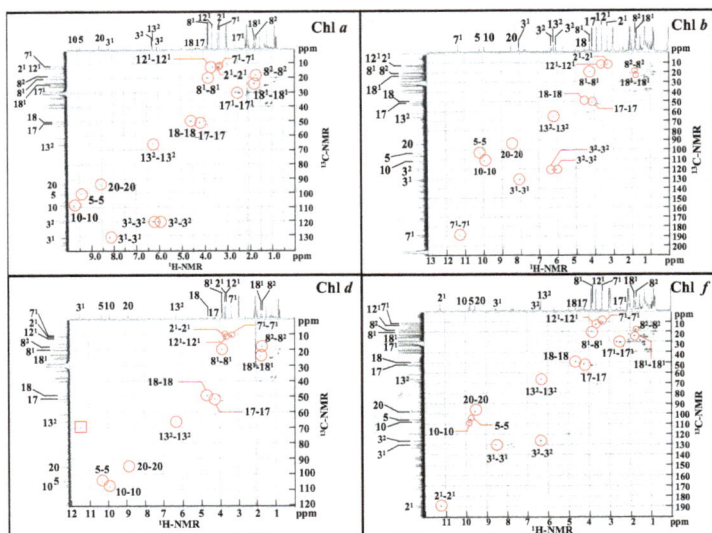

Figure 14. ¹H-¹³C-HSQC spectra of Chls *a*, *b*, *d* and *f* measured in acetone-d_6 at 273 K. □ in Chl *d*: δC189.59 (folded from outside the spectral window)

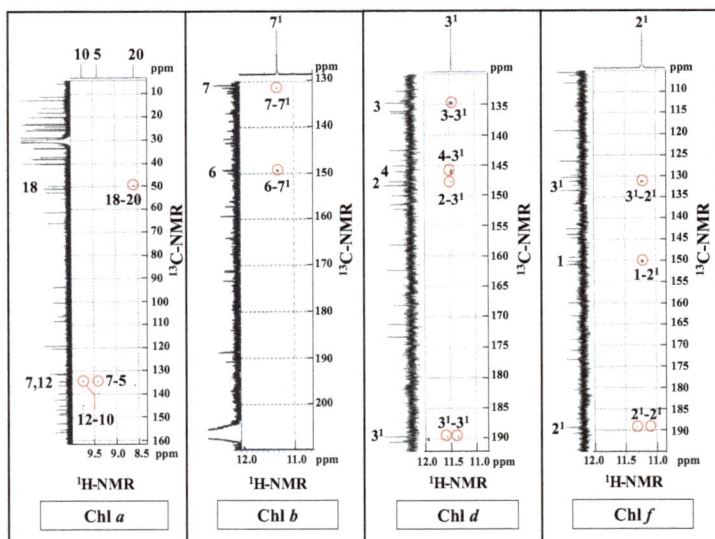

Figure 15. ¹H-¹³C-HMBC spectra of Chls *a*, *b*, *d* and *f* measured in acetone-d_6 at 273 K.

3.6. Redox potentials

To understand the charge separation in the RC, electrochemical characterization of chloro-phylls is of crucial importance. In this section, the redox potentials of Chls and Phes *in vitro* are presented.

Acetonitrile (Aldrich, anhydrous grade: water < 50 ppm) was deoxygenated and dried before use. The solvent was subjected to freeze-pump-thaw cycles at least three times under about 10^{-5} torr. Under a nitrogen atmosphere, the deoxidized solvent was then dried for 24 h with activated molecular sieves (4A 1/16, Wako), pretreated *in vacuo* at 473 K over 24 h. Tetra-*n*-butylammonium perchlorate (Bu_4NClO_4, TBAP) (Aldrich, Electrochemical grade: > 99.0 %), was used as the supporting electrolyte, which had been recrystallized from methanol solution and then dried in vacuo at 333 K over 24 h.

The redox potentials of chlorophylls were measured by square wave voltammetry (SWV). The signal-to-noise ratio of SWV is generally better than that of CV, especially for measuring redox couples at such low concentration (ca. 0.5 mM) as in the present case (Cotton et al. 1979, Wasielewski et al. 1980). Measurements were done with an ALS model 620A electrochemical analyzer. Parameters for SWV were V_{step} = 5.0 mV, AC signal (V_{pulse}) = 25 mV, and p-p at 8 Hz. The measurements were carried out in an air-tight electrochemical cell containing a small compartment for a sample solution (ca. 0.5 mM) equipped with a glass filter that can be degassed and filled with dry N_2. A platinum disk electrode with 1.6 mm in diameter (outer diameter: 3 mm) was used as the working electrode, and a platinum black wire fabricated in the small compartment (internal diameter: 8.9 mm) as the counter electrode. An Ag/AgCl electrode, chosen for good reproducibility despite possibility of junction potential, was connected through a salt bridge to the outer electrolytic solution of the small components. After measurement, the redox potentials of the ferrocene-ferrocinium were measured as +0.45 V vs. Ag/AgCl in acetonitrile.

Typical square wave voltammograms (SWVs) for Chls *a*, *b*, *d* and *f* in acetonitrile are illustrated in Fig. 16. Four peaks are observed in each voltammogram and the potentials in anodic sweep and cathodic sweep (data not shown) are identical to each other, indicating that the four redox reactions are reversible. Similar trends are observed for Phes *a*, *b* and *d* (data not shown). The measurement for Phe *f* is now underway.

In Table 4 are summarized the redox potentials for Chls *a*, *b*, *d*, *f*, Phes *a*, *b* and *d*. Chl *d* shows higher oxidation potentials than Chl *a*, lower than Chl *b*, and much lower than Phe *a*, Phe *b* and Phe *d*. Chl *f* exhibits higher oxidation potentials than Chls *a*, *d*, and lower than Chl *b*. The results can be explained by invoking the inductive effect of substituent groups on the macrocycle, because the redox potentials of chlorophylls are sensibly affected by the nature of substituent groups on the π-electron system (Fuhrhop 1975, Watanabe and Kobayashi 1991, Kobayashi et al. 2007).

The -CHO substituent on Chls *b*, *d* and *f* is an electron-withdrawing group (→CHO), and hence reduces the electronic density in the π-system of chlorophyll. The replacements of -CH₃ at C7 or C2 of Chl *a* by →CHO to yield Chl *b* or Chl *f* cause the macrocycle to be electron poor, thus rendering the molecule less oxidizable (E^1_{ox}: Chl b, f > Chl a). Similarly, replacement of -CH=CH$_2$ at C3 of Chl a by →CHO to yield Chl d makes the first oxidation potential, E^1_{ox}, more positive than that of Chl a (E^1_{ox}: Chl d > Chl a). Therefore, the E^1_{ox} order becomes Chls b, d, f > Chl a (see Figs 16 and 17). When one pays attention to the group of -CH$_3$ at C7 of Chl d and the group of -CH=CH$_2$ at C3 of Chl b or C7 of Chl f, the -CH$_3$ group is more electron-donating (←CH$_3$), thus making the macrocycle of Chl d more electron rich, and hence its oxidation potential less positive (Chls b, f > d); the E^1_{ox} order results in Chls b, f > Chl d > Chl a. As expected from the inductive effect of substituent groups, Chls b and f show the almost the same E^1_{ox} values. Consequently, as seen in Fig. 17, the E^1_{ox} order results in Chl b > Chl f > Chl d > Chl a; a little higher oxidation potential of Chl b than that of Chl f, 20 mV, cannot be explained from the primitive way used here.

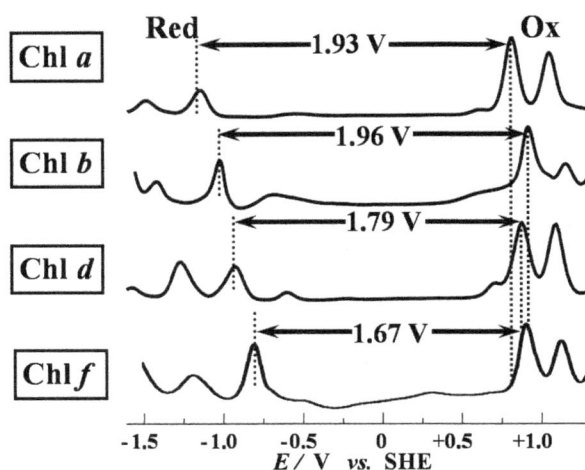

Figure 16. Square wave voltammograms of Chls a, b, d and f in acetonitrile.

	E^2_{red}	E^1_{red}	E^1_{ox}	E^2_{ox}	$E^1_{ox} - E^1_{red}$
			V vs. SHE		
Chl a	-1.46	-1.12	0.81	1.04	1.93
Chl b	-1.41	-1.02	0.94	1.15	1.96
Chl d	-1.27	-0.91	0.88	1.09	1.79
Chl f	-1.12	-0.75	0.92	1.13	1.67
Phe a	-1.00	-0.75	1.14	1.49	1.89
Phe b	-1.05	-0.64	1.25	1.58	1.89
Phe d	-0.87	-0.63	1.21	1.50	1.84

Table 4. Redox potentials of Chls a, b, d, f, Phes a, b and d in acetonitrile

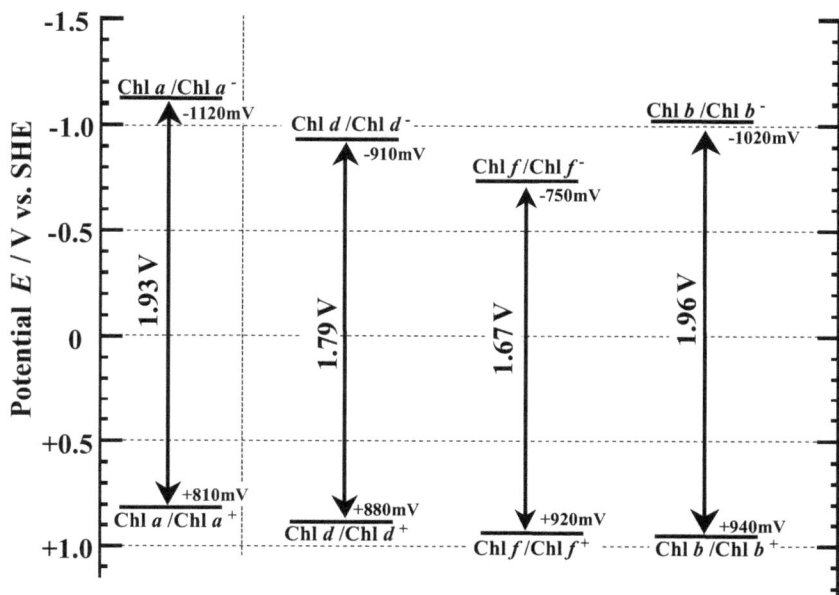

Figure 17. Schematic comparison of redox potentials of Chls *a*, *b*, *d* and *f* in acetonitrile.

The redox behavior of a compound is related to the energy levels of its molecular orbitals: E^1_{ox} is intimately related to the highest occupied molecular orbital (HOMO) and E^1_{red} to the lowest unoccupied molecular orbital (Watanabe and Kobayashi 1991; Hanson 1991). The order of HOMO energy levels well parallels the E^1_{ox} values, while the correlation between the LUMO levels and the E^1_{red} values is less conspicuous (see Fig. 7 in Watanabe and Kobayashi 1991).

As clearly seen in Fig. 17, the order of absolute values of the first reduction potentials, E^1_{red}, of Chls *a*, *d* and *f* is Chl *a* > Chl *d* > Chl *f*, which can be well explained by substituent inductive effect, like E^1_{ox} as mentioned above. This simple rule, however, does not hold for Chls *d* and *f*; their values of E^1_{red} are remarkably more positive than those of Chls *a* and *b*; compared to Chl *a*, Chl *b* is harder to oxidize by 130 mV, and easier to reduce by 100 mV, namely, similar in degree, while Chls *d* and *f* peculiarly easier to reduce by 210 mV and 370 mV. The findings indicate that the inductive effect of substituent groups at ring II conspicuously appears for E^1_{red}. Anyway, this irregularity may come from the fact that the LUMO energy levels are to a lesser extent correlated with the E^1_{red} values.

The primary redox potential difference, $\Delta E = E^1_{ox} - E^1_{red}$, seen in Fig. 17 can be taken as an index for the Q_Y excitation energy (Watanabe and Kobayashi 1991; Hanson 1991). For example, ΔE for Chl *a* is 1.93 eV in Fig. 17, which corresponds to the Q_Y excitation wavelengths to a certain extent, 661-666 nm for Chl *a* in Fig.17. Similarly. ΔE = 1.96 eV, 1.79 eV and 1.67 eV for Chls *b*,

d and f also correlate to the Q_Y wavelengths, 644-662 nm, 686-698 nm. and 695-708 nm. Pheophytins also behave in a similar fashion.

In 1959, the domination of inductive effects of the central metal over a conjugative macrocycle has first been formulated (Gouterman 1959). The redox potential of chlorophyll shows a systematic shifts with the electronegativity of the central metal, and such a trend is rationalized in terms of an electron density decrease in the chlorin π-system by the presence of an electron negative metal in the center of chlorophylls (Watanabe and Kobayashi 1991; Hanson 1991; Noy et al. 1998). Inspection of Table 4 demonstrates that such a trend is essential for the pair of Chls and Phes; the electronegativity of 2.2 for H is significantly higher than that of 1.2 for Mg, which renders Phes more difficult to oxidize than the corresponding Chls.

4. Evolutionary and ecological aspects of chlorophylls

4.1. Diversification of chlorophylls during the evolution of photosynthetic organisms

Chlorophylls are distributed among oxygenic photosynthetic organisms, including cyanobacteria, algae and terrestrial plants (Falkowski et al. 2004). It is generally accepted that plastids first arose by endosymbiosis between photosynthetic prokaryotes (ancestral to present cyanobacteria) and non-photosynthetic eukaryotic hosts (Fig. 18). There are two different types of hypothesis for the evolution of Chl b. One is based on vertical transfer of the gene for Chl b biosynthesis, which includes the lost of the gene. The other is based on the lateral gene transfer for Chl b biosynthesis. In the hypothesis of the former one, the ancestral cyanobacteria contained both Chls a and b, and the gene for Chl b biosynthesis had been lost. In many cyanobacterial linage as well as the linages of the red algae and glaucophytes, the gene had been lost, whereas a few linage of cyanobacteria, (*Prochloron* and *Prochlorothrix*) and the green algae (and terrestrial plants) retained it. This hypothesis is partly supported by the high homology of the enzyme for Chl b biosynthesis (CAO) (Tomitani et al. 1999). In the latter hypothesis, the gene for Chl b biosynthesis was firstly evolved in the ancestor of green algae or cyanobacteria, and the gene was transferred into different linages. This is supported by the fact that one gene transfer was enough for the acquisition of Chl b biosynthesis in cyanobacerium. The secondary endosymbioses of green algae gave rise to euglenophytes, chlorarachniophytes and "green" dinoflagellates, which were containing Chls a and b. The secondary or tertiary plastids of cryptophytes, haptophytes, heterokonts and dinoflagellates contain Chl c in addition to Chl a. Most of present-day cyanobacteria contain only Chl a, and only a few genera are known to keep Chl b (*Prochloron* and *Prochlorothrix*). This suggests that Chl b had been lost in polyphyletic lineage during the evolution of cyanobacteria. Some cyanobacteria contain divinyl chlorophylls (DVChls) a and b (*Prochlorococcus*), Chl d (*Acaryochloris*) and Chl f (strain KC1 and *Halomicronema hongdechloris*). These chlorophylls are only found in cyanobacteria, and it has yet to be revealed when and how they acquired these pigments.

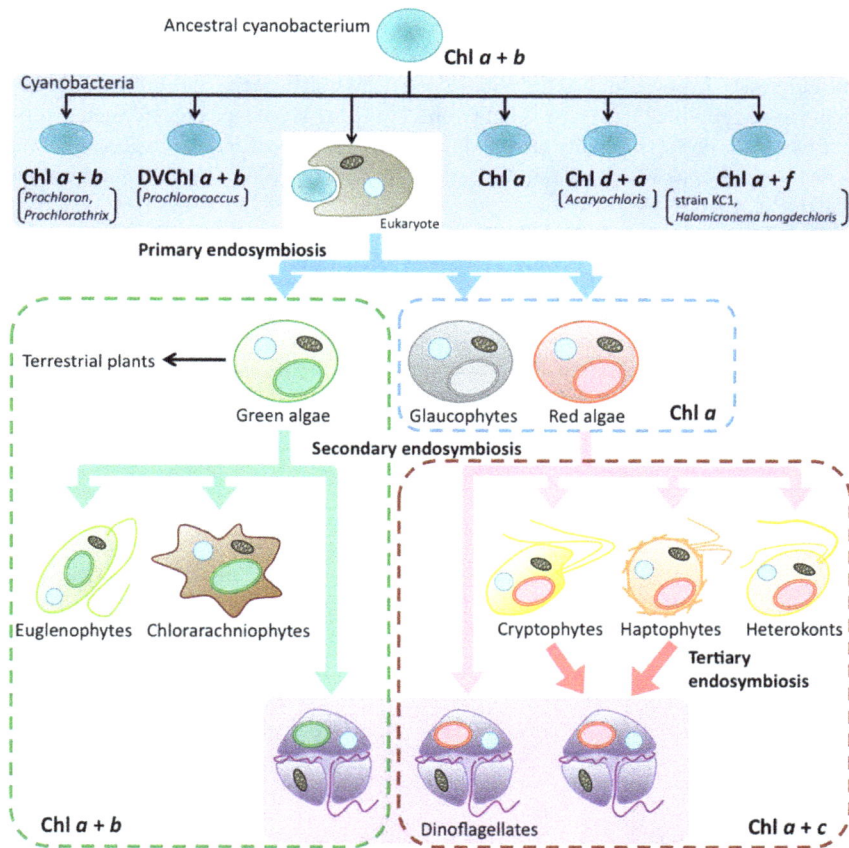

Figure 18. The distribution of various chlorophylls among oxygenic photosynthetic organisms.

4.2. Ecology of the red-shifted chlorophylls

Acquisition of new or additional chlorophylls by photosynthetic organisms is thought to be an adaptation to the light quality of their niches. The two red-shifted chlorophylls, Chls *d* and *f*, are also thought be acquired and selected by a part of cyanobacteria for their growth and survival by sustaining oxygenic photosynthesis using the light available at their niches.

Chl *d*-containing cyanobacterium, *Acaryochloris marina*, was firstly found as a minor symbionts in colonial ascidians which had large numbers of cyanobacterial symbionts, *Prochloron*, which contain Chls *a* and *b* as their photosynthetic chlorophylls (Miyashita et al. 1996). Subsequently, *Acaryochloris* spp. has been found on the surface of coastal macroalgae (Murakami et al. 2004), under those colonial ascidians (Kühl et al. 2005), and in coastal microbial mats at a saline lake (Miller et al. 2005). In all of those niches, *Acaryochloris* spp. competes for light for their oxygenic photosynthesis. However, due to the possession of Chl *d* as primary chlorophyll, *Acaryochloris* spp. can absorb and utilized far-red light from 700 nm to 740 nm for oxygenic photosynthesis (Miyashita et al. 1997). The ability has advantage for their growth and survival, since another oxygenic phototrophs which contain Chl *a* as their primary chlorophyll cannot absorb and utilize the light.

Chl *f*-containing cyanobacterium was firstly reported in microbial mat, where the competition for the light also occurs among the phototrophs (Chen et al. 2010). While the distribution and detailed niche of Chl *f*-containing cyanobacteria have not been elucidated yet, Chl *f* must contribute to the survival of those cyanobacteria. Since Chl *f* in cyanobacteria also absorbed far-red light around 720 nm and Chl *f*-possessing cyanobacteria can grow under far-red LED light as a sole light source (Chen and Blankenship 2011).

Acknowledgements

We thank Dr. Nobuaki Ishida (Ishikawa Prefectural Univ.), Dr. Yoshihiro Shiraiwa and Dr. Koji Iwamoto (Univ. Tsukuba) for their invaluable help. This work was supported in part by Special Project of Organization for the Support and Development of Strategic Initiatives (Green Innovation) (Univ. Tsukuba) to M.K.

Author details

Masami Kobayashi[1], Shinya Akutsu[1], Daiki Fujinuma[1], Hayato Furukawa[1], Hirohisa Komatsu[1], Yuichi Hotota[1], Yuki Kato[2], Yoshinori Kuroiwa[2], Tadashi Watanabe[3], Mayumi Ohnishi-Kameyama[4], Hiroshi Ono[4], Satoshi Ohkubo[5] and Hideaki Miyashita[5]

1 Institute of Materials Science, University of Tsukuba, Tsukuba, Japan

2 Institute of Industrial Science, University of Tokyo, Tokyo, Japan

3 Research Center for Math and Science Education, Organization for Advanced Education, Tokyo University of Science, Tokyo, Japan

4 National Food Research Institute, Tsukuba, Japan

5 Graduate School of Human and Environmental Studies, Kyoto University, Kyoto, Japan

Physicochemical Properties of Chlorophylls in Oxygenic Photosynthesis — Succession of Co-Factors from Anoxygenic to Oxygenic Photosynthesis

83

References

[1] Akiyama M, Miyashita H, Kise H, Watanabe T, Miyachi S and Kobayashi M (2001) Detection of chlorophyll d' and pheophytin a in a chlorophyll d-dominating oxygenic photosynthetic prokaryote Acaryochloris marina. Anal. Sci. 17: 205-208

[2] Akiyama M, Miyashita H, Kise H, Watanabe T, Mimuro M, Miyachi S and Kobayashi M (2002) Quest for minor but key chlorophyll molecules in photosynthetic reaction centers - Unusual pigment composition in the reaction centers of a chlorophyll d-dominated cyanobacterium Acaryochloris marina -. Photosynth. Res. 74: 97-107

[3] Akiyama M, Gotoh T, Kise H, Miyashita H, Mimuro M and Kobayashi M (2004) Stoichiometries of chlorophyll d'/PSI and chlorophyll a/PSII in a chlorophyll d-dominated cyanobacterium Acaryochloris marina. Jpn. J. Phycol. (Supplement for the Proceedings of Algae 2002) 52:67-72

[4] Akutsu S, Fujinuma D, Furukawa H, Watanabe T, Ohnishi-Kameyama M, Ono H, Ohkubo S, Miyashita H and Kobayashi M (2011) Pigment analysis of a chlorophyll f-containing cyanobacterium strain KC1 isolated from Lake Biwa. Photomed. Photobiol. 33: 35-40

[5] Behrendt L, WD Larkum A, Norman A, Qvortrup K, Chen M, Ralph P J, Sørensen S, Trampe E and Ku°hl M (2011) Endolithic chlorophyll d containing phototrophs. ISME J. 5: 1072-1076

[6] Benjamin B, Finazzi G, Benson S, Barber J, Rappaport F and Telfer A (2007) Study of intersystem electron transfer in the chlorophyll d containing cyanobacterium Acaryochloris marina and a reappraisal of the redox properties of P740. Photosynth. Res. 91:155

[7] Boomer SM, Pierson BK, Austinhirst R and Castenholz RW (2000) Characterization of novel bacteriochlorophyll-a-containing red filaments from alkaline hot springs in Yellowstone National Park. Arch. Microbiol. 174: 152-161

[8] Brettel K (1997) Electron transfer and arrangement of the redox cofactors in photosystem I. Biochim. Biophys. Acta 1318: 322-373

[9] Chen M, Schliep M, Willows RD, Cai Z, Neilan BA and Scheer H (2010) A red-shifted chlorophyll. Science 329: 1318-1319

[10] Chen M and Blankenship R E. (2011) Expanding the solar spectrum used by photosynthesis. Trends in Plant Science 16: 427-431

[11] Chen M, Li Y, Birch D and Willows RD (2012) A cyanobacterium that contains chlorophyll f – a red-absorbing photopigment. FEBS Lett. 586: 3249-3254

[12] Cotton TM and Van Duyne RP (1979) Electrochemical investigation of the redox properties of bacteriochlorophyll and bacteriopheophytin in appoticsolvents, J. Am. Chem. Soc. 101: 7605–7612.

[13] Eskins K Scholfield CR and Dutton HJ (1977) High-performance liquid chromatography of plant pigments. J. Chromatogr. 135: 217-220.

[14] Falkowski PG, Katz ME, Knoll AH, Quigg A, Raven JA, Schofield O and Taylor FJR (2004) The evolution of modern eukaryotic phytoplankton, Science 305: 354.

[15] Fajer J, Brune DC, Davis MS, Forman A and Spaulding LD (1975) Primary charge separation in bacterial photosynthesis: Oxidized chlorophylls and reduced pheophytin. Proc. Natl. Acad. Sci. USA 72: 4956-4960

[16] Fuhrhop JH (1975) Reversible reactions of porphyrins and metalloporphyrins and electrochemistry, in: Smith K.M. (ed.), Porphyrins and Metalloporphyrins. Elsevier, Amsterdam p. 14.

[17] Gouterman M (1959) Study of the effects of substitution on the absorption spectra of porphin. J. Chem. Phys. 30: 1139-1161

[18] Gouterman M (1961) Spectra of porphyrins. J. Mol. Spectrosc. 6: 138-163.

[19] Gouterman M, Wagniere G.H and Snyder LC (1963) Spectra of porphyrins part II. Four orbital model, J. Mol. Spectrosc. 11: 108-127

[20] Hanson LK (1991) Molecular orbital theory on monomer pigments. in: H. Scheer (ed.), Chlorophylls, CRC Press, Boca Raton. 993-1014.

[21] Hu Q, Miyashita H, Iwasaki I, Kurano N, Miyachi S, Iwaki M and Itoh S (1998) A photosystem I reaction center driven by chlorophyll d in oxygenic photosynthesis. Proc. Natl. Acad. Sci. USA 95: 13319-13323

[22] Hunt JE and Michalski TJ (1991) Desorption-ionization mass spectrometry of chlorophylls. In: Scheer H (ed) Chlorophylls, pp 835-853. CRC Press, Boca Raton

[23] Iriyama K, Yoshlura M and Shiraki M (1978) Micro-method for the qualitative and quantitative analysis of photosynthetic pigments using high-performance liquid chromatography. J. Chromatogr. 154: 302-305.

[24] Itoh S, Iwaki M, Noguti T, Kawamori A, Mino H, Hu Q, Iwasaki I, Miyashita H, Kurano KN, Miyachi S and Shen R (2001) Photosystem I and II reaction centers of a new oxygenic organism Acaryochloris marina that use chlorophyll. In: PS2001: Proceedings of the 12th International Congress on Photosynthesis, S6-028. CSIRO Publishing, Melbourne (CD-ROM)

[25] Itoh S, Mino H, Itoh K, Shigenaga T, Uzumaki T and Iwaki M (2007) Function of chlorophyll d in reaction centers of photosystems I and II of the oxygenic photosynthesis of Acaryochloris marina. Biochemistry 46:12473-12481.

[26] Jordan P, Fromme P, Witt HT, Klukas O, Saenger W, Krauβ N (2001) Three-dimensional structure of cyanobacterial photosystem I at 2.5 Å resolution. Nature 411: 909-917

[27] Kashiyama Y, Miyashita H, Ogawa NO, Chikaraishi Y, Takano Y, Suga H, Toyofuku T, Nomaki H, Kitazato H, Nagata T, and Ohkouchi N (2008) Evidence of global chlorophyll d. Science 321: 658

[28] Kaufmann KJ, Dutton PL, Netzel TL, Leigh JS and Rentzepis PM (1975) Picosecond kinetics of events leading to reaction center bacteriochlorophyll oxidation. Science 188: 1301-1304

[29] Ke B (2001) The primary electron donor of photosystem I-P700. in Photosynthesis. Photobiochemistry and Photobiophysics. pp. 463-477, Kluwer Academic Publishers, Dordrecht

[30] Kirmaier C, Blankenship RE and Holten D (1986) Formation and decay of radical-pair P+I- in Chloroflexus aurantiacus reaction centers. Biochim. Biophys. Acta 850: 275-285

[31] Klimov VV, Klevanik AV, Shuvalov VA and Krasnovsky AA (1977a) Reduction of pheophytin in the primary light reaction of photosystem II. FEBS Lett. 82: 183-186

[32] Klimov VV, Allkhverdiev SI, Demeter S and Krasnovsky AA (1977b) Photoreduction of pheophytin in photosystem 2 of chloroplasts with respect to the redox potential of the medium. Dokl. Akad. Nauk SSSR 249: 227-230

[33] Klimov VV, Shuvalov VA, Krakhmaleva IN, Klevanik AV and Krasnovskii AA (1977c) Photoreduction of bacteriopheophytin b in the primary light reaction of Rhodopseudomonas viridis chromatophores. Biokhimiya 42: 519-530

[34] Kobayashi M, Watanabe T, Nakazato M, Ikegami I, Hiyama T, Matsunaga T and Murata N (1988) Chlorophyll a'/P700 and pheophytin a/P680 stoichiometries in higher plants and cyanobacteria determined by HPLC analysis. Biochim. Biophys. Acta 936: 81-89

[35] Kobayashi M, van de Meent EJ, Amesz J, Ikegami I and Watanabe T (1991) Bacteriochlorophyll g epimer as a possible reaction center component of heliobacteria. Biochim. Biophys. Acta 1057: 89-96

[36] Kobayashi M, van de Meent EJ, Oh-oka H, Inoue K, Itoh S, Amesz J and Watanabe T (1992) Pigment composition of heliobacteria and green sulfur bacteria. In: Murata N (ed) Research in Photosynthesis, Vol 1, pp 393-396. Kluwer Academic Publishers, Dordrecht

[37] Kobayashi M, Akiyama M, Yamamura M, Kise H, Ishida N, Koizumi M, Kano H and Watanabe T (1998a) Acidiphilium rubrum and zinc-bacteriochlorophyll, part2: Physicochemical comparison zinc-type chlorophylls and other metallochloro-phylls. in

Photosynthesis: Mechanism and Effects, ed. by Garab, G., Kluwer Academic Publishers, Dordrecht, The Netherlands, vol. 2, 735-738

[38] Kobayashi M, Hamano T, Akiyama M, Watanabe T, Inoue K, Oh-oka H, Amesz J, Yamamura M and Kise H (1998b) Light-independent isomerization of bacteriochlorophyll g to chlorophyll a catalyzed by weak acid in vitro. Anal. Chim. Acta 365: 199-203

[39] Kobayashi M, Akiyama M, Kise H, Takaichi S, Watanabe T, Shimada K, Iwaki M, Itoh S, Ishida N, Koizumi M, Kano H, Wakao N and Hiraishi A (1998c) Structural determination of the novel Zn-containing bacteriochlorophyll in Acidiphilium rubrum. Photomed. Photobiol. 20: 75-80

[40] Kobayashi M, Yamamura M, Akutsu S, Miyake J, Hara M, Akiyama M and Kise H (1998d) Successfully controlled isomerization and pheophytinization of bacteriochlorophyll b by weak acid in the dark in vitro. Anal. Chim. Acta 361: 285-290

[41] Kobayashi M, Yamamura M, Akiyama M, Kise H, Inoue K, Hara M, Wakao N, Yahara K and Watanabe T (1998e) Acid resistance of Zn-bacteriochlorophyll a from an acidophilic bacterium Acidiphilium rubrum. Anal. Sci. 14: 1149-1152

[42] Kobayashi M, Oh-oka H, Akutsu S, Akiyama M, Tominaga K, Kise H, Nishida F, Watanabe T, Amesz J, Koizumi M, Ishida N and Kano H (2000) The primary electron acceptor of green sulfur bacteria, bacteriochlorophyll 663, is chlorophyll a esterified with $\Delta 2,6$-phytadienol. Photosynth. Res. 63: 269-280

[43] Kobayashi M, Watanabe S, Gotoh T, Koizumi H, Itoh Y, Akiyama M, Shiraiwa Y, Tsuchiya T, Miyashita H, Mimuro M, Yamashita T and Watanabe T (2005) Minor but key chlorophylls in Photosystem II. Photosynth. Res. 84:201-207

[44] Kobayashi M, Akiyama M, Kise H, Watanabe T (2006) Unusual tetrapyrrole pigments of photosynthetic antennas and reaction centers: Specially-tailored chlorophylls, Chlorophylls and Bacteriochlorophylls: Biochemistry, Biophysics, Functions and Applications, ed. by B. Grimm, R. J. Porra, W. Rüdiger and H. Scheer, Springer, Dordrecht, 55-66

[45] Kobayashi M, Akiyama M, Kano H, Kise H (2006) Spectroscopy and structure determination, in Chlorophylls and Bacteriochlorophylls: Biochemistry, Biophysics, Functions and Applications, ed. by B. Grimm, R. J. Porra, W. Rüdiger and H. Scheer, Springer, Dordrecht, 79-94

[46] Kobayashi M, Ohashi S, Iwamoto K, Shiraiwa Y, Kato Y, Watanabe T (2007) Redox potential of chlorophyll d in vitro. Biochim. Biophys. Acta 1767:596–602.

[47] Koizumi H, Itoh Y, Hosoda S, Akiyama M, Hoshino T, Shiraiwa Y, Kobayashi M (2005) Serendipitous discovery of Chl d formation from Chl a with papain, Sci. Tech. Adv. Material 6: 551–557.

[48] Krabben L, Schlodder E, Jordan R, Carbonera D, Giacometti G, Lee H, Webber A.N, Lubitz W (2000) Influence of the axial ligands on the spectral properties of P700 of photosystem I: a study of site-directed mutants. Biochemistry 39: 13012–13025

[49] Kühl M, Chen M, Ralph PJ, Schreiber U and Larkum AWD (2005) A niche for cyanobacteria containing chlorophyll d. Nature 433: 820

[50] Kumazaki S, Abiko K, Ikegami I, Iwaki M and Itoh S (2002) Energy equilibration and primary charge separation in chlorophyll d-based photosystem I reaction center isolated from Acaryochloris marina. FEBS Lett. 530:153-157

[51] Manning WM and Strain HH (1943) Chlorophyll d, a green pigment of red algae. J. Biol. Chem. 151: 1-19

[52] Miller SM, Augustine S, Olson TL, Blankenship RE, Selker J and Wood AM (2005) Discovery of a free-living chlorophyll d-producing cyanobacterium with a hybrid proteobacterial/cyanobacterial small-subunit rRNA gene. Proc. Natl. Acad. Sci. USA 102: 850–855.

[53] Miyashita H, Ikemoto H, Kurano N, Adachi K, Chihara M and Miyachi S (1996) Chlorophyll d as a major pigment. Nature 383: 402

[54] Miyashita H, Adachi K, Kurano N, Ikemoto H, Chihara M, Miyachi S (1997) Pigment composition of a novel oxygenic photosynthetic prokaryote containing chlorophyll d as the major chlorophyll. Plant Cell Physiol. 38:274-281

[55] Mohr R, Voss B, Schliep M, Kurz T, Maldener I and Adams DG (2010). A new chlorophyll d-containing cyanobacterium: evidence for niche adaptation in the genus Acaryochloris. ISME J. 4: 1456-1469.

[56] Murakami A, Miyashita H, Iseki M, Adachi K and Mimuro M (2004) Chlorophyll d in an epiphytic cyanobacterium of red algae. Science 303:1633.

[57] Nakamura A, Suzawa T and Watanabe T (2004) Spectroelectrochemical determination of the redox potential of P700 in spinach with an optically transparent thin-layer electrode. Chem. Lett. 33: 688-689.

[58] Noy D, Fiedor L, Hartwich G, Scheer H and Scherz A (1998) Metal-substituted bacteriochlorophylls. 2. Changes in redox potentials and electronic transition energies are dominated by intramolecular electrostatic interactions. J. Am. Chem. Soc. 120: 3684-3693

[59] Ohashi S, Miyashita H, Okada N, Iemura T, Watanabe T and Kobayashi M (2008) Unique photosystems in Acaryochloris marina. Photosynth. Res. 98: 141-149

[60] Ohashi S, Iemura T, Okada N, Itoh S, Furukawa H, Okuda M, Ohnishi-Kameyama M, Ogawa T, Miyashita H, Watanabe T, Itoh S, Oh-oka H, Inoue K and Kobayashi M (2010) An overview on

[61] chlorophylls and quinones in the photosystem I-type reaction centers. Photosynth. Res.104: 305-319

[62] Ohkubo S, Usui H and Miyashita H (2011) Unique chromatic adaptation of a unicellular cyanobacterium newly isolated from Lake Biwa. Jpn. J. Phycol. (Sorui) 59: 52(A22) (in Japanese)

[63] Parson WW, Clayton RK and Cogdell RJ (1975) Excited states of photosynthetic reaction centers at low redox potentials. Biochim Biophys Acta 387: 265-278

[64] Pelletier PJ and Caventou JB (1818) Ann. Chim. Phys. 9: 194

[65] Petke JD, Maggiora G, Shipman L and Christoffersen R (1979) Stereoelectronic Properties of Photosynthetic and related systems - v. ab initio configuration interaction calculations on the ground and lower excited singlet and triplet states of ethyl chlorophyllide a and ethyl pheophorbide a. Photochem.Photobiol. 30: 203-223.

[66] Porra RJ and Scheer H (2000) 18O and mass spectrometry in chlorophyll research: Derivation and loss of oxygen atoms at the periphery of the chlorophyll macrocycle during biosynthesis, degradation and adaptation. Photosynth. Res. 66: 159–175

[67] Rockley MG, Windsor MW, Cogdell RJ and Parson WW (1975) Picosecond detection of an intermediate in the photochemical reaction of bacterial photosynthesis. Proc. Natl. Acad. Sci. USA 72: 2251-2255

[68] Schoch S (1978) The esterification of chlorophyllide a in greening bean leaves. Z . Naturforsch., C: Biosci. 33C: 712-714.

[69] Shoaf W T (1978) Rapid method for the separation of chlorophylls a and b by high-pressure liquid chromatography. J. Chromatogr. 152: 247-249.

[70] Shuvalov VA, Amesz J and Duysens LNM (1986a) Picosecond spectroscopy of isolated membranes of the photosynthetic green sulfur bacterium Prosthecochloris aestuarii upon selective excitation of the primary electron donor. Biochim. Biophys. Acta 851: 1-5

[71] Shuvalov VA, Vasmel H, Amesz J and Duysens LNM (1986b) Picosecond spectroscopy of the charge separation in reaction centers of Chloroflexus aurantiacus with selective excitation of the primary electron donor. Biochim. Biophys. Acta 851: 361-368

[72] Sivakumar V, Wang R and Hastings G (2003) Photo-oxidation of P740, the primary electron donor in photosystem I from Acaryochloris marina. Biophys. J. 85:3162–3172

[73] Smith KM (1975) Mass spectrometry of porphyrins and metalloporphyrins. In: Smith KM (ed) Porphyrins and Metalloporphyrins, pp 381-398, Elsevier Scientific Publishing Company, Amsterdam

[74] Smith KM and Unsworth JM (1975) The nuclear magnetic resonance spectra of porphyrins-◦1. Tetrahedon 31: 367-375

[75] Strain HH and Manning WM (1942) Isomerization of chlorophylls a and b. J. Biol. Chem. 146: 275-276

[76] Swingley WD, Chen M, Cheung PC, Conrad AL, Dejesa LC, Hao J, Honchak BM, Karbach LE, Kurdoglu A, Lahiri S, Mastrian SD, Miyashita H, Page L, Ramakrishna P, Satoh S, Sattley WM, Shimada Y, Taylor HL, Tomo T, Tsuchiya T, Wang ZT, Raymond J, Mimuro M, Blankenship RE and Touchman JW (2008) Niche adaptation and genome expansion in the chlorophyll d-producing cyanobacterium Acaryochloris marina. Proc. Natl. Acad. Sci .USA 105:2005–2010.

[77] Telfer A, Pascal A, Barber J, Schenderlein M, Schlodder E and Çetin M (2007) Electron transfer reactions in photosystem I and II of the chlorophyll d containing cyanobacterium, Acaryochloris marina. Photosynth. Res. 91:143

[78] Tomitani A, Okada K, Miyashita H, Matthijs HCP, Ohno T and Tanaka T (1999) Chlorophyll b and phycobilins in the common ancestor of cyanobacteria and chloroplasts. Nature 400: 159-162.

[79] Tomo T, Okubo T, Akimoto A, Yokono M, Miyashita H, Tsuchiya T, Noguchi T and Mimuro M (2007) Identification of the special pair of photosystem II in a chlorophyll d-dominated cyanobacterium. Proc. Natl. Acad. Sci. USA 104: 7283-7288

[80] Tomo T, Kato Y, Suzuki T, Akimoto S, Okubo T, Noguchi T, Hasegawa K, Tsuchiya T, Tanaka K, Fukuya M, Dohmae N, Watanabe T and Mimuro M (2008) Characterization of highly purified photosystem I complexes from the chlorophyll d-dominated cyanobacterium Acaryochloris marina MBIC 11017. J. Biol. Chem. 283:18198-18209

[81] van Gorkom HJ (1974) Identification of the reduces primary electron acceptor of photosystem II as a bound semiquinone anion. Biochim. Biophys. Acta 347: 439-442

[82] Wasielewski MR, Smith RL and Kostka AG (1980) Electrochemical production of chlorophyll a and pheophytin a excited states. J. Am. Chem. Soc. 102: 6923-6928.

[83] Watanabe T, Hongu A, Honda K, Nakazato M, Konno M and Saitoh S (1984) Preparation of chlorophylls and pheophytins by isocratic liquid chromatography. Anal. Chem. 56: 251-256

[84] Watanabe T, Nakazato M, Konno M, Saitoh S and Honda K (1984) Epimerization in the pheophytin a/a' system. Chem. Lett. 1411-1414

[85] Watanabe T and Kobayashi M (1991) Electrochemistry of chlorophylls, in: H. Scheer(ed.), Chlorophylls, CRC Press, Boca Raton pp. 287–315.

[86] Weiss C (1978) Electronic absorption spectra of chlorophylls. In: Dolphin D (ed) The Porphyrins, Vol. III, Physical Chemistry, Part A, pp 211-223. Academic Press, New York

[87] Wolf H and Scheer H (1973) Stereochemistry and chiroptic properties of pheophorbides and related compounds. Ann. N. Y. Acad. Sci. 206: 549-567

[88] Zouni A, Witt HT, Kern J, Fromme P, KrauβN, Saenger W and Orth P (2001) Crystal structure of photosystem II from Synechococcus elongatus at 3.8 Å resolution. Nature 409: 739-743

Chapter 4

Quantum Yields in Aquatic Photosynthesis

David Iluz and Zvy Dubinsky

Additional information is available at the end of the chapter

1. Introduction

1.1. Photosynthetic efficiency

Primary productivity is the product of the light energy absorbed by plants and the efficiency by which this energy is stored as a photosynthate. The quantum yield (ϕ) of photosynthesis is defined (Eq. 1) as the molar ratio between oxygen released in photosynthesis (or carbon assimilated) to photons absorbed in the process (Fig. 1) (Dubinsky, 1980; Dubinsky & Berman, 1976, 1979, 1981; Dubinsky et al., 1984). The quantum yield is, therefore, equal to the ratio of photosynthetically stored radiation (PSR) to the energy absorbed photosynthetically usable radiation (PUR)[for definitions, see Morel (1978)].

$$\phi = \frac{\text{moles oxygen evolved (CO}_2 \text{ absorbed)}}{\text{moles light quanta absorbed}} = \frac{\text{PSR}}{\text{PUR}} \tag{1}$$

To calculate ϕ, we need to know the fraction of incident light that is absorbed by phytoplankton cells (Morel, 1978). This fraction is proportional to the product of the photosynthetic pigment concentration and a* - the specific *in vivo* absorbance constant of pigments (Dubinsky, 1980, 1992; Kirk, 1994). For *in situ* studies, both the spectral absorbance of the cells and the intensity and spectrum of ambient light are taken into account as the k_c parameter (Dubinsky, 1992; Dubinsky et al., 1986; Schanz et al., 1997) whose dimensions are $m^2 \, mg^{-1}$ chl. Thus, the absorbed light is proportional to k_c, the *chlorophyll a* concentration, and light intensity (E).

Early attempts to define and measure the quantum yield were published by several pioneers in photosynthesis research. Some of the symbols for quantum yield used in the past and references are listed in Table 1. The present review focuses on the measurement of the quantum yields of photosynthesis in the aquatic domain.

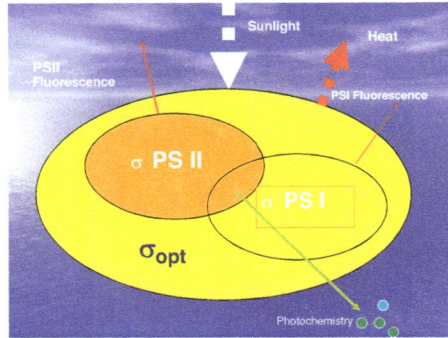

Figure 1. Scheme of quantum yield, which is the ratio between the energy in photochemistry products to photons absorbed from sunlight in the photosynthesis process, whereby the major initial loss is thermal dissipation (see chapter by Pinchasov, this volume), and minor ones are fluorescence emitted primarily by PSII and marginally by PSI.

Source	Symbol	Value
Aruga and Ichimura, 1968	ε	0.0184
Taling, 1970	K_s	0.01-0.12
Megard, 1972	K_s	0.01-0.12
Bannister, 1974	K_c	0.016
Ganf, 1974	ε_s	0.012-0.016
Berman, 1976	ε_s	0.006
Jewson, 1976	ε_s	0.011
Bindloss, 1976	K_s	0.0086
Dubinsky and Berman, 1979	k_c, from η^n_{450}	0.0121
Dubinsky and Berman, 1979	k_c, from η^n_{560}	0.005
Dubinsky and Berman, 1979	k_c, from η^n_{650}	0.0112
Dubinsky and Berman, 1979	k_c, from η^n_{PhAR}	0.0067

Table 1. Specific extinction coefficients of chlorophyll (in units* mg $^{-1}$Chl*m^{-2}) (Dubinsky & Berman, 1979)

1.2. Maximal quantum yield

In principle, ϕ_{max} cannot exceed 0.125 since four electrons are required for the evolution of one molecule of O_2 from water. According to the "Z" diagram of photosynthesis (Emerson & Lewis, 1943), each electron is driven by two photons, each absorbed in PSII and PSI, resulting in a minimal photon requirement ($1/\phi$) of eight (Fig. 2).

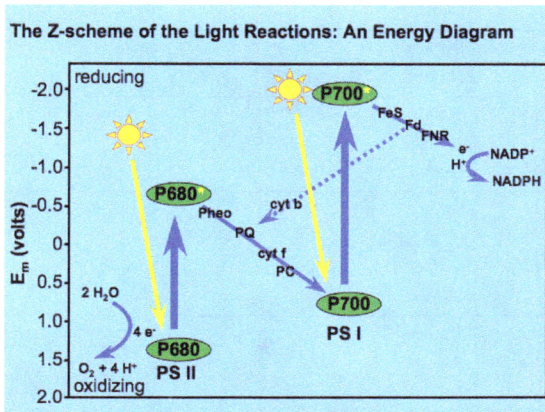

Figure 2. The "Z" scheme of photosynthesis, demonstrating the minimum requirement of 2 x 4 photons for the evolution of one molecule of O_2 and the concomitant assimilation of CO_2.

In cases where the nitrogen source was nitrate, not ammonium, at least two more photons were consumed to provide for its reduction to cell components, reaching ϕ_{max} values <0.1. High values, approaching the theoretical maxima, were observed only under low light. Under high light, the rates of photon absorption by antennae exceed the rates of energy utilization in photochemical processes. These rates were limited by the bottleneck of electrons shuttling through the quinone pool from PSII to PSI. The excess energy had to be dissipated as heat following absorption by photoprotective pigments, such as peridinin and astaxanthine, and by means of the xanthophyll cycle. Table 2 lists some ϕ_{max} values found in different regions.

In all studies, an increase in ϕ_{max} values with depth from the surface to the deepest samples, was observed. Morel (1978) estimated ϕ_{max} values at the surface based on ^{14}C incubation data in the oligotrophic Sargasso Sea and the highly productive Mauritanian upwelling zone. In the eutrophic, green Mauritanian data, ϕ_{max} was considerably higher than in the Sargasso area, as would be expected in a nutrient-replete environment, and reached 0.012. In the oligotrophic Sargasso Sea, ϕ_{max} was found to be only around 0.003, reflecting the oligotrophic conditions in that region. Kishino et al. (1985), working in the Pacific, south of Japan, reported values of 0.004-0.01 at the surface and 0.01-0.026 at the 10-20 m level, where the peak of photosynthesis was found. Values increased further up to 0.026-0.075 at the deep chlorophyll maximum (DCM), at 70 m. During the annual *Peridinium* bloom in the monomictic, mesotrophic Lake Kinneret (Israel), Dubinsky and Berman (1981) observed an increase from 0.025 at the surface to 0.043 at 3 m, below the euphotic zone. During vernal stratification with nutrients in the epilimnion (the nutrients were exhausted in that period), phytoplankton was dominated by minute chlorophytes and ϕ_{max} values were lower as expected: 0.0126 at the surface and a maximum of 0.06 at 5-7 m.

Locality	Depth	φ	Reference
Sargasso Sea, Mauritania	Surface	0.003	Morel (1978)
	Surface	0.012	
Pacific Ocean, south of Japan	Surface	0.004-0.01	Kishino et al. (1985)
	10-20 m	0.01-0.026	
	70 m	0.026-0.075	
Gulf of Eilat, Israel.	Surface	0.00025	Iluz (2008)
winter.	80 m	0.110	
Gulf of Eilat, Israel. summer	Surface	0.00087	
	80 m	0.0266	
Lake Kinneret, Israel.Winter	Surface	0.025	Dubinsky and Berman (1981)
Lake Kinneret, Israel.summer	3 m	0.043	
	Surface	0.0126	
	5-7 m	0.06	
Zooxanthellae in hospice		0.001[1]-0.125[2]	Dubinsky et al. (1984)

[1] From high-light corals

[2] From low-light corals

Table 2. Quantum yield (φ) values found in different regions

In the Gulf of Eilat, we found an increase in ϕ_{max} values with depth for all profiles (Fig. 3). However, it is noteworthy that the correlation coefficient between light and φ was only R^2 = 0.7 for all profiles, whereas in summer it was 0.85 and in winter it reached 0.91. For the pooled data, these differences indicate that additional factors besides light intensity do affect the quantum yields of photosynthesis. The seasonal trend lines clearly point towards the effect of nutrients. In summer, lack of nutrients does restrict photosynthetic efficiency, whereas in winter, due to vertical mixing, no such effect is evident (Iluz et al., 2008).

The trends observed in quantum-yield values are in agreement with their oceanic distributions. For instance, Prezelin et al. (1991) found φ ranging from0.01 to 0.06 in a transect 200 km south of California. These differences were linked to different water masses and depths. Some of these spatial changes were related to the taxonomic differences between phytoplankton assemblages. For instance, diatom-dominated sites had φ twice as high as those consisting of picoplankton at the DCM (Schofield et al., 1991). However, these represent complex differences, not necessarily taxonomic *per se*. Diatoms thrive in nutrient-rich situations, whereas picophytoplankton outcompete all larger eukaryotes in oligotrophic waters due to their higher surface/volume ratios. Furthermore, the DCM is found at the very bottom, or even just below, the euphotic depth at very low light, where higher φ values are always to be expected.

In their study on the Sargasso Sea, Cleveland et al. (1989) reported an inverse correlation between φ values ranging from 0.033 to 0.102, and the distance from the nitrocline. These results are in accord with those of Kolber et al. (1990) from the Gulf of Maine. In both cases,

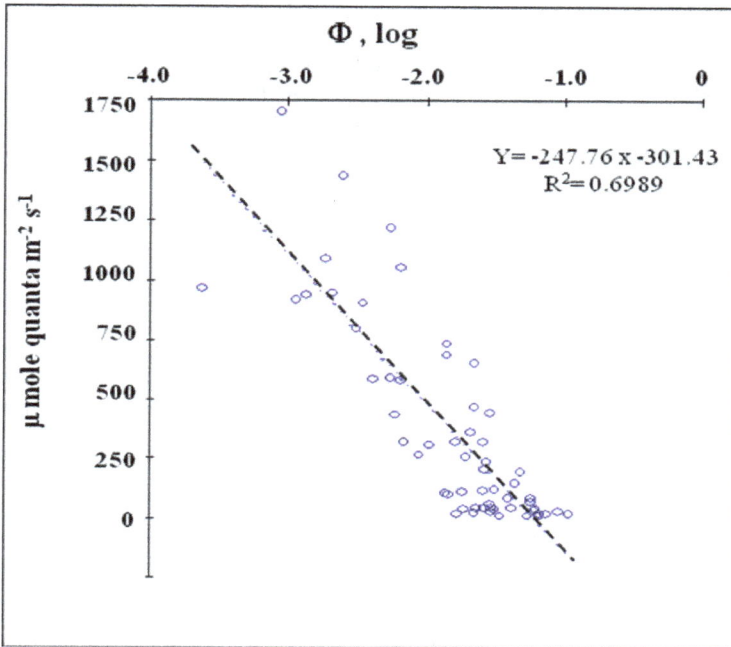

Figure 3. Linear regression of the quantum yield vs. light intensity for all profiles down to 100 m depth, between 24.10.94 and 18.06.96 (n= 56).

the difference in quantum yields was explained by the flux of nitrogen from deep waters into the euphotic zone. A diel pattern superimposed upon the spatial differences in φ, peaking in the morning and declining in the afternoon (Kishino et al., 1985; Prezelin et al., 1987, 1991), was also reported. Such an "afternoon nap" (Schanz & Dubinsky, 1988; Walsby et al., 2001) was also seen by Tilzer (1984) in Lake Konstanz, where threefold diel changes in φ exceeded their seasonal amplitude. Here, too, values decreased from morning to afternoon. In general, other than in their universal bathymetric trend, quantum yields of photosynthesis [e.g., Cleveland et al. (1989)] and the related fluorescence yields (Falkowski, 1991; Kolber et al., 1990)reveal the overriding control of nutrients over oceanic primary productivity.

2. Methods based on measuring light absorption and photosynthesis rates

To calculate the quantum yield most ways, we need to measure the rate of photosynthesis (but see below the sections on photoacoustics and variable fluorescence) and light absorption (Eq. 2).

2.1. Absorbed light

The denominator of that fraction, the absorbed light, was calculated as follows (Dubinsky & Berman, 1981). From the vertical attenuation coefficient, the total light absorbed in the slice was calculated as the difference between light entering the top of the layer and that at the bottom. It was then partitioned into that absorbed by the phytoplankton and all other light-absorbing substances in the water according to:

$$K_d = k_{w'} + k_c chl \text{ and } k_{w'} = k_w + k_g + k_{tr} \tag{2}$$

where K_d is the attenuation coefficient PAR [ln units m^{-1}], $k_c[m^2mg^{-1}chl\ a]$ is the spectrally averaged, in situ, specific extinction coefficient of *chlorophyll a*, and chl is *chlorophyll a* concentration [mg *chl a* m^{-3}]. k_w is attenuation coefficient due to water alone, k_g is gilvin, and k_{tr} is tripton - all of these in ln units m^{-1}.

Thus the absorbed light, given in mole quanta is:

$$\frac{(PARz_1 - PARz_2)k_c^*\ chl}{K_d} \tag{3}$$

From these, using ^{14}C tracer incubations (Steemann-Nielsen, 1952) in order to estimate the enumerator of the quantum yield fraction, the authors were able to calculate the ratio of mole carbon stored as photosynthate in a defined water volume to that of light quanta absorbed by the phytoplankton in the same water volume and at the same time interval.

By dividing primary production rates for the same time interval and volume as the absorbed light, we obtained the values of ϕ for all depths as molar ratios O_2/quanta.

A part of the light impinging on a phototroph cell is absorbed according to the relation between the absorbance spectrum of that cell's pigment assortment and the spectral distribution of the surrounding underwater light field. The fate of harvested light and the losses incurred until the remainder is stored in a generalized algal cell, and until it becomes available as a substrate for life-supporting respiration and building blocks for cell growth and multiplication, are summarized in Figure 1.

Under low-light intensity, photosynthesis is proportional to photon flux. In the light-limited range of the photosynthesis versus energy curve (P vs. E*), the quantum yield remains maximal, ϕ_{max}. This is equal to the ratio between α and k_c, when α is defined as the initial slope of the P vs. E curve and expressed as moles O_2 evolved per mg chlorophyll. Chlorophyll unit *in vivo*:

$$\alpha = \frac{\text{mole } O_2 mg^{-1}chl\ s^{-1}}{\text{mole photons } m^{-2}s^{-1}} = \phi_{max}k_c \tag{4}$$

From here we can release ϕ_{max} without knowing the amount of absorbed light at any given depth (Dubinsky et al., 1986; Falkowski et al., 1990).

The maximum quantum yield (ϕ_{max}) is measured when photosynthesis is light-limited, a situation diagnosed by a linear relationship between photosynthesis and photosynthetic photons, also incorrectly termed photon flux (flux already implies density!) density (PPFD).

Theoretically, ϕ_{max} is 0.125, since 8 moles of photons are required for a mole of oxygen to evolve from water according to the "Z" scheme of photosynthesis (Emerson & Arnold, 1932) in two photoactivations per electron, and concomitantly reduce 1 mole of CO_2 in the absence of photorespiration. Because there is some cyclic photophosphorylation, ϕ_{max} may be closer to 0.112 in most plants (Long et al., 1993). Furthermore, whenever the source of nitrogen is nitrate rather than ammonia, ϕ_{max} is further reduced due to the energy required for its reduction

2.2. Carbon

Primary production is usually measured in terms of the amount of biomass, carbon fixation, or oxygen produced. Gross primary production (Pg) is the rate of photosynthesis, the total amount of fixed energy (sunlight that has been transformed into the chemical energy of organic materials, i.e., photosynthesis):

$$Pg = R+Pn \tag{5}$$

where R is the energy that has been used by the autotrophs themselves in their respiration, and Pn is the energy that was not consumed, and results in growth or food for grazers.

The most common method for measuring aquatic photosynthesis is based on ^{14}C assimilation, which is briefly summarized below. Samples for ^{14}C productivity measurements -P(^{14}C)- are usually incubated in 60-ml polycarbonate bottles. Carbon uptake is measured with a modified ^{14}C uptake technique (Steemann-Nielsen, 1952). A spike of approximately 8 mCi of [^{14}C] bicarbonate is added to each bottle. After incubation, the samples are filtered under light vacuum (about 100 mm Hg) onto 25-mm 0.45-m filters (Millipore), rinsed with 15 ml of filtered lake water, and briefly fumed in HCl vapor to eliminate any remaining traces of inorganic ^{14}C. Control samples poisoned by Lugol's iodine at time zero were run with each experimental series to compensate for nonbiological adsorption to filters. The total added ^{14}C is also checked for each sampling series by counting 0.1-ml portions directly from each of the incubated bottles. Total radioactivity and the radioactivity in the particulate fraction retained on the filters are determined by liquid scintillation with quench correction. From the ratio of ^{14}C added to ^{12}C in the water, ^{14}C assimilation rates are converted to photosynthesis rates, taking into account the isotopic discrimination factor of 1.06.

2.3. Oxygen

Whenever quantum yields are based on oxygen evolution rates, there is an inherent difficulty since measured changes in oxygen concentration are net rates (Eq. 6), where P_G stands for gross photosynthesis, the parameter needed for the calculation of quantum yields, P_N is the measured, net photosynthesis rate, and R is respiration, usually measured in the dark. This assumes that dark respiration is the same as in the light, which is likely to be an underestimate, resulting in too low estimates of P_G and, consequently, of the quantum yield.

Oxygen exchange by photosynthesis and respiration is the largest biogeochemical cycle in aquatic systems, and the major biogeochemical cycle in the biosphere. In order to understand this cycle, it is necessary to know the gross rates of the major processes involved in oxygen production and uptake. The production of O_2 is known to occur in a four-step process in photosystem II, but O_2 consumption in aquatic organisms takes place by several reactions. These include ordinary respiration through the cytochrome oxidase pathway, respiration by the alternative oxidase pathway, Mehler reaction, and photorespiration. The first two processes take place in the light as well as under dark conditions, whereas the latter two occur only under illumination. Although the presence of the above-mentioned mechanisms has been established in different studies, their quantitative importance in the overall O_2 uptake in aquatic systems is not well known, and it is necessary to assess their role in natural environments. In this respect, ordinary O_2 incubation methods, which are very useful for the assessment of photosynthetic production from light- and dark-incubation experiments [e.g., Williams &Purdie (1991)], do not provide the necessary information.

Clark electrodes were used to measure the effect of intermittent light on photosynthetic oxygen evolution and on dark respiration rates. The main parameters of photosynthesis will be derived from the generated photosynthesis versus energy (P vs. E) curves (Fig. 4), including α, the initial slope of light-limited photosynthesis, and φ, the quantum yield of that process. E_k ($=I_k$), the irradiance level of incipient light saturation, the light saturated rate of photosynthesis, P_{max}, the light compensation point E_c ($=I_c$), and dark respiration R (Fisher et al., 1996), will also be obtained. The enhanced post-illumination respiration (EPIR) rates will be determined according to Falkowski et al. (1985) and Beardall et al. (1994). The Arnold and Emerson number will be determined from the evolution of oxygen per short, saturating flash, as chlorophyll molecules/O_2 molecules evolved (Emerson & Arnold, 1932).

$E_k=I_k$, $I_c=E_c$.

2.3.1. Stable isotopes

The main drawback of the above methods is their inability to measure the rate of respiratory O_2 uptake in the light, and it is assumed that the rates of dark and light uptakes are equal. The rates of gross production as well as light O_2 uptake can be estimated in field incubation experiments using $H_2^{18}O$ as a spike (e.g., (Bender et al., 1987, 1999; Luz & Barkan, 2000). However, this method alone cannot help to characterize the type of the respiratory mechanisms involved in aquatic O_2 uptake.

The discrimination against ^{18}O associated with the cytochrome oxidase pathway is 18‰ (Guy et al., 1992), but with the cyanide-resistant alternative oxidase pathway (AOX) it is much greater: 31‰ in green tissues and26‰ in nongreen tissues (Robinson et al., 1992). The discrimination in the Mehler reaction is 15‰ and in photorespiration – 21‰ (Berry, 1992; Guy

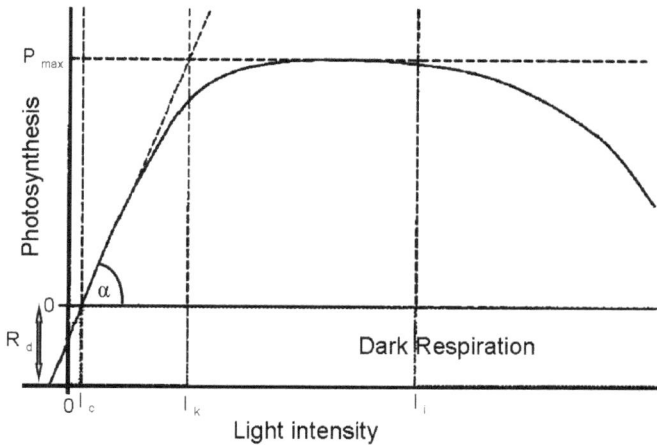

Figure 4. Light response curve of photosynthesis versus light intensity (Grobbelaar, 2006).

et al., 1993). d^{18}O of O$_2$ in a purified oxygen–argon mixture determined by dual inlet mass spectrometry.

3. Methods that do not require light absorption and photosynthesis rates

3.1. Variable fluorescence

Using a custom-built fluorescence induction and relaxation (FIRe) instrument, as described in Gorbunov and Falkowski (2004). The FIRe technique is based on recording fluorescent transients (called "variable fluorescence") induced by a sequence of excitation flashes of light with precisely controlled intensity, duration, and intervals between flashes. Analysis of the fluorescence signals provides a comprehensive suite of photosynthetic characteristics of the organism, including the minimum (F$_o$) and maximum (F$_m$) fluorescence yields corresponding to open and closed reaction centers of PSII, respectively, variable fluorescence component (F$_v$), the quantum yield of photochemistry in PSII (simply put - photosynthetic efficiency), the functional absorption cross section of PSII, and the rates of photosynthetic electron transport down to PSII (Gorbunov & Falkowski, 2004; Kolber et al., 1998) (Fig. 5). The size of the plastoquinone (PQ) pool (i.e., the number of PQ molecules per reaction center) can be determined from the comparative analysis of the fluorescence induction on the millisecond time scale in the absence and presence of (3-(3,4-dichlorophenyl)-1,1-dimethylurea) (DCMU). The redox state of the PQ pool is also assessed from the shape of the fluorescence induction curve. Variable fluorescence measurements under ambient light provide information about the efficiency of non-photochemical quenching (NPQ) and the rates of photosynthetic electron transport as a function of light intensity. A computer-controlled ambient light source is integrated into the FIRe instrument for fully automatic measurements of the above variables. The photoinhibition of PSII by supra-optimal light will be estimated from the reduction in dark-adapted values of Fv/Fm compared to their night values (Long et al., 1994).

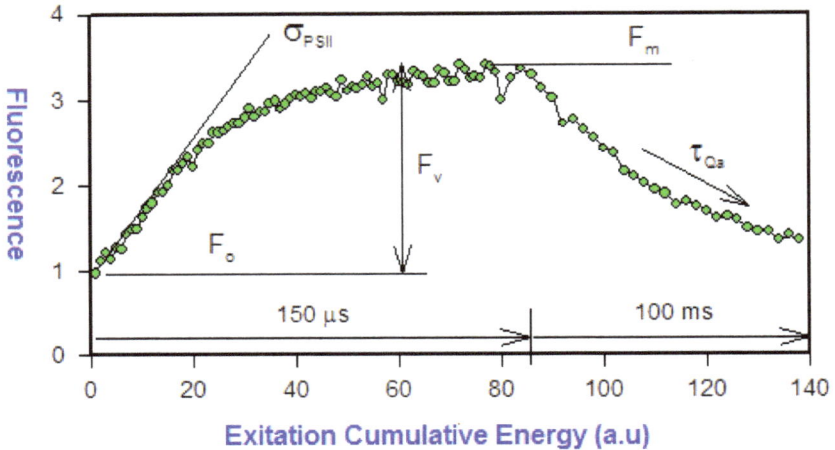

Figure 5. The parameters obtainable from FIRe measurements: the minimum (F_o) and maximum (F_m) fluorescence yields corresponding to open and closed reaction centers of PSII, respectively, variable fluorescence component (F_v), the quantum yield of photochemistry in PSII (simply put - photosynthetic efficiency), and the functional absorption cross section of PSII.

3.2. Fast Repetition Rate (FRR) fluorescence

The simultaneous response of σ_{PSII}(Å^2 quanta^{-1}) and τ(μs) reveals important information about the photosynthetic response to the growth environment. Specifically, the light saturation parameter (E_K, μmol photons m^{-2} s^{-1}) was estimated as $[1/(\tau\,\sigma)]\,1.66 \times 10^8$ (Falkowski & Raven, 2007), where the factor 1.66×10^8 accounts for the conversion of Å^2 to m^2, quanta to μmol quanta (photons), and μs to s (e.g., Moore et al. (2006). The actual value of E_K is dependent upon both the wavelength used to generate σ_{PSII} as well as which time constant associated with the FRR relaxation phase is used to describe τ (Kolber et al., 1998).

3.3. Photoacoustics

Energetics of photosynthesis determined by pulsed photoacoustics. This methodology directly determines the light energy not stored in photosynthesis. It is, thus, ideal for determining the changes in stored energy on the microsecond, i.e., the photochemical time scale, caused by the differing light regimes. The methodology has been described and protocols given for measurements of both reaction centers and whole cells (Boichenko et al., 2001; Hou et al., 2001a, 2001b). The method has been successfully applied to the measurement of biomass (Dubinsky et al., 1998), to discriminate between taxa of phytoplankton (Mauzerall et al., 1998), and to study the physiological state of phytoplankton (Pinchasov et al., 2005). The efficiency of energy storage will be determined in the sample before and after a light regime that affects the growth rate and/or oxygen production rate of the organism. The variable light regime is continued until a steady state is reached and the photoacoustic measurements can be made in a shorter time compared to that of the light regime. For the slower intermittent light-dark regimes, one may be able to measure the state of the system in each phase. The results will indicate if the variable light effect is caused by a change in the efficiency of light utilization at the photochemical level. If a change is seen then, energy utilization has changed and the xanthophyll cycle will be implicated. If no change is seen, then the change has occurred on the long time scale, such as in the case of CO_2 fixation.

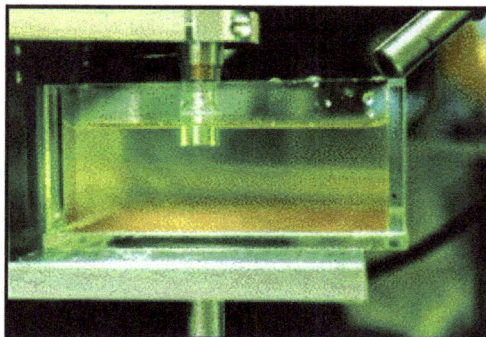

Figure 6. Photoacoustic phytoplankton cell.

4. Factors affecting quantum yields

4.1. Photoacclimation and photoinhibition

Phytoplankton photosynthesis at any depth depends principally on the intensity and spectral quality of the ambient light. However, the amount of light harvested by phytoplankton also depends on the quantity of chlorophyll present and on the variable, average, spectral in vivo attenuation coefficient, k_c (Dubinsky & Berman, 1981; Schanz et al., 1997). Moreover, as available light changes, so does the fraction of the harvested light that can be transduced by the cell into photochemical products - this fraction is ϕ. Contrary to what has been discussed above, happening under limiting light, at high irradiances the photon flux harvested by the photosynthetic pigments exceeds the rate at which these photons can be utilized by the photochemical reaction centers. In such light-saturated situations, an increased fraction of the harvested light will be dissipated by nonradiative decay as heat, and emitted as fluorescence. Under such conditions, the photosynthetic apparatus may be temporarily or irreversibly damaged, leading to photoinhibition and ensuing reduction in quantum yields. In a bathymetric profile, in any water body (Fig. 7) exposed to full sunshine, the photosynthesis of phytoplankton is inhibited at the surface due to supra-optimal irradiance levels, photosynthesis is inhibited, and quantum yields are low, as light reaches limiting levels at (E_k), and below that depth throughout the photic zone, photosynthesis is light limited and quantum yields are constant (Fig. 7). All these depth-related changes in irradiance, chlorophyll, k_c, and ϕ are essential inputs for modeling and predicting the depth-distribution of photosynthesis in aquatic environments.

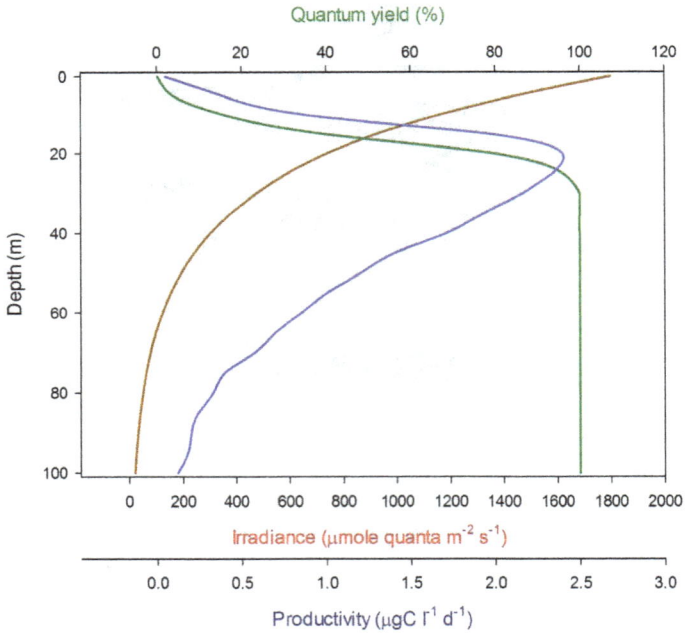

Figure 7. Profile of irradiance, photosynthesis and quantum yield.

In the laboratory, Dubinsky et al. (1986) exposed different cultures to different light levels and found changes in quantum yield (ϕ) vs. light intensity (Fig. 8).

Additional features emerge when summer (June 1995, June 1996) data are compared to winter (February 1995, December 1996) data (Fig. 9). Winter ϕ values decreased steeply at high

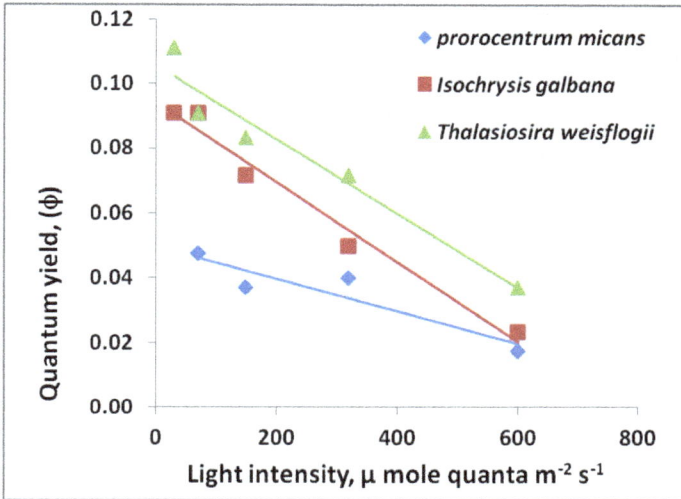

Figure 8. Quantum yield (φ) values found in cultures, after Dubinsky et al. (1986).

irradiance levels, reaching undetectably low ones at I>400 μmole quanta $m^{-2}s^{-1}$, whereas in summer samples, values were measurable even at irradiances twice as high. We attribute the high sensitivity of summer phytoplankton to light intensity to the fast vertical mixing characteristic of that period, which did not allow sufficient time for photoacclimation (Falkowski & Wirick, 1981). The phytoplankton organisms acclimate to a light intensity that is the average over the entire mixing depth of ~400 m, resulting in cells with far too much pigmentation not to be damaged or at least strongly photoinhibited upon exposure to near-surface irradiances. That also explains why they fare better under dim light, under which the nearly optically black cells maximize light harvesting and its efficient utilization. Conversely, during summertime, stratification cells acclimate to the light intensity at each depth, thereby mitigating the effects of light gradients. However, the shortage of nutrients limits the ability of phytoplankton to fully exploit the advantages of photoacclimation.

At all times, the quantum yields of photosynthesis are strongly reduced under high light since non-photochemical quenching (NPQ) excess-energy–dissipating processes avert photodynamic damage. Furthermore, not only is there, under high light, a mismatch between light harvesting and end-electron flow rates, but also between the fast light-driven carbon assimilation and the Redfield rate supply of nitrogen and phosphorus [see (Dubinsky & Berman-Frank, 2001)], all depressing the quantum yields of photosynthesis. Under low light approaching ~1% of the subsurface, light-harvesting rates, being in step with τ values, allow quantum yields to reach their theoretical upper boundaries of 8-10 photons per mole O_2 evolved.

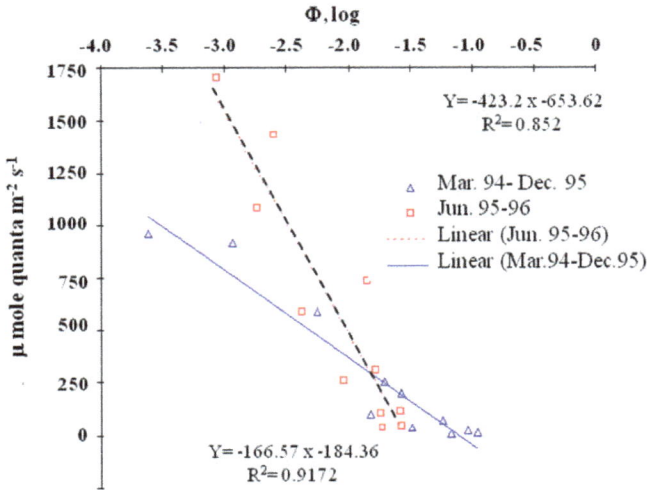

Figure 9. Linear regression of the quantum yield vs. light intensity, comparison between summer (June 1995, June 1996) and winter profiles (February 1995, December 1996) down to 100 m depth.

4.2. Nutrient status

Al Qutob et al. (2002) showed the co-limitation of phytoplankton photosynthesis in the gulf by both nitrogen and phosphorus. During thermal summer stratification, nutrient depletion was severe, and no nitrite could be detected in the upper 70 m. Their field data suggest that the accumulation of nitrite is associated with nutrient-stimulated phytoplankton growth. This hypothesis was supported by nutrient-enrichment bioassays performed concomitantly: only when phytoplankton growth was stimulated by nutrient additions, did nitrite accumulate in the water., Using photoacoustics, Pinchasov et al. (2005) showed the depression of photosynthetic efficiency of several phytoplankton species under iron deficiency (Fig. 10) and under nitrogen and phosphorus starvation (Fig. 11) (Pinchasov et al., 2005).

There is a dependence of quantum yields on light intensity (Dubinsky, 1992; Dubinsky & Berman, 1981; Morel, 1978). However, quantum yields depend not only on ambient light. It was also reported that in laboratory experiments, nutrient limitation lowers quantum yields (Cleveland et al., 1989; Falkowski, 1991; Kolber et al., 1990). Where the specific importance of an individual nutrient is concerned, Kolber et al. (1988) showed the effects of lack of nitrogen, Greene et al. (1991) showed the effects of lack of iron, and Falkowski (1991) reported the effects of phosphorus limitation. The impact of these shortages depends on their cellular requirements for balanced growth, N>P>Fe, which differ among phytoplankton taxa, but also in their amounts stored in cells [*sensu* cell quota, Droop (1983)]. In the case of the Gulf of Eilat, during

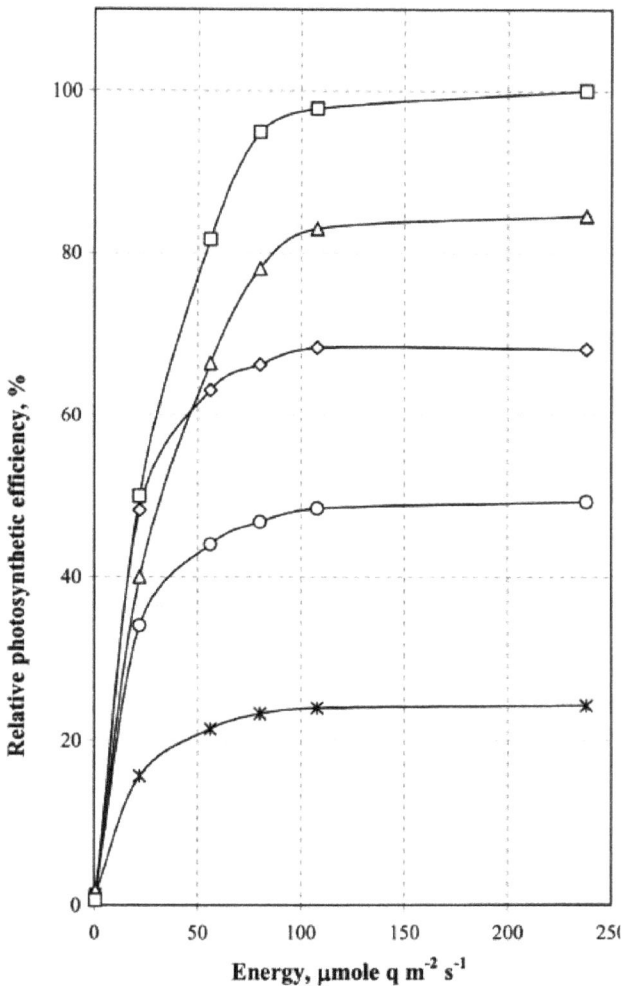

Figure 10. The effect of changing iron concentration on photosynthetic energy storage for *Isochrysis galbana*. The iron concentrations in the cultures were 0 mg L⁻¹, (.); 0.03 mg L⁻¹, (o); 0.09 mg L⁻¹, (◊); 0.18 mg L⁻¹ (Δ); and for the control, 0.6 mg L⁻¹ (*). The maximal storage in the nutrient-replete control was taken as 100%.

summertime stratification, there is a concomitant shortage of both nitrogen and phosphorus as their winter concentrations of $NO_3^- = 0.2$ μmol l⁻¹ and $PO_4^{-3} = 0.1$ μmol l⁻¹ in winter, drop to $PO_4^{-3} = 0.02$ μmol l⁻¹ and $NO_3^- = 0.04$ μmol l⁻¹ in summer, respectively (Al-Qutob, 2001; Al-Qutob et al., 2002; Genin et al., 1995; Iluz et al., 2008; Labiosa et al., 2003; Lazar & Erez, 1992; Levanon-

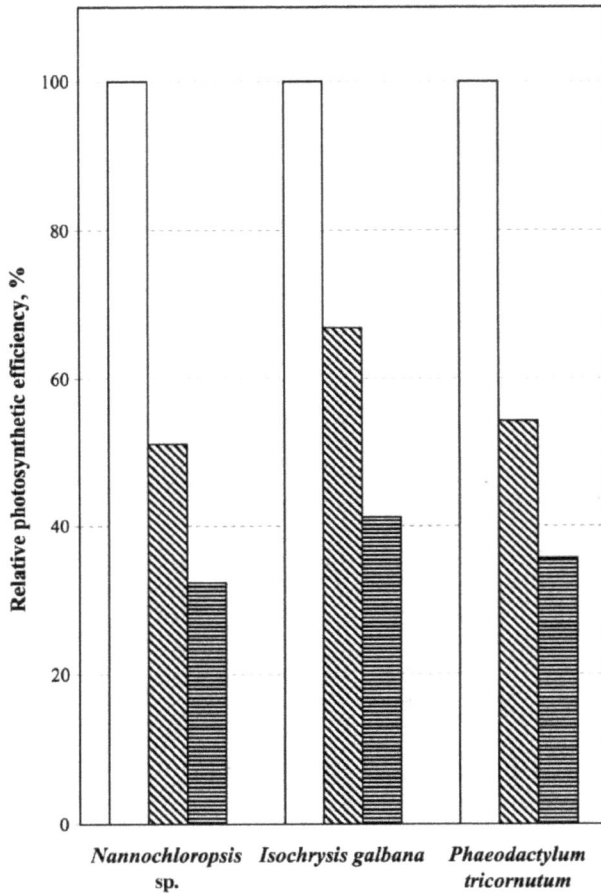

Figure 11. The effect of nutrient limitation on the relative photosynthetic energy storage in the three algae, *Nanno-chloropsis* sp., *Phaeodactylu mtricornutum*, and *Isochrysis galbana*. For each species, photosynthetic energy storage of the nutrient-replete control was taken as 100%. Controls (clear columns) were grown in nutrient-replete media, whereas in the P (oblique hatch) and N (horizontal hatch) cultures, phosphorus and nitrogen were omitted from the medium.

Spanier et al., 1979; Lindell & Post, 1995; Reiss et al., 1984). It seems that due to the limited availability of nitrogen and phosphorus in the gulf, it is these two nutrients that control the annual phytoplankton cycle in the gulf (Al-Qutob et al., 2002). Thus, the changes in iron availability (Chase et al., 2006) themselves may play, at most, a minor role in that cycle, even though iron does affect the supply of the two major nutrients.

In the Gulf of Eilat, the quantum yields of photosynthesis are constrained in summer by low nutrient supply and in winter by vertical mixing rates being too fast to allow for effective photoacclimation (Iluz et al., 2008).

4.3. Pollutants

Like exposure to excessive light intensity and nutrient limitation, pollutants may also affect the normal function of various components of the photosynthetic apparatus, leading to decreased quantum yields. Pinchasov et al. (2006) reported a biphasic decrease in the quantum yield of the cyanobacterium *Synechococcusleopoliensis.* A rapid decrease in the very first minutes of exposure was followed by a slower decline over the next few hours, probably due to harm to different components of the photosynthetic apparatus (Fig. 12).

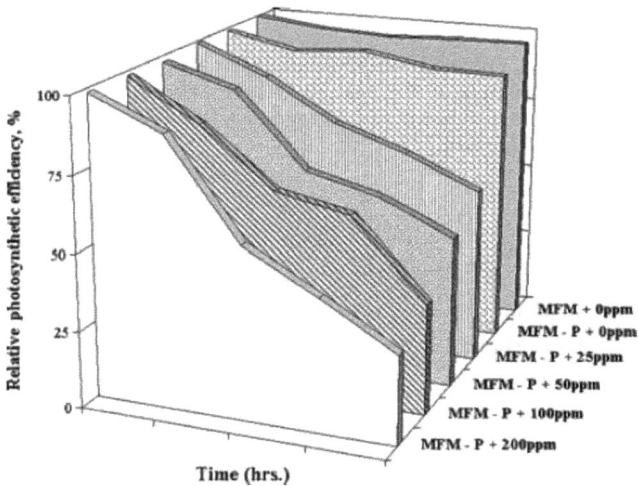

Figure 12. Relative photosynthetic efficiency following lead application of *Synecococcus leopoliensis* [after Pinchasov et al. (2006)].

5. Importance

5.1. Ecology

Hyperspectral satellite-image–based global oceanic primary productivity estimates use a universal factor of ϕ for converting absorbed light to primary productivity (Kolber et al., 1990). These values should be determined for each oceanic province and, other than in the tropics, are adjusted for seasonal differences. Parallel approaches are also applied in terrestrial ecology.

In ecological studies, in many cases the preferred efficiency parameter is not the quantum yield, where the energy absorbed by the ecosystem's primary producers is difficult to estimate. In such cases, the denominator used is the total light energy impinging on the area studied. This ecological efficiency parameter is always considerably smaller than the quantum yield, since usually all of the solar energy is absorbed by phototrophs, except in dense plant cover or hypertrophic water bodies. The ecological efficiency parameter allows the comparison of natural and man-made ecosystems, and of aquatic and terrestrial ecosystems, whereas the quantum yield reports the physiological status of the plants. High ecological light utilization efficiencies form the basis for high-biomass ecosystems, such as tropical rainforests, upwelling oceanic regions, and coral reefs, whereas desert and steppe biomes are biomass-limited by low ecological light-utilization efficiencies.

5.2. Biotechnology

The economic viability of photobioreactors and algal mass culture ponds depends, to a large extent, on the quantum yield of the algae, since that translates into harvested product per insolation both for the same time interval and area. That easily determines into annual income per surface devoted to that crop.

Acknowledgements

This study was supported by NATO SfP 981883. We thank Sharon Victor for editorial assistance. Gal Dishon and Yulia Pinhasov for their help with graphs

Author details

David Iluz[1,2] and Zvy Dubinsky[1]

1 The Mina and Everard Goodman Faculty of Life Sciences, Bar-Ilan University, Ramat-Gan, Israel

2 Dept. of Environmental Science and Agriculture, Beit Berl College, Kfar-Saba, Israel

References

[1] Al-qutob, M. (2001). Nutrient distributions and dynamics in the Gulf of Eilat (Aqaba), Red Sea. Ph.D. thesis, Bar-Ilan University, Ramat-Gan, Israel

[2] Al-qutob, M, Hase, C, Tilzer, M. M, & Lazar, B. (2002). Phytoplankton drives nitrite dynamics in the Gulf of Aqaba, Red Sea. Marine Ecology Progress Series, , 239, 233-239.

[3] Beardall, J, Burgerwiersma, T, Rijkeboer, M, Sukenik, A, Lemoalle, J, Dubinsky, Z, & Fontvielle, D. (1994). Studies on enhanced post-illumination respiration in microalgae.Journal of Plankton Research, 0142-7873, 16(10), 1401-1410.

[4] Bender, M, Grande, K, Johnson, K, Marra, J, Williams, P. J. L, Sieburth, J, Pilson, M, Langdon, C, Hitchcock, G, Orchardo, J, Hunt, C, Donaghay, P, & Heinemann, K. (1987). A comparison of four methods for determining planktonic community production Limnology and Oceanography, 0024-3590, 32(5), 1085-1098.

[5] Bender, M, Orchardo, J, Dickson, M. L, Barber, R, & Lindley, S. (1999). In vitro O-2 fluxes compared with C-14 production and other rate terms during the JGOFS Equatorial Pacific experiment. Deep-Sea Research Part I-Oceanographic Research Papers, 0967-0637, 46(4), 637-654.

[6] Berry, J. A. (1992). Biosphere, atmosphere, ocean interactions- a plant physiologist's perspective. In: Primary productivity and biogeochemical cycles in the sea, P.G. Falkowski & A.D. Woodhead (Eds.), Plenum Press, New York, 441-454.

[7] Boichenko, V. A, Hou, J. M, & Mauzerall, D. (2001). Thermodynamics of electron transfer in oxygenic photosynthetic reaction centers: volume change, enthalpy, and entropy of electron-transfer reactions in the intact cells of the cyanobacterium Synechocystis PCC 6803. Biochemistry, , 40, 7126-7132.

[8] Chase, Z, Paytan, A, Johnson, K. S, Street, J, & Chen, Y. (2006). Input and cycling of iron in the Gulf of Aqaba, Red Sea. Global Biogeochemical Cycles, Art., 20(GB3017)

[9] Cleveland, J. S, Perry, M. J, Kiefer, D. A, & Talbot, M. C. (1989). Maximal quantum yield of photosynthesis in the northwestern Sargasso Sea.Journal of Marine Research, , 47, 869-886.

[10] Droop, M. R. (1983). years of algal growth kinetics, a personal view. Botanica Marina, , 26, 99-112.

[11] Dubinsky, Z. (1980). Light utilization efficiency in natural phytoplankton communities. In: Primary productivity in the sea, P.G. Falkowski (Ed.), Plenum Press, New York, 83-97.

[12] Dubinsky, Z. (1992). The optical and functional cross-sections of phytoplankton photosynthesis. In: Primary productivity and biogeochemical cycles in the sea, P.G. Falkowski & A.D. Woodhead (Eds.), Plenum Press, New York, 31-45.

[13] Dubinsky, Z, & Berman, T. (1976). Light utilization efficiencies of phytoplankton in Lake Kinneret (sea of Galilee). Limnology and Oceanography, , 21, 226-230.

[14] Dubinsky, Z, & Berman, T. (1979). Seasonal changes in the spectral composition of downwelling irradiance in Lake Kinneret (Israel) Limnology and Oceanography, 0024-3590, 24(4), 652-663.

[15] Dubinsky, Z, & Berman, T. (1981). Light utilization by phytoplankton in Lake Kinneret (Israel). Limnology and Oceanography, , 26, 660-670.

[16] Dubinsky, Z, & Berman-frank, I. (2001). Uncoupling primary production from population growth in photosynthesizing organisms in aquatic ecosystems.Aquatic Sciences, , 63(1), 4-17.

[17] Dubinsky, Z, Berman, T, & Schanz, F. (1984). Field experiments for in situ measurement of photosynthetic efficiency and quantum yield. Journal of Plankton Research, , 6, 339-349.

[18] Dubinsky, Z, Falkowski, P. G, & Wyman, K. (1986). Light harvesting and utilization by phytoplankton. Plant and Cell Physiology, 0032-0781, 27(7), 1335-1349.

[19] Dubinsky, Z, Feitelson, J, & Mauzerall, D. C. (1998). Listening to phytoplankton: Measuring biomass and photosynthesis by photoacoustics.Journal of Phycology, , 34(5), 888-892.

[20] Emerson, R, & Arnold, W. J. (1932). A separation of the reactions in photosynthesis by means of intermittent light.General Physiology, , 15, 391-420.

[21] Emerson, R, & Lewis, C. M. (1943). The dependence of the quantum yield of Chlorella photosynthesis on wavelength of light.American Journal of Botany, , 30, 165-178.

[22] Falkowski, P. (1991). Species variability in the fractionation of 13C and 12C by marine phytoplankton.Journal of Plankton Research, , 13, 21-28.

[23] Falkowski, P. G, & Raven, J. A. (2007). Aquatic Photosynthesis, 2nd edition.Princeton University Press, Princeton, NJ.

[24] Falkowski, P. G, & Wirick, C. D. (1981). A simulation model of the effects of vertical mixing on primary production.Marine Biology, , 65, 69-75.

[25] Falkowski, P. G, Dubinsky, Z, & Wyman, K. (1985). Growth-irradiance relationships in phytoplankton.Limnology and Oceanography, , 30, 311-321.

[26] Falkowski, P. G, Jokiel, P. L, & Kinzie, R. A. (1990). Irradiance and corals. In: Coral reefs. Ecosystems of the world, Z. Dubinsky (Ed.), Elsevier Science Publishers, Amsterdam, 89-107.

[27] Fisher, T, Minnaard, J, & Dubinsky, Z. (1996). Photoacclimation in the marine alga Nannochloropsis sp. (Eustigmatophyte): a kinetic study. Journal of Plankton Research, , 18, 1797-1818.

[28] Genin, A, Lazar, B, & Brenner, S. (1995). Vertical mixing and coral death in the Red Sea following the eruption of Mount Pinatubo.Nature, , 377, 507-510.

[29] Gorbunov, M. Y, & Falkowski, P. G. (2004). Fluorescence Induction and Relaxation (FIRe) technique and instrumentation for monitoring photosynthetic processes and primary production in aquatic ecosystems. In: 13th International Congress of Photosynthesis, A. van der Est & D. Bruce (Eds.), 1029-1031, Allen Press, Montreal, 2

[30] Greene, R. M, Geider, R. J, & Falkowski, P. G. (1991). Effect of iron limitation on photosynthesis in a marine diatom.Limnology and Oceanography, , 36, 1772-1782.

[31] Grobbelaar, J. U. (2006). Photosynthetic response and acclimation of microalgae to light fluctuations In: Algal cultures analogues of blooms and applications, D.V. Subba Rao (Ed.), Science Publishers, Enfield (NH), USA, Plymouth, UK, 671-683.

[32] Guy, R. D, Berry, J. A, Fogel, M. L, Turpin, D. H, & Weger, H. G. (1992). Fractionation of the stable isotopes of oxygen during respiration by plants- the basis of a new technique to estimate partitioning to the alternative path. In: Molecular, biochemical and physiological aspects of plant respiration, H. Lambers & L.H.W. van der Plas (Eds.), SPB Academic Publishing, The Hague, The Netherlands, 443-453.

[33] Guy, R. D, Fogel, M. L, & Berry, J. A. (1993). Photosynthetic fractionation of the stable isotopes of oxygen and carbon.Plant Physiology, 0032-0889, 101(1), 37-47.

[34] Hou, J. M, Boichenko, V. A, Diner, B. A, & Mauzerall, D. (2001a). Thermodynamics of electron transfer in oxygenic photosynthetic reaction centers: volume change, enthalpy, and entropy of electron-transfer reactions in manganese-depleted photosystem II core complexes. Biochemistry, , 40, 7117-7125.

[35] Hou, J. M, Boichenko, V. A, Wang, Y. C, Chitnis, P. R, & Mauzerall, D. (2001b). Thermodynamics of electron transfer in oxygenic photosynthetic reaction centers: a pulsed photoacoustic study of electron transfer in photosystem I reveals a similarity to bacterial reaction centers in both volume change and entropy. Biochemistry, , 40, 7109-7116.

[36] Iluz, D, Yehoshua, Y, & Dubinsky, Z. (2008). Quantum yields of phytoplankton photosynthesis in the Gulf of Aqaba (Elat), Northern Red Sea. Israel Journal of Plant Sciences, 0792-9978, 56(1-2), 29-36.

[37] Kirk, J. T. O. (1994). Light and photosynthesis in aquatic ecosystems, 2nd ed., Cambridge University Press, London, New York

[38] Kishino, M, Takahashi, M, Okami, N, & Ichimura, S. (1985). Estimation of the spectral absorption coefficients of phytoplankton in the sea.Bulletin of Marine Sciences, , 37, 634-642.

[39] Kolber, Z, Zehr, J, & Falkowski, P. G. (1988). Effect of growth irradiance and nitrogen limitation on photosynthetic energy conversion in photosystem II.Plant Physiology, , 88, 923-929.

[40] Kolber, Z, Wyman, K. D, & Falkowski, P. G. (1990). Natural variability in photosynthetic energy conversion efficiency: a study in the Gulf of Maine. Limnology and Oceanography, , 35(1), 72-79.

[41] Kolber, Z. S, Prasil, O, & Falkowski, P. G. (1998). Measurements of variable chlorophyll fluorescence using fast repetition rate techniques: defining methodology and experimental protocols. Biochimica et Biophysica Acta-Bioenergetics, , 1367(1-3), 88-106.

[42] Labiosa, R. G, Arrigo, K. R, Genin, A, Monismith, S. G, & Van Dijken, G. (2003). The interplay between upwelling and deep convective mixing in determining the seasonal phytoplankton dynamics in the Gulf of Aqaba: Evidence from SeaWiFS and MODIS. Limnology and Oceanography, , 48, 2355-2368.

[43] Lazar, B, & Erez, J. (1992). Carbon geochemistry of marine derived brines: I. ^{13}C depletions due to intense photosynthesis. Geochimica et Cosmochimica Acta, , 56, 335-345.

[44] Levanon-spanier, L, Padan, E, & Reiss, Z. (1979). Primary production in desert-enclosed sea the Gulf of Elat (Aqaba), Red Sea.Deep-Sea Research, , 26, 673-685.

[45] Lindell, D, & Post, A. F. (1995). Ultraphytoplankton succession is triggered by deep winter mixing in the Gulf of Aqaba (Eilat), Red Sea. Limnology and Oceanography, , 40, 1130-1141.

[46] Long, S. P, & Postl, W. F. Bolhár Nordenkampf, H.R. ((1993). Quantum yields for uptake of carbon dioxide in C3 vascular plants of contrasting habitats and taxonomic groupings. Planta, 0032-0935, 189(2), 226-234.

[47] Long, S. P, Humphries, S, & Falkowski, P. G. (1994). Photoinhibition of photosynthesis in nature.Annual Review of Plant Physiology and Plant Molecular Biology, , 45, 633-662.

[48] Luz, B, & Barkan, E. (2000). Assessment of oceanic productivity with the triple-isotope composition of dissolved oxygen. Science, 0036-8075, 288(5473), 2028-2031.

[49] Mauzerall, D. C, Feitelson, J, & Dubinsky, Z. (1998). Discriminating between phytoplankton taxa by photoacoustics.Israel Journal of Chemistry, , 38(3), 257-260.

[50] Moore, C. M, Suggett, D. J, Hickman, A. E, Kim, Y. N, Tweddle, J. F, Sharples, J, Geider, R. J, & Holligan, P. M. (2006). Phytoplankton photoacclimation and photoadaptation in response to environmental gradients in a shelf sea.Limnology and Oceanography, 0024-3590, 51(2), 936-949.

[51] Morel, A. (1978). Available, usable, and stored radiant energy in relation to marine photosynthesis.Deep-Sea Research, , 25, 673-688.

[52] Pinchasov, Y, Kotliarevsky, D, Dubinsky, Z, Mauzerall, D. C, & Feitelson, J. (2005). Photoacoustics as a diagnostic tool for probing the physiological status of phytoplankton.Israel Journal of Plant Sciences, , 53(1), 1-10.

[53] Pinchasov, Y, Berner, T, & Dubinsky, Z. (2006). The effect of lead on photosynthesis, as determined by photoacoustics in Synechococcus leopoliensis (Cyanobacteria).Water Air and Soil Pollution, , 175, 117-125.

[54] Prezelin, B. B, Bidigare, R. R, Matlick, H. A, Putt, M, & Hoven, B. V. (1987). Diurnal patterns of size fractioned primary productivity across a coastal front.Marine Biology, , 96, 563-574.

[55] Prezelin, B. B, Tilzer, M. M, Schofield, O, & Haese, C. (1991). The control of the production process of phytoplankton by the physical structure of the aquatic environment with special reference to its optical properties.Aquatic Sciences, , 53, 136-186.

[56] Reiss, Z. S, & Hottinger, L. eds. ((1984). The Gulf of Aqaba.Ecological micropaleontology. Springer Verlag, Berlin

[57] Robinson, S. A, Yakir, D, Ribas-carbo, M, Giles, L, Osmond, C. B, Siedow, J. N, & Berry, J. A. (1992). Measurements of the engagement of cyanide-resistant respiration in the Crassulacean acid metabolism plant Kalanchoe daigremontiana with the use of online oxygen isotope discrimination. Plant Physiology, 0032-0889, 100(3), 1087-1091.

[58] Schanz, F, & Dubinsky, Z. (1988). The afternoon depression in primary productivity in a high rate xxidation pond (Hrop).Journal of Plankton Research, , 10(3), 373-383.

[59] Schanz, F, Senn, P, & Dubinsky, Z. (1997). Light absorption by phytoplankton and the vertical light attenuation: ecological and physsiological significance. Oceanography and Marine Biology: An Annual Review, , 35, 71-95.

[60] Schofield, O, Prezelin, B. B, Smith, R. C, Stegmann, P. M, Nelson, N. B, Lewis, M. R, & Baker, K. S. (1991). Variability in spectral and nonspectral measurements of photosynthetic light utilization efficiencies.Marine Ecology Progress Series, , 78, 253-271.

[61] Steemann-nielsen, E. (1952). The use of radioactive carbon (^{14}C) for measuring organic production in the sea.Journal du Conseil Perm. Internationl pour l Exploration de la Mer, , 18, 117-140.

[62] Tilzer, M. M. (1984). Seasonal and diurnal of photosynthetic quantum yields in the phytoplankton of Lake Constance.Verhandlungen des Internationalen Verein Limnologie, , 22, 958-962.

[63] Walsby, A. E, Dubinsky, Z, Kromkamp, J. C, Lehmann, C, & Schanz, F. (2001). The effects of diel changes in photosynthetic coefficients and depth of Planktothrix rubescens on the daily integral of photosynthesis in Lake Zurich.Aquatic Sciences, , 63(3), 326-349.

[64] Williams, P. J. B, & Purdie, D. A. (1991). In vitro and in situ derived rates of gross production, net community production and respiration of oxygen in the oligotrophic

subtropical gyre of the North Pacific Ocean.Deep-Sea Research Part A-Oceanographic Research Papers, 0198-0149, 38(7), 891-910.

Chapter 5

CAM Photosynthesis in Bromeliads and Agaves: What Can We Learn from These Plants?

Alejandra Matiz, Paulo Tamaso Mioto,
Adriana Yepes Mayorga, Luciano Freschi and
Helenice Mercier

Additional information is available at the end of the chapter

1. Introduction

The term Crassulacean Acid Metabolism (CAM) was introduced in the 1940s as a result of observations in *Bryophyllum calycinum*, a crassulacean plant, which showed prominent diel variations in leaf acid content, with increases at night followed by daytime deacidification [1, 2]. Nowadays, we know that CAM is present in plant families other than Crassulaceae, being found in about 20,000 terrestrial and aquatic species, with representatives in 343 genera of 35 families [3, 4].

In general, CAM photosynthesis consists of the nocturnal carboxylation of phosphoenolpyruvate (PEP) by using atmospheric or respiratory CO_2, giving rise to oxaloacetate (OAA), a reaction mediated by the enzyme phosphoenolpyruvate carboxylase (PEPC). OAA is then reduced by malate dehydrogenase (MDH) to malate, which is subsequently transported into the vacuole and stocked in the form of malic acid, generating the typical nocturnal acidification of CAM plants. This transport into the vacuole is mediated by an active process of proton pumping through H^+-V-ATPases in the tonoplast and an organic acid anion channel. In the following light period, the stomata are maintained closed, and vacuolar malic acid is remobilized into the cytoplasm (returning to the malate form) and decarboxylated, releasing CO_2 (a process mediated by malic enzyme, ME-type, or phosphoenolpyruvate carboxykinase, PEPCK-type, enzymes, depending on the plant species) and causing the deacidification of the cells. The liberated CO_2 is refixed in the chloroplasts by the bifunctional enzyme ribulose-bisphostate carboxylase/oxygenase (Rubisco) [4-8].

Given the complex set of overlapping phenomena in CAM plants, including the diel patterns of acid synthesis and degradation, the presence of two carboxylases (PEPC and Rubisco) and the peculiar stomatal opening behaviors, the concept of CAM phases was proposed, for didactic purposes. Basically, CAM can be divided into four phases: 1) Phase I consists of the nocturnal fixation of CO_2 via open stomata and its storage in the form of organic acids in the vacuole; 2) at the start of Phase II in the early morning, when PEPC is becoming inactive while Rubisco is progressively being activated, the stomata still remain open, and the fixation of CO_2 can occur via both enzymes; 3) during Phase III, the stomata remain closed, while the organic acids are remobilized from the vacuoles and decarboxylated, generating CO_2 to Rubisco in the Calvin-cycle (C_3); and finally 4) in Phase IV, which occurs at the transition from the light to the dark period, the storage of organic acid is already exhausted and the stomata reopen again, allowing the CO_2 to be assimilated directly in carbohydrates via the Calvin-cycle [8-11] (Figure 1).

Considering the fact that CAM photosynthesis continuously offers CO_2 to Rubisco (during Phases II and III), in essence, this photosynthetic adaptation consists of a CO_2-concentrating mechanism. The internal storage of carbon in the form of organic acids and its subsequent decarboxylation generates an internal increase of CO_2, which largely reduces the oxygenase activity of Rubisco, therefore alleviating photorespiration [12]. In fact, reduced photorespiratory rates in CAM plants are expected to occur mainly during the first part of the light period when prominent organic acid decarboxylation fluxes significantly elevate internal CO_2 concentration. On the other hand, depending on the duration of the light period, the photosynthetic active radiation (PAR) received and the amount of organic acid accumulated during the previous night, CAM plants might also face the completely opposite scenario in terms of internal CO_2 and O_2 availability during the last part of Phase III since the eventual depletion of organic acid reserves and the daytime accumulation of O_2 produced due to the photosynthetically-driven water photolysis behind closed stomata would lead to O_2/CO_2 rates greatly favoring the oxygenase activity of Rubisco. In addition, under severe water stress conditions, stomatal opening at Phase IV might not occur, leading to high levels of photorespiration during the last hours of the light period [12].

In a similar way to CAM, C_4 photosynthesis also represents an important CO_2-concentrating mechanism for terrestrial plants. Although sharing many biochemical similarities and evolutionary driving forces, C_4 and CAM also display marked differences. For instance, C_4 plants spatially concentrate a small pool of transitory C_4 acids in mesophyll cells that turnover rapidly in the bundle sheath cells, while the concentrating mechanism of CAM plants is based on the temporal storage of a larger pool of end products of C_4 acids (mainly malic acid) in the vacuoles at night, which slowly turnover in the cytoplasm of the same photosynthetic cells during the day [3]. Another significant difference between C_4 and CAM photosynthesis involves the mechanism used to separate the two carboxylase activities, *i.e.*, a spatial PEPC-Rubisco separation in C_4 (PEPC in the mesophyll and Rubisco in the bundle sheath cells) and a temporal separation between these enzymes in CAM plants (PEPC at night and Rubisco during the day). C_4 and CAM are not only advantageous CO_2 concentrating systems but are also mechanisms capable of providing elevated water use efficiency (WUE). In C_4, the concentration of CO_2 at the site of Rubisco activity usually reduces stomatal conductance and, therefore, the general transpirational water losses through the light period. On the other hand, the ability to restrict stomatal opening during the periods in which higher air humidity is available (*e.g.*, at night, during dawn and dusk) allows CAM plants to fix higher amounts of CO_2 with very low rates of water loss through transpiration.

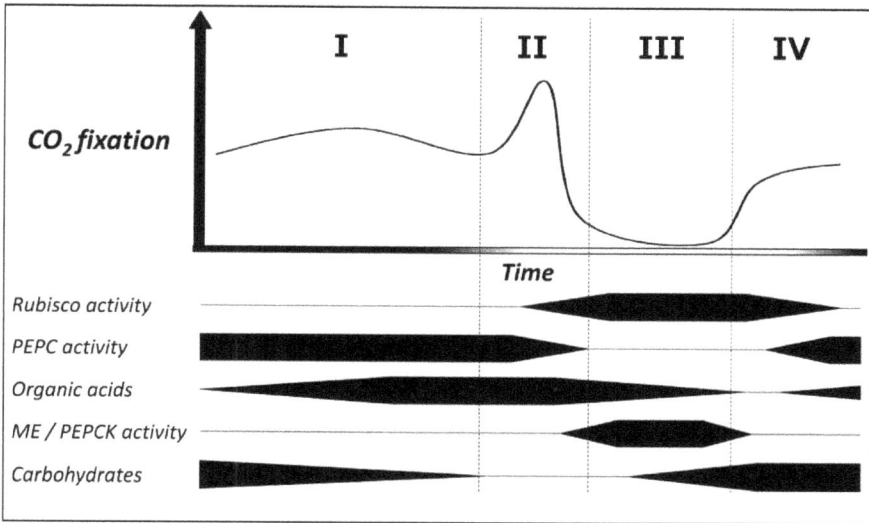

Figure 1. Four temporal Phases of CAM (I, II, III and IV) indicated over a 24-hour photoperiod by the main features of CAM: CO_2 fixation, Rubisco, PEPC and NAD(P)-ME-type or PEPCK-type activities and organic acid and carbohydrate accumulation. The black shapes below the x-axis indicate when the processes described above were happening. The black and white bars in the x-axis indicate night and day, respectively.

The presence of two pathways of carboxylation in CAM plants, one mediated by PEPC and the other by Rubisco, determines the differences in the ratio of carbon isotope discrimination ($\delta^{13}C$). Those differences are given by the fractionation characteristics of PEPC and Rubisco. Therefore, $\delta^{13}C$ values found in the plant are indicative of how much CO_2 was fixed by Rubisco (during the day) or PEPC (at night). Consequently, growth dependent on dark CO_2 fixation generates less negative values of $\delta^{13}C$ (CAM plants) while growth dependent on direct atmospheric CO_2 fixation during the day results in more negative $\delta^{13}C$ values (C_3 plants) [13]. For this reason, the values of $\delta^{13}C$ found in the tissues of plants leads to their classification as either CAM or C_3.

In the early 1970s, Winter and von Willert [14] discovered the capacity of *Mesembryanthemum crystallinum*, a plant until that time considered C_3, of expressing CAM (fixing CO_2 at night) when treated with NaCl, showing that plants did not uniquely express CAM or C_3. After this discovery, Osmond and cols. [13] analyzed the $\delta^{13}C$, CO_2 fixation and malate synthesis when regulating day/night length and temperature during growth of *Kalanchoë daigremontiana* and *K. blossfeldiana*. The results led to the conclusion that, depending on the environmental conditions, CAM plants are capable of using both C_3 (Rubisco) and C_4 (PEPC) fixation mechanisms during their development. Thus, the environmental conditions regulate the proportion of carbon acquired through each pathway.

1.1. Types of CAM

With more information in the literature regarding CAM, we now know that not all CAM plants just fix CO_2 at night while the stomata remain closed during the day (*e.g.* classical CAM); CAM photosynthesis is now considered a more flexible phenomenon in plants. It has been proposed that there is a continuum of photosynthetic expression from C_3 to CAM and in this progression several types of CAM photosynthesis can be found, ranging from a weak to a strong degree of expression [4], which is determined by the stage of plant development, environmental conditions and/or the species. Although classical CAM was characterized as traditionally showing a four-phase pattern, the flexibility of CAM and the diversity of CAM species illustrate the existence of a wide range of responses; thus, within CAM, three main types can be identified: classical CAM, CAM-cycling and CAM-idling.

The term CAM-cycling is used to explain the photosynthetic condition represented by a diel gas exchange pattern similar to that found in C_3 plants, combined with nocturnal organic acid accumulation [7, 14, 15], which results from the internal respiratory CO_2 refixation via PEPC at night. During the day, these acids are decarboxylated, releasing CO_2 to Rubisco, while the stomata remain open [8,16]. CAM-cycling can be expressed in some species at the early stages of shifting from C_3 to CAM in response to water limitation (facultative CAM plants) [16], maintaining a positive carbon balance during repeated drought events [17], reducing respiratory CO_2 losses, allowing water retention and extending the life cycle [7]. On the other hand, CAM-idling exhibits a null CO_2 gas exchange over the entire 24-hour period, while a small nocturnal acid accumulation still continues [7,15], in which the respiratory CO_2 is recaptured and used for synthesizing organic acids at night, which are decarboxylated the following day, recovering the carbohydrate used during the refixation of respiratory CO_2 [6]. Thus, under extended drought conditions, for example, CAM-idling allows recycling internal CO_2, avoiding a negative balance of carbon at the expense of growth and maintaining photosynthetic competence [16].

1.2. Constitutive and facultative CAM

Some species possess the capacity to exhibit facultative CAM, in which the degree of CAM expression greatly varies depending on internal or environmental cues. Facultative CAM may be part of the ontogenetic plant program, in which environment factors could accelerate or delay the preprogrammed C_3-CAM transition [17-20]. Thus, it is possible to find variable degrees of CAM modulation in facultative species, ranging from exclusively environmental to strictly developmental control of CAM, passing through an intermediate state in which both environmental and developmental cues influence CAM expression [19, 20] (Figure 2). Some well-characterized examples of facultative CAM plants are *Mesembryanthemum crystallinum* and some *Clusia* species.

As above mentioned, *Mesembryanthemum crystallinum* was initially considered an exclusive C_3 plant; however, after the discovery of its capacity to switch from C_3 to CAM [14], this plant became a common model species for facultative CAM photosynthesis research. *M. crystallinum* is an annual halophyte widely distributed in places with hot and dry summers, wet and cold winters and increased salinity, as found in south and east Africa, along the coasts of the Mediterranean and west United States [21, 22]. Thus, after a short cold and rainy season (winter), *M. crystallinum* germinates and the juvenile plants perform exclusively C_3 photosynthesis. Then, during the summer, in which the temperature and salinity increase and water availability decreases, the mature plants produce smaller and succulent leaves with epidermal bladders coinciding with the developmental CAM induction [21, 23]. Since young plants cannot perform CAM, ontogenetic factors become visibly important for this plant [24], and stressful conditions, like high salinity, would only accelerate the rate of CAM induction already preprogrammed by the plant development [8, 9, 17, 21]. Nevertheless, Winter and Holtum [19] observed that *M. crystallinum* is capable of performing C_3 photosynthesis during its entire life cycle if maintained in non-saline and well-watered conditions, demonstrating that CAM induction in *M. crystallinum* is mainly controlled by environmental conditions rather than by preprogrammed developmental processes. Once *M. crystallinum* shifts from C_3 to CAM, it never returns to C_3 mode [17, 21, 22], even if the adverse conditions are removed. Curiously, some degree of reversibility in CAM-induced *M. crystallinum* plants has already been reported [25], but more studies are necessary to confirm this. On the other hand, some species of the genus *Clusia* clearly show their capacity to switch from C_3 to CAM and back again in response to different environmental conditions.

In fact, the genus *Clusia* represents a magnificent example of diversity and inducibility of CAM. In *Clusia minor*, for example, opposite leaves on the same node can perform either C_3 or CAM in response to different temperatures and leaf-air vapor pressures, as only the leaves maintained in dry air conditions were capable of expressing CAM [26]. Among *Clusia* species, *C. minor* is currently the most widely used model for C_3-CAM photosynthesis studies [18, 20, 27, 28]. It has already been shown that juvenile plants of *C. minor* perform C_3 photosynthesis and switch their photosynthetic behavior to CAM when mature, increasing dark CO_2 fixation [18, 20], activity of PEPC and PEPCK (key enzymes in CAM photosynthesis), and nocturnal accumulation of organic acids [18]. Thus, apparently, *C. minor* plants are controlled by programmed development. However, these plants when exposed to drought (or dry season in nature) up-regulate CAM activity [18, 20, 29, 30], indicating that CAM can also be controlled by the environment, even though reversion to the C_3 state never occurs. Therefore, both development and environment are capable of regulating CAM in *C. minor*. In another species of the same genus, *C. pratensis*, shows that, in this case, CAM responds almost exclusively to environmental cues and the switch from C_3 to CAM is fully reversible [20].

Plants that always perform CAM in mature tissues independently of the environmental conditions (stressful or not) are classified as constitutive (or obligate) CAM species. Examples of constitutive CAM are some species from the Cactaceae family (*e.g.*, some *Opuntia* species),

Crassulaceae (*e.g.*, some *Kalanchoë* species) [16, 20] and Clusiaceae (*C. rosea*) [20, 31]. However, despite being classified as constitutive CAM plants, some of these species might also show some degree of plasticity in CAM expression in response to environmental conditions (Figure 2). *Kalanchoë pinnata, K. daigremontiana* and *Opuntia ficus-indica*, for example, are capable of up-regulating CAM when maintained under drought conditions [16, 20, 32]. Also, *Opuntia basilaris*, another constitutive CAM species, enhances its nocturnal CO_2 fixation when exposed to favorable conditions of watering [33]. It should be noted that neither favorable nor unfavorable conditions are capable of changing the CAM mode to C_3 mode because the ontogenetic program of the plant does not allow such modification [20, 32].

Nowadays, there is an important debate going on about establishing clear differences between facultative CAM and constitutive CAM plants since it has already been demonstrated that, in some constitutive CAM plants, when juvenile (*e.g.*; *K. daigremontiana, K. pinnata, Opuntia ficus-indica* and *Clusia rosea*), C_3 photosynthesis is the main pathway to uptake CO_2 and CAM only becomes the dominant pathway when they are mature [20, 29]. For example, *C. rosea* seems to be strongly controlled by ontogenetic factors, as variations in the environment did not affect the degree of CAM in this species; nevertheless, when juvenile, facultative and fully reversible CAM can be observed in this species [20]. Thus, due to the facultative component that exists in constitutive CAM, it would be very difficult to strictly define a species as facultative or constitutive CAM. We, therefore, propose that facultative CAM species should be those that are capable of going from typically C_3 photosynthesis to CAM and back, even in adulthood. However, species that are capable of expressing CAM at some moment in their life cycle, regardless of whether they are influenced by either environmental or developmental factors, but cannot return to a C_3 after they become adults should be considered simply CAM plants.

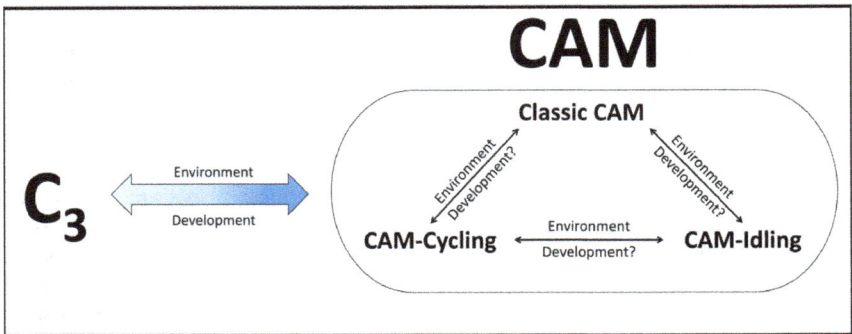

Figure 2. Capacity of plants to transition from C_3 to CAM modes is controlled by environmental and developmental cues (blue arrow). CAM and its flexibility to transit between different modes of CAM (classic-CAM, CAM-cycling and CAM-idling) is under environmental control; however, it is unknown if it could also be regulated by developmental factors.

2. Biochemistry of CAM

2.1. Importance of carbohydrates in CAM plants

Due to its biochemical nature, the CAM cycle is inexorably associated with a serious metabolic constraint: at the end of the Phase IV (end of the light period – Figure 1) there must be an adequate pool of carbohydrates to generate PEP, required for nocturnal CO_2 fixation via PEPC during Phase I. That pool of carbohydrates must differ from the pool of carbohydrates needed to ensure other metabolic processes including dark respiration, export and growth [6, 9, 17, 34, 35]. In several CAM species, the net flux of carbon for regenerating PEP is mainly based on the pool of starch; thus, these plants show a large diel change in transitory starch [34, 36, 37]. Therefore, when CAM is induced, an increase in starch degradative enzymes is required [38, 39], and specific transporters of intermediate products across the chloroplast membrane must also be present [38, 40]. The degradation of transitory starch can be very important in CAM plants providing PEP through glycolysis, as demonstrated for *Mesembryanthemum crystallinum,* in which mutant plants deficient in starch synthesis due to a lack of the enzyme phosphoglucomutase were incapable of operating in CAM mode and had lower fecundity than the wild type. However, when these mutants were fed with glucose, the nocturnal acidification was recovered [41].

Interestingly, it was observed that *M. crystallinum,* when switching from C_3 to CAM, changes the metabolic pathway of starch degradation from hydrolytic to phosphorolytic [42]. In the latter, after the starch is degraded, the main product exported from the chloroplast is glucose-6-phosphate, while in the hydrolytic pathway the main product exported is maltose [43]. Thus, when in C_3 mode, *M. crystallinum* exports maltose from the chloroplast as a result of starch degradation [42], but when it operates in CAM mode, the export switches from maltose to glucose-6-phosphate [42, 44]. Interestingly, it was observed that in several so-called facultative CAM species exposed to drought or salinity stress, PEPC increased its sensitivity to activation by glucose-6-phosphate [45, 46]. This may suggest a link between the metabolic pathway of phosphorolytic starch degradation and PEPC activation. It has been proposed that starch degradation and its flux through PEPC is under circadian regulation in CAM plants because it was observed that the activity of some enzymes that participate in the starch degradation (*e.g.* β-amylase and starch phosphorylase) are coordinated in time with the phosphorylation of PEPC at night [9].

As mentioned above, transitory starch plays an important role in the transition from C_3 to CAM in plants of *Mesembryanthemum crystallinum* [41] because starch is the main carbohydrate used to generate PEP either in primary or axillary leaves [34]. The partitioning of assimilates in different pools originating from C_3-carboxylation or C_4-carboxylation have been suggested as a possible regulatory mechanism of carbohydrate metabolism in CAM plants [47]. It has also been demonstrated that this carbohydrate partitioning has a crucial ecophysiological function besides allowing CAM to function; in C_3 primary leaves of *M. crystallinum,* the carbohydrate partitioning works mainly to facilitate the development of the whole plant. Then, when the dry season arrives, the growth of axillary leaves accelerates and the development of the plant is mainly directed towards reproduction. Therefore, when the plant switches from C_3 to CAM, the axillary leaves are capable of exporting sugars during the day and the night derived from C_3 and C_4-carboxylation, in contrast to primary leaves, which possess a limited export of sugars [34]. Thus, the axillary leaves ensure the reproductive success by exporting sugars, while starch guarantees the CAM cycle [17, 34]. Interestingly, measurements of $\delta^{13}C$ in seeds of this species showed a value of -16.4‰, indicating an important contribution of the nocturnal CO_2 fixated (CAM) to seedset [48].

Although starch represents the main carbohydrate to provide PEP in many CAM plants, there are also CAM plants that show a smaller diel change in starch levels because they are also capable of storing carbohydrates in the form of hexose inside the vacuole [49], as observed in *Ananas comosus* (pineapple) [50, 51]. Therefore, there are differences among CAM plants in their diel changes of energy-rich compounds used for nocturnal organic acid synthesis [52].

In addition, it was observed that species of *Clusia* display differences between their carbohydrate pools derived from C_3 and C_4-carboxylation and the partitioning of each pool to storage or exportation. In *Clusia minor*, in which CAM expression is environmentally and ontogenetically controlled, there was a doubling of its pool of carbohydrates after the switch from C_3 to CAM. Interestingly, two pools of soluble sugars were identified, one enriched in ^{13}C and the other depleted in ^{13}C, the latter destined to be transported, indicating the existence of a regulated partitioning of carbohydrates in this species [47]. Surveys comparing *Clusia minor* (performing CAM) and *Clusia rosea* (constitutive CAM) showed that the carbohydrate pool in leaves of *C. rosea* is mainly derived from PEPC carboxylation, while the carbohydrate pool of *C. minor* is derived from Rubisco carboxylation. In both species, leaf soluble sugars were enriched in ^{13}C when compared with the leaf starch, indicating that C_3 assimilates were preferably redirected into starch [34]. In addition, in the same work, it was demonstrated that regardless of the degree of CAM the pool of carbohydrates exported from the leaves to reproductive sinks (fruits) was mainly derived from C_3-carboxylation (showing values of $\delta^{13}C$ of -25.6 and -25.8 in *Clusia minor* and *Clusia rosea*, respectively). Thus, those results confirm the existence of different partitioning of assimilates derived from C_3 or C_4-carboxylation between reproductive and vegetative growth. As observed for most CAM plants, growth and productivity are maximized when direct CO_2 fixation via Rubisco in Phase IV predominates [17, 34, 35]. Therefore, the fact that assimilates are formed during this phase and exported from the leaves to the sink shows the importance of Rubisco in the reproductive success and growth of CAM plants.

2.2. An overview of PEPC and PPCK regulation

PEPC is a homotetrameric enzyme that participates in a broad range of functions (both photosynthetic and non-photosynthetic) in the plant cell [53], not found in fungi nor animals [54]. In C_4 and CAM plants, PEPC irreversibly carboxylates PEP, using HCO_3^- and generating oxaloacetate (OAA). Both the capacity to use HCO_3^- instead of gaseous CO_2 (different from Rubisco) and the high affinity for HCO_3^- make CAM plants more efficient in terms of nitrogen usage because less nitrogen allocation is required to form adequate quantities of the carboxylating enzymes (PEPC and Rubisco) than in C_3 plants, in which a larger pool of Rubisco is required to fixate CO_2.

Due to the activity of two carboxylase enzymes (PEPC and Rubisco) temporally separated in CAM plants, a tight metabolic control to avoid futile cycles of carbon must exist [6]. PEPC enzyme can be regulated by environment and endogenous circadian signals, resulting in the reversible activation of the enzyme by phosphorylation. This regulation of PEPC through phosphorylation is mediated by a PEPC kinase enzyme (PPCK- a specific Ca^{2+} -independent serine/threonine kinase), which phosphorylates PEPC on its serine residue (Figure 3), reducing its sensitivity to malate inhibition at night [55] and increasing its allosteric activation by glucose 6-phosphate and affinity for PEP [53]. During the day, PEPC is highly inhibited by malate because this enzyme is dephosphorylated. Thus, PEPC activity is restricted to Phase I, early Phase II and late Phase IV [6, 55] (Figure 1), minimizing the futile cycling of simultaneous malate synthesis (mainly at night) and breakdown (only during daytime).

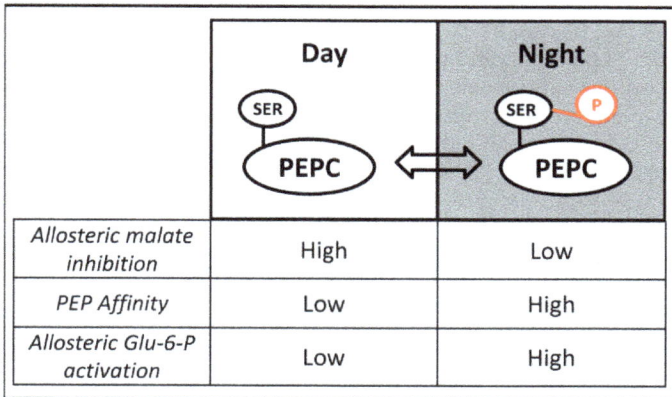

	Day	Night
Allosteric malate inhibition	High	Low
PEP Affinity	Low	High
Allosteric Glu-6-P activation	Low	High

Figure 3. Daytime inactivation and nighttime posttranslational activation of PEPC enzyme. Activation of PEPC at night depends on the phosphorylation of its serine residue by PPCK enzyme. Different allosteric malate inhibition and Glu-6-P activation and affinity for PEP in light and dark periods is shown.

PEPC kinase enzyme is mainly regulated by transcriptional level abundance, showing the peculiar characteristic among kinase enzymes of requiring *de novo* synthesis to be active [55]. In C_3 and C_4 plants, PPCK seems to be activated by light [56-59], while in CAM plants this enzyme responds mostly to a circadian oscillator and is active during the night [55, 60, 62]. Taybi and cols. [63] showed that the transcript accumulation of PPCK in *M. crystallinum* is largely subjected to circadian control under continuous light condition. However, that circadian control is itself regulated by other factors, such as cytosolic malate level [55, 64]. In *Zea mays*, a C_4 species, it was found that ZmPPCK2, a specific PPCK isoform, is expressed preferentially during the night [59]. These findings point toward a flexibility in the expression of this enzyme and indicate that a simple step could lead to its expression during nighttime, as happens in CAM plants.

Differences in regulation of PEPC would also require changes in Rubisco activity to avoid futile cycling of CO_2. In a C_3 plant, Rubisco apparently needs to be activated by Rubisco activase (RCA), which responds positively to irradiance and ATP supply and negatively to high concentrations of sugar phosphate [65]. The same authors found that in CAM-induced M. crystallinum, the activation of Rubisco followed a rhythmicity of activation-deactivation. Whether this is due to circadian or metabolite control, the authors could not confirm.

3. CAM origins

Phylogenetic reconstructions from comparative physiology and taxonomy confirm that C_3 photosynthesis is ancestral to both CAM and C_4, with CAM appearing multiple times in the taxa and earlier than C_4 photosynthesis [54, 66]. Since all enzymes related to CAM functions are also found in C_3 plants, several aspects of CAM appear to have evolved as minor modifications of processes already present in ancestral C_3, suggesting that CAM may have originated from the reorganization of ancient metabolic pathways (co-option) [67]. Thus, processes such as gene duplication, processing of mRNA and gene expression control by cis-regulatory elements or enhancers can maintain essential ancestral functions along with new ones related to CAM [67].

The majority of CAM plants is found in monocot and eudicot taxa, which had high diversification by the early and middle Miocene [68]. It has been hypothesized that atmospheric CO_2 levels and the CO_2/O_2 ratio decreased sufficiently some time after the Cretaceous [69] and during the Miocene, allowing the evolution of CAM in terrestrial and aquatic environments [66, 70]. Thus, atmospheric CO_2 reductions, coupled with warm climates in subtropical latitudes, were factors with negative impacts on C_3 photosynthesis, due to a depletion of CO_2 diffusion gradient that resulted in a decrease in the carboxylation efficiency and an increase in photorespiration rates [71]. Therefore, the restricted daytime CO_2 availability in that environment may have been the most important driving force in the evolution of CAM. As a result, plants that rely on the C_4 and CAM pathway would have advantages compared to C_3 plants because of the higher carboxylation efficiency [70]. It is also proposed that a photosynthetic mechanism similar to aquatic CAM arose during the late Paleozoic, also a period of low atmospheric CO_2 concentrations [72]. Thus, CAM photosynthesis may have emerged several times as a means of improving carbon economy.

Currently, another evidence of the multiple origins of CAM is its occurrence in 35 taxonomically diverse families [16], of which 32 belong to Magnoliophyta division. Among these 32 families, eight are monocots (Agavaceae, Alismataceae, Araceae, Asphodelaceae, Bromeliaceae, Hydrocharitaceae, Orchidaceae and Ruscaceae) [4].

Nowadays, there is little fossil evidence about CAM origins. Isoetes species are at present-day a monophyletic CAM taxon which represents the oldest clade of known CAM plants since there is fossil evidence of the existence of that taxon by the early Cretaceous [70]. More recent CAM plant fossils are found from the middle Miocene and late Pleistocene to the Holocene. The plant fossil from the middle Miocene, *Protoyucca shadishii*, is very likely an early member of dry communities of present *Yucca* CAM plants (Agavaceae) [73]. Another survey analyzed $\delta^{13}C$ values of samples of CAM plant *Opuntia polyacantha* (Cactaceae) found in old pack rat middens in the southwestern United States more than 40,000 years old and others 10,000 years old, in which a shift in the $\delta^{13}C$ value from -21.9‰ (in the 40,000-year-old sample) to - 13.9 ‰ (10,000-year-old sample) was observed. These results provided physiological evidence of drier climates in the late Pleistocene in that region [74], which apparently favored CAM expression. Recently, the fossil *Karatophyllum bromelioides* from the late Pleistocene to Holocene was studied, which shares excellently correlated morphological characteristics with *Aechmea magdalenae*, a CAM bromeliad [75]. Since succulence within Bromeliaceae seems to be related to CAM (leaf thickness values greater than 1 mm showed carbon-isotope ratios lower than -20‰ [76], both the similar morphology of this fossil plant to *A. magdalenae* and the leaf thickness of 1.6 mm showed that, indeed, *Karatophyllum bromelioides* could be a fossil CAM [75].

4. CAM in Bromeliaceae

Bromeliaceae is considered a monophyletic family inside the order Poales [77]. It is one of the largest and most widespread plant families in the neotropics, occupying a wide variety of niches with different conditions [68, 78]. Recently, of the three classic subfamilies, only Bromelioideae and Tillandsioideae are considered monophyletic, while Pitcairnioideae was subdivided into five new subfamilies: Pitcairnioideae, Puyoideae, Navioideae, Hechtioideae, Lindmanioideae and Brocchinioideae [79].

In the Bromeliaceae, CAM has arisen multiple times in a seemingly independent way. Crayn and cols. [68] identified at least three independent origins for CAM in the 51 species analyzed, and this pattern was maintained when more species were taken into account. Silvera and cols. [4] also found multiple origins for CAM in the Orchidaceae family, which appeared to be linked with the colonization of the epiphytic habitat. In contrast to the Orchidaceae family, however, among bromeliads a strong correlation was not found between the occurrence of CAM and epiphytism [68]. In fact, a more recent work suggested that CAM could be more common in terrestrial, rather than epiphytic, bromeliad species [80]. Although very enlightening, these studies focused on the phylogeny and took into account only whether the plants were epiphytes or not. Some extra details could arise by observing how each species couples with its environment. For example, one of the major morphologic features of bromeliads is the rosette conformation of the leaves. This morphology is very important for some species in which the overlapping of its leaves forms a structure capable of accumulating water and organic matter – the so-called tank-bromeliads [78]. Based on the presence of the tank and how the plant acquires nutrients, Pittendrigh [81] separates bromeliads into four classes (Types I, II, III and IV - Figure 4). Table 1 shows a list of species grouped by type with their respective habitat (simplified only as moist or dry) and photosynthetic pathway.

4.1. Type I Bromeliads

Type I bromeliads are the ones that obtain their nutrients from the soil through the roots. The main differentiating feature of Type I bromeliads is that they do not have a structure capable of accumulating water. For this reason it would be expected that, when growing in dry habitats, these species would need other mechanisms to preserve water, such as CAM. Among this group are a few species that have already been studied regarding CAM. Cristopher and Holtum [50] studied some Type I bromeliads, such as *Pitcairnia paniculata, Fosterella schidosperma, Cryptanthus zonatus, Ortophytum vagans* and *Dyckia sp,* along with species belonging to other types. All Type I bromeliads showed some degree of CAM, except *P. paniculata,* indicating that this particular species might be better adapted to water-abundant environments, while the others may be more successful in drier habitats. In the same paper, the authors detailed the carbohydrate profile for these species but found no correlation between the type of accumulated carbohydrate and habit or CAM expression. Among the subfamilies, however, there were some similarities; for example, Tillandsioideae and Bromelioideae species accumulated starch, while plants belonging to the former Pitcairnioideae subfamily accumulated mainly soluble sugars.

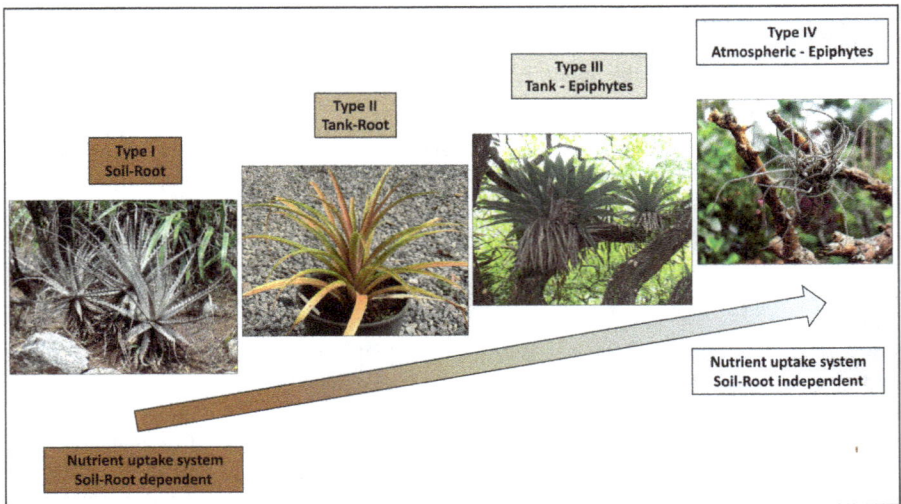

Figure 4. Four main bromeliad types proposed by Pittendrigh [81] and their dependency to acquire nutrients from the soil through a root system (upward arrow). As examples, types are illustrated by different bromeliad species. Type I –

Soil-Root: *Dyckia* sp.; Type II - Tank-Root: *Ananas* sp.; Type III –Tank-Epiphyte: *Vriesea* sp.; Type IV – Atmospheric-Epiphyte: *Tillandsia* sp.

Species	Type	Habitat	Photosynthetic patway	Reference
Aechmea allenii	III	moist	CAM[1,2]	[76]
Aechmea dactylina	III	moist	CAM[1,2]	[76]
Aechmea fasciata	III	dry	CAM[3]	[82]
Aechmea fendleri	III	moist	CAM[2]	[83]
Aechmea floribunda	III	dry	CAM[3]	[82]
Aechmea haltonii	III	moist	CAM[3]	[68]
Aechmea magdalenae	I	moist	CAM[1]	[84]
Aechmea nudicaulis	III	dry	CAM[3]	[82]
Aechmea sphaerocephala	III	dry	CAM[3]	[82]
Ananas ananassoides	I	dry	CAM[1,2,3]	[68, 85]
Ananas comosus	II	dry	CAM[1,2,3]	[50, 85]
Araeococcus pectinatus	III	moist	CAM[3]	[68]
Billbergia amoena	III	dry	CAM[3]	[82]
Billbergia pyramidalis	III	dry	CAM[3]	[82]
Brocchinia acuminata	I	moist	C3[3]	[68, 86]
Brocchinia micrantha	I	moist	C3[3]	[68, 86]
Bromelia chrysantha	II	dry	CAM[3]	[68]
Bromelia humilis	II	dry	CAM[1,2,3]	[83, 87-88]
Bromelia plumieri	I	dry	CAM[1]	[84]
Catopsis floribunda	III	moist	C3[3]	[68]
Catopsis wangerinii	III	moist?	C3[3]	[68]
Catopsis micrantha	III	moist	C3[1,2]	[76]
Cottendorfia florida	I	dry	C3[3]	[68]
Cryptanthus zonatus	I	dry	CAM[2]	[50]
Cryptanthus beuckeri	I	dry	CAM[3]	[68]
Deuterocohnia meziana	I	dry	CAM[3]	[68, 86]
Deuterocohnia longipetala	I	dry	CAM[3]	[68, 86]
Deuterocohnia lotteae	I	dry	CAM[3]	[68, 86]
Dyckia dawsonii	I	dry	CAM[3]	[86]
Dyckia ferox	I	dry	CAM[3]	[68, 86]

Dyckia hilaireana	I	dry	CAM^3	[68]
Dyckia sp.	I	dry	CAM^2	[50]
Encholirium inerme	I	dry	CAM^3	[68, 86]
Encholirium irwinii	I	dry	CAM^3	[68, 86]
Edmundoa perplexa	III	moist	CAM^3	[68]
Fosterella schidosperma	I	dry	CAM^2	[50]
Fosterella elata	I	moist	C3^3	[68, 86]
Fosterella penduliflora	I	moist	C3^3	[68, 86]
Fosterella petiolata	I	moist?	C3^3	[68, 86]
Guzmania calamifolia	I	moist	C3^{1,2}	[76]
Guzmania circinnata	III	moist	C3^{1,2}	[76]
Guzmania coriostachya	III	moist	C3^{1,2}	[76]
Guzmania desautelsii	III	moist	C3^{1,2}	[76]
Guzmania filiorum	III	moist	C3^{1,2}	[76]
Guzmania glomerata	III	moist	C3^{1,2}	[76]
Guzmania lingulata	III	moist	C3^{1,2,3}	[76, 89]
Guzmania macropoda	III	moist	C3^{1,2}	[76]
Guzmania monostachia	III	moist	C3/CAM^{1,2,3}	[68,89-90]
Guzmania mucronata	III	moist	C3^2	[83]
Guzmania musaica	III	moist	C3^{1,2}	[76]
Guzmania scherzeriana	III	moist	C3^{1,2}	[76]
Guzmania sprucei	III	moist	C3^{1,2}	[76]
Guzmania subcorymbosa	III	moist	C3^{1,2}	[76]
Guzmania wittmackii	III	moist	C3^3	[68]
Hechtia glabra	I	dry	CAM^3	[68, 86]
Hechtia glomerata	I	dry	CAM^3	[68, 86]
Hechtia guatemalensis	I	dry	CAM^3	[68, 86]
Hechtia lindmanioides	I	dry	CAM^3	[68, 86]
Lymania alvimii	III	moist	CAM^3	[68]
Mezobromelia pleiosticha	III	moist?	C3^3	[68]
Navia igneosicola	I	moist	C3^3	[68, 86]
Navia phelpsiae	I	moist	C3^3	[68]
Neoregelia eltoniana	III	dry	CAM^3	[82]
Neoregelia pineliana	III	moist	CAM^3	[68]

Neoregelia spectabilis	III	moist	CAM[2]	[50]
Nidularium bilbergioides	III	moist	CAM[2]	[50]
Orthophytum vagans	I	moist	CAM[2]	[50]
Pepinia beachiae	I	-	C3[3]	[68, 86]
Pitcairnia arcuata	I	moist	C3[1,2]	[76]
Pitcairnia burle-marxii	I	moist	C3[3]	[68]
Pitcairnia carinata	I	moist	C3[3]?	[68]
Pitcairnia corallina	I	moist	C3[3]	[86]
Pitcairnia heterophylla	I	moist	C3[3]	[68, 86]
Pitcairnia hirtzii	I	moist	C3[3]	[68]
Pitcairnia orchidifolia	I	moist	C3[3]	[68,86]
Pitcairnia paniculata	I	moist	C3[2]	[50]
Pitcairnia poortmanii	I	moist	C3[3]	[68]
Pitcairnia recurvata	I	moist	C3[3]	[68, 86]
Pitcairnia rubronigriflora	I	-	C3[3]	[68, 86]
Pitcairnia squarrosa	I	moist	C3[3]	[68, 86]
Pitcairnia smithiorum	I	moist	C3[3]	[86]
Pitcairnia sprucei	I	moist	C3[3]	[86]
Pitcairnia valerii	I	moist	C3[1,2]	[76]
Pitcairnia wendlandii	I	moist	C3[3]	[68]
Portea petropolitana	I	moist	CAM[2]	[50]
Puya aequatorialis	I	dry	C3/CAM[3]	[68, 86]
Puya floccosa	I	dry	C3/CAM[2,3]	[92]
Puya humilis	I	dry	C3/CAM[3]	[68, 86]
Puya laxa	I	dry	C3/CAM[3]	[68, 86]
Puya werdermannii	I	dry	C3/CAM[3]	[86]
Racinaea fraseri	III	moist	C3[3]	[68]
Racinaea spiculosa	III	moist	C3[1,2]	[76]
Ronnbergia explodens	I	moist	C3[1,2]	[76]
Tillandsia anceps	III	moist	C3[1,2]	[76]
Tillandsia balbisiana	IV	moist	CAM[1,2]	[93]
Tillandsia bulbosa	IV	moist	CAM[1,2,3]	[76]
Tillandsia circinnata	IV	moist	CAM[1,2]	[93]
Tillandsia dodsonii	III	moist	C3[3]	[68]

Tillandsia fasciculata	III	moist	CAM[1,2]	[93]
Tillandsia fendleri	III	moist	C3[2]	[83]
Tillandsia flexuosa	III	dry	CAM[2]	[83]
Tillandsia gardneri	IV	dry	CAM[3]	[82]
Tillandsia ionantha	IV	moist	CAM[1,2]	[94]
Tillandsia monadelpha	III	moist	C3[1,2]	[76]
Tillandsia pohliana	IV	moist	CAM[1,2]	[91]
Tillandsia recurvata	IV	moist	CAM[1,2]	[93]
Tillandsia schiedeana	IV	moist	CAM[1,2]	[93, 95]
Tillandsia setacea	IV	moist	CAM[1,2]	[93]
Tillandsia stricta	IV	dry	CAM[3]	[82]
Tillandsia tricolor	IV	moist	CAM[2]	[50]
Tillandsia usneoides	IV	moist	CAM[1,2,3]	[82, 93, 96]
Tillandsia utriculata	III	moist	CAM[1,2]	[93, 97]
Tillandsia valenzuelana	III	moist	CAM[1,2]	[93]
Vriesea carinata	III	moist	C3[2]	[50]
Vriesea espinosae	III?	moist?	CAM[3]	[68]
Vriesea espinosae	III?	moist?	CAM[3]	[68]
Vriesea monstrum	III	moist	C3[1,2]	[76]
Vriesea procera	III	dry	C3[3]	[82]
Vriesea sucrei	III	dry	CAM[3]	[82]
Werauhia viridifolia	III	moist	C3[3]	[68]
Werauhia capitata	III	moist	C3[1,2]	[76]
Werauhia greenbergii	III	moist	C3[1,2]	[76]
Werauhia hygrometrica	III	moist	C3[1,2]	[76]
Werauhia jenii	III	moist	C3[1,2]	[76]
Werauhia kupperiana	III	moist	C3[1,2]	[76]
Werauhia lutheri	III	moist	C3[1,2]	[76]
Werauhia milennia	III	moist	C3[1,2]	[76]
Werauhia panamensis	III	moist	C3[1,2]	[76]
Werauhia vittata	III	moist	C3[1,2]	[76]

Table 1. Habitat (moist or dry), type classification (according to Pittendrigh [81]) and photosynthetic pathway of 129 bromeliad species. Photosynthetic pathways were defined by [1] Gas exchange, [2] Biochemical parameters, and [3] δ^{13}C values.

4.2. Type II Bromeliads

The leaves of Type II bromeliads are disposed in such a way that their bases overlap, forming a space where water and nutrients can accumulate, called "tank". The absorption of the contents inside the tank can be performed in some cases by a root system that grows through the overlapping rosette leaves (called tank-roots) or even directly by the leaves, through epidermal structures called absorbing trichomes [78]. In this group, however, absorption of water and nutrients is still mostly performed by the roots. Some studies regarding CAM have been performed in this type of bromeliad. For instance, in *Bromelia humilis*, Medina and cols. [87] found differences in nocturnal CO_2 fixation and acid accumulation that increased in the wet season and with irrigation. Similar results were reported by Lee and cols. [98], indicating that in this possibly constitutive CAM plant, a more favorable condition would lead to higher stomatal conductance during the night, thereby increasing acid accumulation and plant growth. Partially confirming this, Fetene and cols. [88] showed more nocturnal CO_2 fixation by *Bromelia humilis* in the presence of nitrogen and higher irradiance. On the other hand, it was also demonstrated that another species, *Puya floccosa*, is capable of increasing nocturnal acid accumulation in response to drought and/or other unfavorable microclimatic cues [92]. Working with *Achmea magdalenae*, *Bromelia plumieri* and *Ananas comosus*, Skillman and cols. [84] showed that these plants performed well in a shaded understory, presenting higher photo-synthetic capacity when compared to C_3 plants growing in the same conditions. The growth rate, however, was inferior, when the same comparison was made, possibly because of differences in the partitioning of carbohydrates produced. Also in this work, it was noted that CAM plants grew more during the dry season, different from the C_3 plants, which grew more in the wet season. Therefore, CAM seems to be advantageous over C_3 photosynthesis in conditions with water shortage. This can be easily observed in facultative CAM species, in which CAM is often induced by drought.

Recent studies on *Ananas comosus* provided some insight into the signaling of CAM up-regulation in the Bromeliaceae [99]. Although considered a constitutive CAM species, this work indicated that *A. comosus* is capable of performing C_3 photosynthesis when young plants are grown under *in vitro* conditions. In this same study, by investigating the signaling events controlling pineapple CAM expression in response to water deficit, the authors characterized the existence of at least three signaling pathways: one inhibitory, mediated by cytokinins, and two stimulatory, one dependent and one independent of ABA. Furthermore, both stimulatory pathways converged on cytosolic calcium signaling, while the ABA-dependent pathway also involved the free radical nitric oxide. Another intriguing observation was made on *A. comosus*, along with other species, regarding the longitudinal distribution of metabolites along the leaves. Popp and cols. [83] showed that in this species, along with six other bromeliads, the nocturnal accumulation of acids and the amount of carbohydrates showed an increase from the leaf base to the tip. This interesting longitudinal gradient along the leaves of bromeliads, and possibly other rosette plants, will be addressed and further discussed later in this chapter.

Apparently, CAM occurs in Type I and Type II bromeliads (both terrestrial) when required by the environment (*e.g*, dry or exposed habitats). Types III and IV, however, are epiphytic.

4.3. Type III Bromeliads

Epiphytes notably enrich the tropical forest ecosystems by providing new niches for a great number of organisms [100, 101] but are subjected to several environmental limitations. For example, nutrients and water are only available sporadically or seasonally through rainfall [102]. Therefore, epiphytes may face water shortage even if living in a moist environment like tropical forests.

One of the main characteristics of Type III bromeliads is the presence of the tank. In these plants, the acquisition of water and nutrients comes mainly from the solution impounded there [78]. These species present a large number of absorptive trichomes on the base of the leaves, which allow them to directly absorb water and nutrients from the tank [78]. Their roots play a minor role in nutrition, being more restricted to mechanical support [103]. In the epiphytic habitat the tank is a very important structure, which allows the accumulation of water and/or nutrients during drier periods. Therefore, Type III bromeliads have a reservoir of water even when rain is absent. Another remarkable feature of epiphytic tank bromeliads is their inter-action with other organisms. The rosette conformation provides different compartments with distinct ecological conditions, serving as favorable habitats for a wide variety of organisms that, besides shelter, also need a supply of water to conclude their life cycles [104]. In addition, the organisms living in the tank may provide important nutrients to the plant [101, 105].

Among Type III epiphytes it is possible to find C_3 and CAM photosynthesis (Table 1). In fact, some Type III bromeliads have been intensively studied in terms of photosynthetic plasticity. For instance, *Aechmea* 'Maya', which is a cross between *A. fasciata* and *A. tessmanii*, expresses CAM photosynthesis and has been used as a model for analyzing the impacts of several environmental factors (*e.g.*, light, nutrition, CO_2 availability) on the CAM behavior and carbohydrate partitioning [106-108]. Interestingly, when maintained under elevated concentrations of CO_2, this species increased the CO_2 uptake during Phases II and IV, but the nocturnal uptake of CO_2 remained similar to control conditions [106]. The extra CO_2 absorbed in those phases was used for the synthesis of hexoses that were not exported from the leaves. A later work on the same species showed that when these plants were transferred to low luminosity conditions, CAM was strongly dampened in the short term [107]. In fact, there is an acidifica-tion of the cytosol, which seems to be the result of an incapacity of the cells to degrade the malic acid formed during the previous night. The acid concentrations remained high for at least the first two days, resulting in serious damage to the cells. Also, CO_2 assimilation ceased and remained null at least until the sixth day of treatment. In the long term, the plant recovered a small part of the capacity to assimilate CO_2, when compared to the levels observed for control plants [107]. Accordingly, this species had a similar response when the four seasons were taken into account, with a higher level of carbon assimilation in more illuminated seasons (summer and spring) than in darker ones (winter and autumn - [108]).

Another Type III species that is receiving more and more attention in CAM studies is *Guzmania monostachia*. Maxwell and cols. [109] noted that plants of this species accumulated less acid when the photosynthetic active radiation decreased due to the season of the year or shading. Later, Maxwell and cols. [110] verified that high light and drought had a positive effect on CAM expression in this species, along with a powerful photoprotective mechanism, which could explain how *G. monostachia* couples with changing light and water supply in its environment. In fact, it was further described that when exposed to full sunlight, this species increases its accumulation of acids along with its photoprotective mechanisms within five days of the start of the dry season [111]. More recently, Freschi and cols. [90] studied the upregulation of CAM in response to drought in *G. monostachia*. The authors found that the plants presented CAM-idling after seven days of drought. Another interesting feature brought forth by these authors is the differences in CAM expression along the length of the leaves, relating them to those observed in other bromeliads by Popp and cols. [83]. In the case of *G. monostachia*, there was an up-regulation of PEPC activity, followed by an increase in nocturnal acid accumulation only in the apical portion of the leaves. This observation agrees with other data regarding Type III bromeliads, showing that there is a division of functions along the leaf length (Figure 5). This is easily understandable, since the leaves of these species must perform both absorption of water/nutrients and photosynthesis. The base of the leaf, which receives less light and is in direct contact with tank contents, may be more specialized in absorption, while the apex, which intercepts more light, may be specialized in photosynthesis.

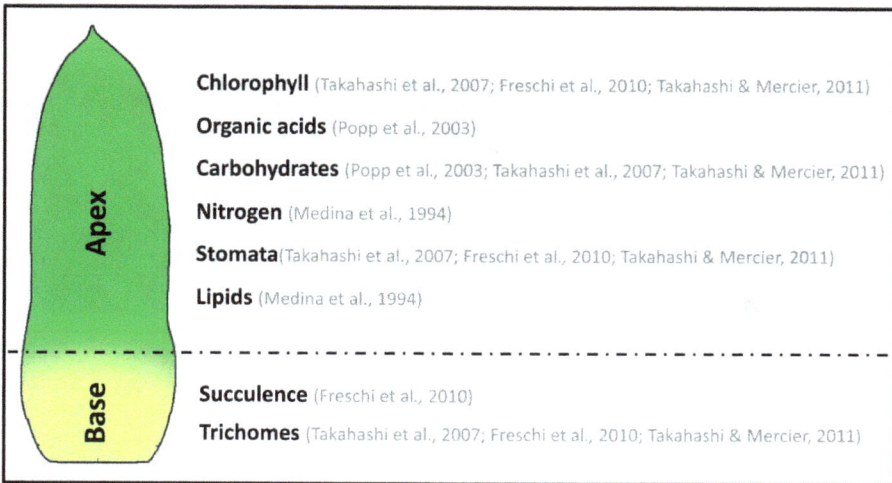

Figure 5. Comparative characteristics between apex and base of Type III bromeliad leaves. Characteristics described in the different leaf portions illustrate that they are more abundant in that region than the other.

However, some differences in gas exchange along the leaf were also observed, even if to a lesser extent, in other species that do not form the tank, like the Type I bromeliad *Ananas comosus* [112] and agaves such as *Agave sisalana* (James Hartwell, personal communication). This leads to the possibility that other factors besides functional specialization may be at work. In fact, both agaves and bromeliads have a similar structure: they are monocarpic monocots with the leaves disposed in a rosette (Figure 6). Therefore, these plants have a structure that always places younger leaves inside the rosette and at a greater angle than those of the periphery [83]. Thus, light interception changes with leaf age [112]. Each individual leaf also has an age gradient along its length, due to the action of an intercalary meristem present in the leaf bases. Consequently, the apices of the leaves are older than their bases, and since CAM can be developmentally regulated, this would at least partially account for this base-apex gradient pattern. If this is true, then, this phenomenon should be found in all rosette monocots, regardless of whether they have a tank or not. It is probable, however, that the presence of the tank enhances the differences observed in the basal and apical portions of the leaf. The apparent occurrence of a base-apex CAM expression gradient along the leaves of the other non-bromeliad rosette plants, such as agaves, is discussed later in this chapter.

In Type III bromeliads, recycling of internal CO_2 seems to be an important resistance mechanism which allows the plant to keep stomata closed for a longer time and, therefore, save more water. In fact, Stiles and Martin [97] demonstrated that, when *Tillandsia utriculata* remained without water for several days, the stomatal conductance diminished, causing an increase in the proportion of refixed CO_2. This extreme response to drought can be even more important for Type IV bromeliads, which are presented next.

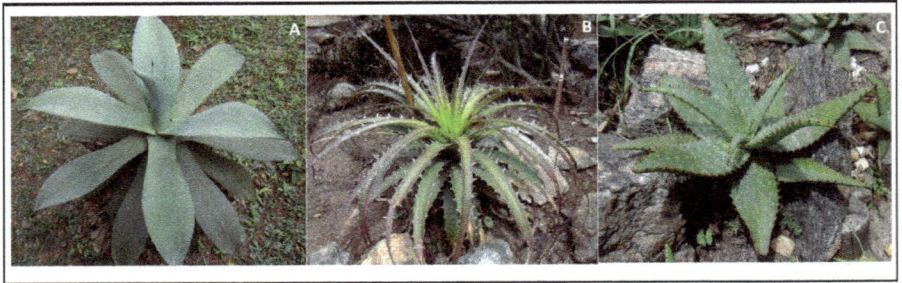

Figure 6. Monocot rosette plants. A. *Agave* sp. B. *Dyckia* sp. C. *Aloe* sp.

4.4. Type IV Bromeliads

Type IV bromeliads are the so-called atmospheric ones. In the absence of a tank, their leaves are covered by absorbing trichomes, which allow the uptake of water and nutrients that are suspended in the atmosphere or during rainfall [78]. The trichomes also seem to be useful in reflecting light to avoid photoinhibition [113, 114]. Therefore, atmospheric bromeliads must have a very efficient mechanism to capture water and nutrients as quickly as possible when these resources are available [102]. These species also need ways to save the absorbed water. In fact, all atmospheric *Tillandsia* species present CAM [68]. Martin and Adams III [95] measured the 24h gas exchange pattern in the CAM plant *Tillandsia schideana*, using it to estimate the amount of acid that should accumulate during the night, based on the amount of CO_2 fixed. When they measured nighttime acid accumulation, however, the values were much higher than the estimate. The authors concluded that this excess in acidity originated from recycling of the respired CO_2. Moreover, drought exposure caused a drop in CO_2 assimilation, and this difference between estimated and measured acidity values increased even more [95]. It is possible, then, that this species was on its way to CAM-idling as a resistance mechanism to withstand the drought period.

By comparing nighttime atmospheric CO_2 uptake and malate accumulation in 12 *Tillandsia* species, Loeschen and cols. [93] verified that respiratory CO_2 recycling significantly occurred only in *T. schideana*. One possibility is that these were well-hydrated greenhouse plants, so the rates of CO_2 recycling would be low when compared to CO_2 assimilated during night. In another study, Nowak and Martin [94] analyzed the CAM responses of *Tillandsia ionantha* under drought for 60 days. From the 30th day on, the CO_2 uptake started to drop until the 60th day. When comparing it to the acid accumulation, the authors observed that in response to drought the percentage of recycled CO_2 increased. Based on these data, it seems that *T. ionthana* was also on its way to a CAM-idling state in order to withstand drought. Later, Haslam and cols. [96] demonstrated that increasing irradiance was also capable of enhancing CO_2 recycling, along with water shortage in the atmospheric *T. usneoides*. A recent work with the atmospheric *T. pohliana* found that this plant accumulates almost equimolar concentrations of malate and citrate [91]. This is interesting because citrate accumulation does not result in a net carbon gain [115]; therefore, this accumulation may serve other purposes. Freschi and cols. [91] suggest that, since MDH is inhibited by OAA, in this species citrate formation would be a way to avoid excessive OAA accumulation. Moreover, citrate formation generates NADH, which could be consumed during the night by MDH and other enzymes. In fact, the same authors found that nitrate reductase, a great consumer of NAD(P)H, is mainly active at night in this species. Therefore, Type IV bromeliads are very good examples of plants that are capable of performing distinct CAM modes.

In conclusion, CAM can be found in Type I bromeliads when they inhabit xeromorphic areas, such as deserts, following a distribution similar to other plants, including other non-bromeliad terrestrial monocot rosette plants (*e.g.*, agaves, aloes). Type II would allow some accumulation of water, but since the contents of the tank are not so well absorbed and they have access to soil water, they seem to follow a similar CAM expression pattern to that found in Type I bromeliads. Among Types III and IV, the presence of a tank is perhaps a major difference that permits C_3, or at least facultative CAM, in the epiphytic habitat since water accumulation and absorption through this structure is possible. Without the tank, atmospheric bromeliads (Type IV) have to perform CAM or they will not survive. Therefore, the presence of the tank in some epiphytic bromeliads could be the reason why a clear correlation between CAM and epiphytism was not found in Bromeliaceae, different from other families with epiphytic species [4, 68, 80].

5. CAM in Agavaceae

When compared with the relatively abundant literature about CAM photosynthesis in bromeliads, much less is known about the photosynthetic biochemistry, regulation and plasticity in Agavaceae. As many bromeliads, most Agavaceae representatives are terrestrial monocarpic rosette plants, inhabiting predominantly arid and semiarid regions. Similar to other xeric plants, many species of Agavaceae exhibit relatively low growth rates under both natural and optimal environmental conditions. On the other hand, sometimes impressive rates of biomass productivity can be found in *Agave* plants [116], which might be associated with their wide range of anatomical and physiological adaptations to survive and thrive under water-limited conditions. Among the physiological adaptations exhibited by Agavaceae representatives, the capacity to express CAM photosynthesis naturally deserves to be highlighted and, therefore, will be the main focus of this chapter section. The current and ancient uses of agaves for food, beverages and diverse natural products have already been recently covered by excellent reviews [116-118] and will not be emphasized in this chapter.

According to APG III in 2009 [77], Agavaceae belongs to the expanded family Asparagaceae, and *Agave* is currently treated as one of the 18 genera of the subfamily Agavoideae [119]. The genus *Agave sensu stricto* (166 species), which is divided into the two subgenera *Littaea* (53 species) and *Agave* (113 species), is principally monocarpic and covers the most succulent and dry-adapted members of the family. Among other adaptations, large succulent rosettes, which funnel water, and shallow root systems, which allow a rapid uptake of sudden precipitation, are traits commonly found among *Agave* species. Diversification and speciation of *Agave*, the largest genus of the family, were significantly elevated between 8-6 million years ago (late Miocene and early Pliocene), coinciding with an increase in dry conditions in central Mexico [120, 121], and then again between 3-2.5 million years ago (late Pliocene), coinciding with the distribution of nectarivorous bats, the main pollinators of current *Agave* species. The diversification of *Agaves* in North America during the phenomenon of reduced precipitation and atmospheric CO_2 availability in the late Miocene-Pliocene was simultaneous with the diversification of several other succulent plant lineages across the world, such as the ice plants in south Africa [122].

In contrast to observations in the Bromeliaceae, virtually all Agavaceae genera are believed to have at least some species capable of expressing CAM photosynthesis [123, 124]. Naturally, a certain number of Agavaceae representatives, such as some *Yucca* species, have been clearly demonstrated to be C_3 plants [125, 126]. For succulent agaves, CAM seems to be a ubiquitous trait, expressed many times in a quite rigid pattern with reduced gas exchange at Phases II and IV. For instance, it is known that species such as *Agave deserti, A. fourcroydes, A. tequilana, A. angustifolia, A. lecheguilla, A. lurida, A. murpheyi, A. parryi, A. salmiana, A. scabra, A. schottii, A. shawii, A. sisalana, A. utahensis, A. vilmoriniana, A. virginica* and *A. weberii* typically perform CAM photosynthesis under natural conditions [127-131]. Nevertheless, it was reported that *Agave deserti* is clearly able to change from CAM to C_3 photosynthesis when maintained under well-watered laboratory conditions [132]. In fact, a complete switch from CAM to C_3 diel gas exchange pattern with almost all net CO_2 uptake occurring during daytime and virtually no day/night acid fluctuations was observed in *A. deserti* plants maintained under well-watered greenhouse conditions, reinforcing the notion that a possible C_3-CAM facultative behaviour might occur in this particular agave [132]. Moreover, a certain level of photosynthetic plasticity has even been described for young and adult plants of the CAM constitutive *A. tequilana*, which allows them to modulate the contribution of daytime (Phases II and III) and nighttime (Phase I) carbon acquisition when facing different environmental conditions [133-134]. In fact, although most CO_2 uptake in *A. tequilana* takes place at night (Phase I) [133-135], the relative contribution of daytime carbon gain (especially in Phase IV) can be modulated throughout the year [134]. Curiously, in both adults and young individuals of this species, at least some Phase IV CO_2 acquisition can be maintained even during the driest months of the year, which is a phenomenon not very commonly observed among other CAM plants growing under arid conditions [133, 134].

In addition to optimizing carbon gain and growth, the occurrence of some daytime gas exchange might also contribute as a transpiration cooling mechanism, which would significantly benefit these plants as long as enough water is available in the soil or inside their succulent leaves and stems [136]. This daytime transpiration cooling system can help tropical plants minimize excessive increases in leaf temperature and is logically not present when stomatal opening is restricted to the night period (Phase I). Interestingly, in a recent study, it was verified that the spike (young folded leaves in the centre of the rosette) of *A. tequilana* is the most thermotolerant part of the plant, presenting the highest stomatal density and elevated levels of HSPs (heat-shock proteins) [136]. Considering the fact that the central spike is the youngest part of the agave, it seems plausible to suggest that these young tissues might exhibit lower levels of CAM expression and perhaps higher daytime gas exchange, which would inexorably lead to at least some transpiration cooling during the light period.

Whether young tissues of agaves present lower levels of CAM photosynthesis still remains to be investigated, but some current observations seem to indicate a possible influence of tissue age on CAM expression in these plants. For instance, as observed for rosette bromeliads [90], some base-apex longitudinal gradient in CAM expression might also be present along the leaves of *Agave sisalana*, in which CO_2 uptake in the more basal and younger leaf portion occurred almost exclusively during daytime, whereas a gas exchange pattern typical of CAM photosynthesis was observed in the more mature leaf tip (J. Hartwell, personal communication). In this study, conducted in detached leaves, all leaf portions received the same amount of PAR incidence, ensuring that these changes in CO_2 uptake pattern would not simply be a result of distinct light availability along the leaf blade. Further suggesting some influence of plant ontogeny on CAM photosynthesis in Agavaceae species, Olivares and Medina [112] observed that nocturnal changes in titratable acidity were also dependent on leaf age in *Fourcroya humboldiana* since this parameter increased with the distance from the leaf bases and also from the younger to more mature leaves, reaching maximum values at the 7th leaf (counting from the rosette centre). Moreover, it has also been demonstrated that late-afternoon CO_2 uptake (Phase IV) in *A. deserti* decreases as the seedlings age, being virtually absent in adult plants of this species [137], which might indicate a transition to a CAM mode more strictly dependent on nighttime (Phase I) CO_2 uptake. As in other CAM plants, factors like leaf arrangement, total daytime PAR, daytime/nighttime temperature and drought seems to influence CO_2 uptake in *Agave* plants [129, 135, 138]. For instance, it has been suggested that *Agave* required more than 4 mol photons m^{-2} d^{-1} (93 μmol m^{-2} s^{-1} for 12-photoperiod) to fix CO_2 at night and, therefore, to accumulate organic acids [129]. In *A. fourcroydes*, nearly 90% of

PAR saturation of nocturnal CO_2 uptake happened near a total daily PAR of 20 mol m^{-2} [138], values comparable to other CAM plants. Also in this *Agave* species, studies have demonstrated that when drought conditions were imposed, a higher fraction of daytime CO_2 uptake was lost compared to nighttime CO_2 uptake [138]. In fact, adult *A. fourcroydes* plants exposed to 11 days of drought exhibited a reduction of 99% in net daytime CO_2 uptake and 76% in nighttime CO_2 uptake (Nobel, 1985b). In addition, it has been demonstrated that low leaf temperatures at night are quite beneficial for nocturnal gas exchange and organic acid accumulation in different *Agave* species [127, 135, 138]. Based on the lack of information about CAM in Agavaceae, additional studies are required to determinate whether CAM photosynthesis in *Agave* plants really depends on the plant and/or leaf ontogenetic stage and whether it can be facultatively expressed in some species in response to environmental stimuli. Naturally, the regulatory mechanisms controlling CAM expression in these plants remains even more elusive.

As commonly observed in many other CAM xerophytes, leaf succulence is a widespread feature of agaves. The internal water storage provided by a prominent leaf hydrenchyma might be, at least in part, responsible for the relatively high biomass productivity observed in some *Agave* species living in severely water-limited habitats. This prominent succulence would serve to buffer abrupt and long-term changes in water availability, helping to maximize nocturnal CO_2 uptake and extend the duration of atmospheric CO_2 acquisition beyond the night period [134]. As in other plants displaying the "succulent syndrome", the presence of CAM photosynthesis and leaf succulence in agaves is also correlated to features that reduce water loss, like thick cuticles, reduced stomatal size and/or frequency, and other water-conserving characteristics. As a result, remarkably high water use efficiency (WUE) is usually observed in *Agave*, allowing these plants to colonize dry heterogeneous environments, sometimes even achieving elevated productivity values at these locations. Naturally, precise adjustments in photosynthetic biochemistry are clearly needed to obtain the highest day- and nighttime CO_2 uptake possible with the usually scant and erratic water supply of arid and semi-arid regions.

Studies have demonstrated that the prominent leaf succulence of several adult *Agave* species is key for allowing the occurrence of substantial net CO_2 uptake even when soil water content reaches relatively low levels, reinforcing the importance of this internal water supply to ensure high photosynthetic performance during the entire year [133]. Interestingly, though, young plants of *Agave tequilana*, even while presenting lower succulence than adult individuals and, therefore, comparatively lower internal water storage, were able to obtain carbon during day and night under field conditions. These plants exhibited almost the same carbon gain of adult individuals and maintained relatively high photosynthetic assimilation rates during both dry and wet seasons [134]. Naturally, the continuous water movement from the medullar hydrenchyma to the marginal chlorenchyma during the dry season, when soil water availability can decrease to critical values, might in fact be a critical factor for allowing the occurrence of relatively high levels of CO_2 assimilation year-round, even in these young agave plants [134]. Under the severely dry conditions normally faced by many agaves, internal water storage tissues such as hydrenchyma are obviously much more appropriate than an external water reservoir such as the tank of certain bromeliads. Besides the scarce and sporadic rain events, the air humidity of arid and semi-arid regions would inexorably lead to a fast evaporation of any external water reservoir, whereas internal water storage would last much longer by benefiting from additional morphological, anatomical and physiological adaptations, such as thick cuticles, highly regulated stomata control, accumulation of osmotically active compounds in the water storage tissues, among others.

In this sense, another typical feature of Agavaceae species is the production of fructans, which are polymers of β-fructofuranosyl residues synthesized from sucrose and accumulated in the vacuoles of succulent parenchymatic cells of leaves and stems. Fructans are believed to contribute in several ways to plant metabolism and development, including osmoregulation, cryoprotection, and drought tolerance [139-141], and in mature agave plants fructans become an energy source for flowering. In general, the type and structure of fructans can be indicative of the species, within the limits of effects triggered by the environment and growing stage of the plant [116, 142]. Being water soluble and, therefore, osmotically active, fructans can influence the osmotic potential of parenchymatic cells. This osmotic impact of fructans might depend, among other factors, on their degree of polymerization and relative concentration inside the vacuoles of each cell. Since fructans are not as highly polymerized as glucans (*e.g.*, starch), their use as main storage carbohydrate might be of significance for determining the osmotic pressure in agave cells [112]. Although some evidence suggests that fructans are not broken down during the dark period to provide PEP as substrate for the nocturnal CO_2 fixation in agaves [143], other data indicate a possible use of these carbohydrates as the main source of nighttime PEP production [112]. For instance, it has been observed that diel fluctuations in sucrose could account for more than 83% of the carbon needed for nocturnal PEP regeneration in *Agave americana* [143]. On the other hand, in *Agave guadalajarana* diel fluctuation in the leaf starch, glucose, fructose and sucrose could not account for the carbon required for nighttime PEP production, and a possible use of alternative extrachloroplastic carbohydrates (such as fructans) for PEP generation has been suggested [144]. This suggestion is in agreement with the relatively low content of starch normally observed in *Agave* tissues and with the inverse relationship between fructans and malic acid observed in some Agavaceae species such as *Fourcroya humboldiana* [112]. In fact, the leaf levels of soluble fructans (including sucrose and fructose) in *F. humboldiana* clearly decreased during the night coinciding with the period of malic acid accumulation. The amount of carbon involved in these reciprocating fluxes indicated that fructans apparently represent the exclusive source of PEP for dark CO_2 fixation in this Agavaceae species [112]. Diel changes in the degree of polymerization of *F. humboldiana* fructans were also observed, which might be associated with the hydrolysis of fructose molecules from the fructans during the night for PEP regeneration and a reverse process during the day when CO_2 from malate would be incorporated into new fructose units of the pre-existing fructan molecules [112].

In summary, the impacts of environment and development on CAM expression capacity and carbohydrate metabolism in *Agave* species are still poorly understood. This extensive lack of knowledge regarding the relevant traits that account for the capacity of these plants to productively grow under arid and semiarid conditions contrasts diametrically with the enormous economical potential that some *Agave* species possess for biomass and renewable material production in marginal lands. This situation might possibly change in a near future, especially considering the increasing interest in using these and other CAM species as alternative crops in a context of global climate change and increased desertification.

6. Bromeliads and agaves CAM plants in a climate change and desertification context

Climate change involves elevated CO_2 concentrations, increasing temperatures and/or changes in precipitation patterns. Therefore, the perspective of climate changes around the planet has stimulated extensive research into assessing the impacts of elevated CO_2, elevated temperature and drought on different vegetation types. The atmospheric CO_2 concentration has been increasing rapidly during the last century, now reaching about 390 ppm. Promoted by deforestation, land-use, and the burning of fossil fuels, CO_2 is predicted to double by the middle of this century. Besides being an important change in the environmental conditions in and of itself, this progressive increase in atmospheric CO_2 concentration might indirectly impact climate by leading to increases in global temperature and perturbations in precipitation patterns.

The effects of elevated atmospheric CO_2 on carbon gain, plant growth and physiological performance depends on the CO_2 assimilation pathway, the exposure duration and the environmental conditions, among other factors [106]. For instance, growth under CO_2 enrichment might impact the relative contribution of C_3 and C_4 carboxylation pathways to net carbon gain, which could affect WUE over the day/night cycle and carbohydrate fractioning for growth and export [106]. As CAM plants use Rubisco and PEPC to take up CO_2, different conclusions have been reached. For instance, it has been proposed that PEPC in CAM plants might be saturated at the current atmospheric CO_2 concentration [145, 146]. However, divergent results about the influence of elevated CO_2 in CAM plants have been reported during the last decades [147-153], demonstrating that the enrichment of CO_2 into the atmosphere can trigger complex responses in CAM plants. For example, elevated CO_2 had no effect on diel CO_2 uptake by *Kalanchoë daigremontiana* [147] nor on nighttime CO_2 uptake in *Clusia uvitana* and *Portulacaria afra* [154, 155], whereas in several other CAM species more significant impacts of elevated CO_2 on the daytime, nighttime and/or diel carbon acquisition have been reported [148, 150, 151, 153]. Changes in morphology, anatomy and biochemistry driven by modifications in atmospheric CO_2 concentration have also been observed in some CAM plants, commonly associated with concomitant alterations in their growth rates and biomass accumulation. Moreover, in plants maintained under elevated atmospheric CO_2, some researchers have observed instantaneous net CO_2 uptake increasing over time, which suggests that the response to high levels of CO_2 in the atmosphere is maximized by physiological and morphological changes, such as chlorenchyma thickening [149, 152, 156]. Since leaf succulence limits diffusion of CO_2 and optimizes the accumulation photosynthetic products, changes in this morphological trait over time might contribute to hamper the acclimation of CAM plants under elevated CO_2 in the atmosphere.

Among bromeliads, perhaps the first study with the goal of evaluating the influence of elevated CO_2 on CAM photosynthesis was carried out by Nowak and Martin [157] in the atmospheric epiphyte *Tillandsia ionantha*. In this study, the authors demonstrated for the first time that a CAM plant could respond to high atmospheric CO_2 with significant increases in nocturnal malate accumulation, which would potentially lead to increases in productivity. Subsequent studies conducted with pineapple (*Ananas comosus*) plants have demonstrated that under elevated CO_2 concentration, this species responds with increases in both morning and nighttime CO_2 uptake [150, 151], associated with higher values of WUE [151], productivity, root:shoot ratio and leaf thickness [149]. Interestingly, when the pineapple plants were heavily irrigated, CAM activity and biomass response to elevated CO_2 were significantly reduced [151, 158].

The responses of bromeliads of the genus *Aechmea* have also been investigated under elevated atmospheric CO_2 concentrations [106, 158, 159]. Long-term exposure of *Aechmea magdalenae* to double CO_2 concentration (~ 700ppm) resulted in improved growth, although non-significant increases in daytime and dark CO_2 fixation were observed [158]. For *Aechmea* 'Maya' it was evidenced that an atmosphere of 700 ppm CO_2 promoted a 60% increase in carbon gain, through promotion of diurnal C_3 pathway and C_4 carboxylation during Phases II and IV, where WUE was equivalent to night periods [106]. Elevated CO_2 promoted the accumulation of hexose during Phase IV, stopping neither net daytime carbohydrates export nor the stimulation of dark carboxylation and nocturnal export [106]. These authors point out respiration as the major carbohydrate sink in *A. Maya* and recognized discrete pools of carbohydrates for CAM and for export. They observed a two-fold increase in water use efficiency under elevated CO_2, suggesting this as the major physiological advantage of CO_2 enriched atmosphere, which can be favourable for growth during drought stress events [106]. Curiously, studies conducted on *Aechmea* plants with ornamental value, like *Aechmea* 'Maya' and *Aechmea fasciata* 'Primera', showed different responses under elevated CO_2 atmosphere [159]. For example, during 34 weeks of growth under CO_2 enriched atmosphere, *A.* 'Maya' biomass and leaf micromorphology showed no significant changes, whereas elevated CO_2 promoted a reduction of total leaf area and thickness in *A. fasciata* 'Primera', which led to a reduction in fresh and dry biomass. Curiously, during these changes driven by elevated CO_2, some ornamental traits were lost in *A. fasciata* 'Primera', especially due to the reduction in total chlorophyll and the changes in leaf allometric length/width ratios, producing paleness and more compact plants [159].

Thus far, studies on the impacts of elevated CO_2 on Agavaceae species have been mainly conducted in the genus *Agave* and *Yucca* [126, 156, 160-162]. *Agave deserti*, *A. salmiana* and *A. vilmoriniana* plants grown under elevated CO_2 concentrations showed higher nighttime and/or afternoon CO_2 uptake [156, 160-162] as well as increased WUE and productivity [156]. In *A. deserti*, elevated CO_2 treatment for 17 months resulted in longer and thicker leaves, thicker chlorenchyma and increased root cell length [156]. Moreover, in *A. deserti* and *A. salmiana*, long-term treatments with elevated CO_2 resulted in decreases in total PEPC and Rubisco activity associated with increases in Rubisco *in vivo* activation status [156, 160], which can be interpreted as a strategy for maintaining photosynthetic performance since increases in the activated *vs.* total ratio for Rubisco would compensate for decreases in total activity of this enzyme. Also in *A. deserti* and *A. salmiana,* increases of up to 30% in dry weight gain have been reported after long-term treatments under elevated CO_2 [161]. On the other hand, significant differences in dry weight accumulation in *A. vilmoriniana* grown for 6 months under 350 or 675 ppm of CO_2 were only observed when these plants received water just once a week, but not when they were watered twice a week [163].

Taken together, these reports, sometimes presenting contrasting results, may reflect the inherent plasticity of CAM in terms of optimizing carbon gain and WUE in a changing scenario [106]. The impressive absence of acclimatization in CAM plants to elevated CO_2 enrichment can be perhaps correlated with the prominent succulence observed in many of these species, which might proportionate a large space for accumulating photosynthates without feedback inhibition [152]. Naturally, more research about carbohydrate partitioning in CAM species is clearly needed for a better understanding of the lack of acclimatization of these plants under elevated CO_2. It is also important to keep in mind that all the conclusions currently available about the impacts of high atmospheric CO_2 on CAM photosynthesis are still based in relatively few studies and species.

Since most predictions about the future global climate scenario describe intensification in aridity, with possibly increased desertification around the world, the adaptive success of CAM plants in these challenging environments might represent an important alternative for carbon sequestration and use as crops for production of biofuels and other renewable materials in regions not suitable for cultivation of agronomic species performing C_3 and C_4 photosynthesis [11, 49]. The capacity of CAM plants to survive and productively grow in environments subjected to frequent and/or long-lasting dry periods, especially due to their ability to improve net CO_2 assimilation and WUE by carefully adjusting the period for capturing CO_2 during the 24-hour daily cycle, is certainly a feature that justifies even more intensive research on these plants in both the short and long term. Considering the fact that some CAM agaves (*e.g., Agave salmiana, A. mapisaga, A. tequilana, A. americana, A. sisalana*) and bromeliads (*e.g., Ananas comosus*) can achieve productivity levels only slightly lower than those found in C_3 and C_4 species [164], these plants might represent special targets for studies aimed to optimize their use both at the current and future global climate scenarios.

7. Conclusions

It has been hypothesized that the evolution of CAM in terrestrial and aquatic environment was favored by decreasing atmospheric CO_2 levels and CO_2/O_2 ratio during the Miocene. Thus, atmospheric CO_2 reductions, coupled with warm climates in subtropical latitudes, were the most important driving forces in the evolution of CAM. Then, CAM photosynthesis may have originated as a means of improving carbon economy. Phylogenetic reconstructions from comparative physiology and taxonomy have shown that CAM appeared multiple times and originated from a C_3 ancestor since all enzymes related to CAM functions are also found in C_3 plants. These facts suggest that metabolic pathways related to CAM have arisen from co-option processes.

Remarkable differences between C_3 and CAM metabolisms continue to intrigue many researchers, and the importance of many of these differences on CAM functioning is still not completely understood, for example, the importance of the contribution of C_4-carboxylation or C_3-carboxylation in reproductive development of CAM plants; in addition, nowadays very few explanations about the existence of different starch breakdown pathways in C_3 and CAM plants are currently available. It is possible that Glu-6-P exportation at night from the chloroplast (phosphorolytic pathway) in CAM plants could be related with the allosteric activation of PEPC, but this issue has not been directly studied yet.

Nowadays, it is possible to observe that Crassulacean acid metabolism is a remarkable adaptation to environments with low abundance of water, showing a great plasticity in its expression. This plasticity resulted in a classification of several types of CAM (*e.g.*, CAM cycling, CAM idling, *etc.*), and it is important to keep in mind that even constitutive CAM plants do not appear to perform one of these types exclusively, instead transiting between them in response to environmental or developmental conditions. C_3-CAM plants show an even greater plasticity, being capable of going from the C_3 mode to CAM mode, also as a result of developmental or environmental cues. Nevertheless, it is currently almost impossible to establish clear differences between facultative CAM and constitutive CAM plants since it has already been demonstrated that even some constitutive CAM plants, when juvenile, are capable of switching from C_3 to CAM and return to C_3 mode, in a fully reversible way. Therefore, due to the facultative component that exists in constitutive CAM, we propose that facultative CAM species should be those capable of going from C_3 to CAM and back even in adulthood and not just during the juvenile stage. On the other hand, species capable of irreversibly expressing CAM at some moment in their life cycle, regardless of whether CAM expression is influenced by either environmental or developmental factors, should simply be considered as CAM plants.

Bromeliaceae is a very important neotropical plant family with several CAM-performing species. This family has a variety of life forms, which allows the individuals to cope with different kinds of environments, from semi-arid to rainforests, terrestrial to epiphytic habitats. As a water saving mechanism, CAM is undeniably linked to environments in which water is limited for some reason, including semi-arid regions or the epiphytic niche. Thus, a higher abundance of CAM plants is expected to be found in these niches. This is true for semi-arid environments, yet in the epiphytic habitat this observation is controversial: CAM does not seem to be linked with epiphytism in Bromeliaceae. The response to this apparent contradiction could depend on how each species minimizes the lack of water in the environment. For example, the presence of the tank could provide water for the plant in a longer term, thereby allowing tank bromeliads to be C_3-CAM plants or even exclusively C_3, when almost all epiphytes without the tank could be CAM-performing species.

In tank bromeliads, another issue that draws attention is the division of functions inside a single leaf. The basal portion of the leaf seems to be responsible mainly for absorption of water and nutrients, while the actual photosynthetic activity is more restricted to the apex. This difference of functions could be, at least in part, the result of an intercalary meristem, present in monocot leaves. The activity of this meristem generates a gradient of juvenility along the leaf, the apex being the oldest portion. This may be one of the reasons why other species with an architecture similar to that of bromeliads (like agaves, for example) also show some differences in photosynthetic behavior along their leaves.

Besides sharing similarities with bromeliads in terms of general architecture and photosynthetic activity along the leaves, Agavaceae species also seem to present some degree of flexibility in CAM expression. For instance, the ability to switch from C_3 to CAM metabolism in response to environmental cues has already been suggested for at least one agave, *A. deserti*, and more research is needed to clarify whether CAM expression in other species of Agavaceae family would also be controlled by environmental cues and/or ontogenetic program. On the other hand, while many bromeliads usually rely on external water reservoir (*i.e.*, tank) for coping with periods of water shortage, agaves normally display large internal water storage (*i.e.*, medular hydrenchyma) to survive during the harsh xeric conditions where they usually inhabit. In addition, particular water soluble carbohydrates (*i.e.*, fructans) are usually stored in the medullar hydrenchyma of agaves, possibly helping these plants to osmotically adjust the pressure in their cells under variable conditions of water supply.

Finally, the impacts of elevated CO_2 on CAM bromeliads and agaves seems to be relatively variable, possibly indicating that these plastic plants might display distinct strategies to adapt their photosynthetic activity to changes in this environmental factor. The combinatory effects of high CO_2 and other environmental changes predicted for future global climate scenarios (e.g., high temperatures, intensified aridity, etc.) on the physiology of these plants still need to be better defined through more extensive studies involving a wider range of bromeliad and agave species. The possible economical use of these and other CAM plants as alternative crops in a future scenario of increased temperature and aridity is currently a topic of great interest in the research community.

Acknowledgements

The authors would like to thank Paulo Marcelo Rayner Oliveira for the photos used in the figures and Fundação de Amparo à Pesquisa do Estado de São Paulo (FAPESP - 2011/50637-0) for financial support.

Author details

Alejandra Matiz, Paulo Tamaso Mioto, Adriana Yepes Mayorga, Luciano Freschi and Helenice Mercier

*Address all correspondence to: hmercier@usp.br

Botany Department, University of São Paulo, São Paulo, Brazil

References

[1] Ranson, S. L, & Thomas, M. Crassulacean Acid Metabolism. Annual Review of Plant Physiology (1960). , 11(1), 81-110.

[2] Thomas, M, & Beevers, H. Physiological Studies on Acid Metabolism in Green Plants. II. Evidence of CO_2 Fixation in *Bryophyllum* and the Study of Diurnal Variation of Acidity in this Genus. New Phytologist (1949). , 48(3), 421-447.

[3] Osmond, C. B. Crassulacean Acid Metabolism: Now and Then. In: Esser K., Lüttge U., Beyschlag W., Jin M. (eds.) Progress in Botany. New York: Springer-Verlag Berlin Heilderbeg; (2007). , 3-32.

[4] Silvera, K, Neubig, K. M, Whitten, M, Williams, N. H, Winter, K, & Cushman, J. C. Evolution Along the Crassulacean Acid Metabolism Continuum. Functional Plant Biology (2010). , 37(11), 995-1010.

[5] Cushman, J. C, & Bohnert, H. J. Crassulacean Acid Metabolism: Molecular Genetics. Annual Review of Plant Physiology and Plant Molecular Biology (1999). , 50-305.

[6] Dodd, A. N, Borland, A. M, Haslam, R. P, Griffiths, H, & Maxwell, K. Crassulacean Acid Metabolism: Plastic, Fantastic. Journal of Experimental Botany (2002). , 53(369), 569-580.

[7] Herrera, A. Crassulacean Acid Metabolism and Fitness under Water Deficit Stress: If Not for Carbon Gain, What Is Facultative CAM Good for? Annals of Botany (2009). , 103(4), 645-653.

[8] Lüttge, U. Ecophysiology of Crassulacean Acid Metabolism (CAM). Annals of Botany (2004). , 93(6), 629-652.

[9] Dodd, A. N, Griffiths, H, Taybi, T, Cushman, J. C, & Borland, A. M. Integrating Diel Starch Metabolism with the Circadian and Environmental Regulation of Crassulacean Acid Metabolism in *Mesembryanthemum crystallinum*. Planta (2003). , 216(5), 789-797.

[10] Griffiths, H, Robe, W. E, Girnus, J, & Maxwell, K. Leaf Succulence Determines the Interplay between Carboxylase Systems and Light Use During Crassulacean Acid Metabolism in *Kalanchoë* Species. Journal of Experimental Botany (2008). , 59(7), 1851-1861.

[11] Osmond, C. B. Crassulacean Acid Metabolism: A Curiosity in Context. Annual Review of Plant Physiology (1978). , 29-379.

[12] Lüttge, U. CO_2-Concentrating: Consequences in Crassulacean Acid Metabolism. Journal of Experimental Botany (2002). , 53(378), 2131-2142.

[13] Osmond, C. B, Allaway, W. G, Sutton, B. G, Troughton, J. H, Queiroz, O, Lüttge, U, & Winter, K. Carbon Isotope Discrimination in Photosynthesis of CAM Plants. Nature (1973). , 246-41.

[14] Winter, K, & Von Willert, D. J. NaCl-Induzierter Crassulaceen-Säurestoffwechselbei *Mesembryanthemum crystallinum*. Zeitschriftfür Pflanzenphysiologie (1972). , 67(2), 166-170.

[15] Sipes, D. L, & Ting, I. P. Crassulacean Acid Metabolism Modifications in *Peperomia camptotricha*. Plant Physiology (1985). , 77(1), 59-63.

[16] Borland, A. M. Zambrano VAB., Ceusters J., Shorrock K. The Photosynthetic Plasticity of Crassulacean Acid Metabolism: An Evolutionary Innovation for Sustainable Productivity in a Changing World. New Phytologist (2011). , 191(3), 619-633.

[17] Cushman, J. C, & Borland, A. M. Induction of Crassulacean Acid Metabolism by Water Limitation. Plant, Cell & Environment (2002). , 25(2), 295-310.

[18] Borland, A. M, Técsi, L. I, Leegood, R. C, & Walker, R. P. Inducibility of Crassulacean Acid Metabolism (CAM) in *Clusia* Species; Physiological/Biochemical Characterization and Intercellular Localization of Carboxylation and Decarboxylation Processes in three Species which Exhibit Different Degrees of CAM. Planta (1998). , 205(3), 342-351.

[19] Winter, K. Holtum JAM. Environment or Development? Lifetime Net CO_2 Exchange and Control of the Expression of Crassulacean Acid Metabolism in *Mesembryanthemum crystallinum*. Plant Physiology (2007). , 143(1), 98-107.

[20] Winter, K, & Garcia, M. Holtum JAM. On the Nature of Facultative and Constitutive CAM: Environmental and Developmental Control of CAM Expression During Early Growth of *Clusia, Kalanchoë*, and *Opuntia*. Journal of Experimental Botany (2008). , 59(7), 1829-1840.

[21] Adams, P, Nelson, D. E, Yamada, S, Chmara, W, Jensen, R. G, Bohnert, H. J, & Griffiths, H. Growth and Development of *Mesembryanthemum crystallinum* (Aizoaceae). New Phytologist (1998). , 138(2), 171-190.

[22] Bohnert, H. J, & Cushman, J. C. The Ice Plant Cometh: Lessons in Abiotic Stress Tolerance. Journal of Plant Growth Regulation (2000). , 19-334.

[23] Lin, C. C. The Effects of Environmental Factors in the Induction of Crassulacean Acid Metabolism (CAM) Expression in Facultative CAM Plants. Journal of Undergraduate Life Sciences (2009). , 3(1), 64-66.

[24] Cushman, J. C, Michalowski, C. B, & Bohnert, H. J. Developmental Control of Crassulacean Acid Metabolism Inducibility by Salt Stress in the Common Ice Plant. Plant Physiology (1990). , 94(3), 1137-1142.

[25] Vernon, D. M, Ostrem, J. A, Schmitt, J. M, & Bohnert, H. J. PEPCase Transcript Levels in *Mesembryanthemum crystallinum* Decline Rapidly upon Relief from Salt Stress. Plant Physiology (1988). , 86(4), 1002-1004.

[26] Schmitt, A. K. Lee HSJ., Lüttge U. The Response of the C_3-CAM Tree, *Clusia rosea*, to Light and Water Stress. I: Gas Exchange Characteristics. Journal of Experimental Botany (1988). , 39(208), 1581-1590.

[27] Lüttge, U. One Morphotype, three Physiotypes: Sympatric Species of *Clusia* with Obligate C_3 Photosynthesis, Obligate CAM and C_3-CAM Intermediate Behaviour. Plant Biology (1999). , 1(2), 138-148.

[28] Lüttge, U. *Clusia*: Holy Grail and Enigma. Journal of Experimental Botany (2008). , 59(7), 1503-1514.

[29] Borland, A. M, Griffiths, H, Maxwell, C, Broadmeadow, M. S, Griffiths, N. M, & Barnes, J. On the Ecophysiology of the Clusiaceae in Trinidad: Expression of CAM in *Clusia minor* L. During the Transition from Wet to Dry Season and Characterization of three Endemic Species. New Phytologist (1992). , 122(2), 349-357.

[30] Roberts, A, Borland, A. M, Maxwell, K, & Griffiths, H. Ecophysiology of the C_3-CAM Intermediate *Clusia minor* L. in Trinidad: Seasonal and Short-Term Photosynthetic Characteristics of Sun and Shade Leaves. Journal of Experimental Botany (1998). , 49(326), 1563-1573.

[31] Vats, S. K, Kumar, S, & Ahuja, P. S. CO_2 Sequestration in Plants: Lesson from Divergent Strategies. Photosynthetica (2011). , 49(4), 481-496.

[32] Griffiths, H, Helliker, B, Roberts, A, Haslam, R. P, Girnus, J, Robe, W. E, Borland, A. M, & Maxwell, K. Regulation of Rubisco Activity in Crassulacean Acid Metabolism Plants: Better Late than Never. Functional Plant Biology (2002). , 29(6), 689-696.

[33] Hanscom, Z, & Ting, I. P. Irrigation Magnifies CAM-Photosynthesis in *Opuntia basilaris* (Cactaceae). Oecologia (1978). , 33(1), 1-15.

[34] Borland, A. M, & Dodd, A. N. Carbohydrate Partitioning in Crassulacean Acid Metabolism Plants: Reconciling Potential Conflicts of Interest. Functional Plant Biology (2002). , 29(6), 707-716.

[35] Borland, A. M, & Taybi, T. Synchronization of Metabolic Processes in Plants with Crassulacean Acid Metabolism. Journal of Experimental Botany (2004). , 55(400), 1255-1265.

[36] Black, C. C, & Osmond, C. B. Crassulacean Acid Metabolism Photosynthesis:'Working the Night Shift'. Photosynthesis Research (2003). , 76-329.

[37] Borland, A. M. A Model for the Partitioning of Photosynthetically Fixed Carbon During the C_3-CAM Transition in *Sedum telephium*. New Phytologist (1996). , 134(3), 433-444.

[38] Cushman, J. C, Tillett, R. L, Wood, J. A, Branco, J. M, & Schlauch, K. A. Large-Scale mRNA Expression Profiling in the Common Ice Plant, *Mesembryanthemum crystallinum*, Performing C_3 Photosynthesis and Crassulacean Acid Metabolism (CAM). Journal of Experimental Botany (2008). , 59(7), 1875-1894.

[39] Paul, M, Loos, K, Stitt, M, & Ziegler, P. Starch-Degrading Enzymes During the Induction of CAM in *Mesembryanthemum crystallinum*. Plant, Cell & Environment (1993). , 16(5), 531-538.

[40] Häusler, R. E, Baur, B, Scharte, J, Teichmann, T, Elcks, M, Fischer, K. L, Flügge, U-I, Schubert, S, Weber, A, & Fischer, K. Plastidic Metabolite Transporters and their Physiological Functions in the Inducible Crassulacean Acid Metabolism Plant *Mesembryanthemum crystallinum*. The Plant Journal (2000). , 24(3), 285-296.

[41] Cushman, J. C, Agarie, S, Albion, R. L, Elliot, S. M, Taybi, T, & Borland, A. M. Isolation and Characterization of Mutants of Common Ice Plant Deficient in Crassulacean Acid Metabolism. Plant Physiology (2008). , 147(1), 228-238.

[42] Neuhaus, H. E, & Schulte, N. Starch Degradation in Chloroplasts Isolated from C_3 or CAM (Crassulacean Acid Metabolism)-Induced *Mesembryanthemum crystallinum* L. Biochemical Journal (1996). , 318(3), 945-953.

[43] Weise, S. E, Van Wijk, K. J, & Sharkey, T. D. The Role of Transitory Starch in C_3, CAM, and C_4 Metabolism and Opportunities for Engineering Leaf Starch Accumulation. Journal of Experimental Botany (2011). , 62(9), 109-3118.

[44] Kore-eda, S, & Kanai, R. Induction of Glucose-6-Phosphate Transport Activity in Chloroplasts of *Mesembryanthemum crystallinum* by the C_3-CAM Transition. Plant and Cell Physiology (1997). , 38(8), 895-901.

[45] Borland, A. M, & Griffiths, H. Properties of Phosphoenolpyruvate Carboxylase and Carbohydrate Accumulation in the C_3-CAM Intermediate *Sedum telephium* L. Grown under Different Light and Watering Regimes. Journal of Experimental Botany (1992). , 43(248), 353-361.

[46] Winter, K. Properties of Phosphoenolpyruvate Carboxylase in Rapidly Prepared, Desalted Leaf Extracts of the Crassulacean Acid Metabolism Plant *Mesembryanthemum crystallinum* L. Planta (1982). , 154(4), 298-308.

[47] Borland, A. M, & Griffiths, H. Broadmeadow MSJ., Fordham MC., Maxwell C. Carbon-isotope Composition of Biochemical Fractions and the Regulation of Carbon Balance in Leaves of the C_3-Crassulacean Acid Metabolism Intermediate *Clusia minor* L. Growing in Trinidad. Plant Physiology (1994). , 106(2), 493-501.

[48] Winter, K, & Ziegler, H. Induction of Crassulacean Acid Metabolism in *Mesembryanthemum crystallinum* Increases Reproductive Success under Conditions of Drought and Salinity Stress. Oecologia (1992). , 92(4), 475-479.

[49]   Borland, A. M, Griffiths, H, & Hartwell, J. Smith JAC. Exploiting the Potential of Plants with Crassulacean Acid Metabolism for Bioenergy Production on Marginal Lands. Journal of Experimental Botany (2009). , 60(1), 2879-2896.

[50]   Christopher, J. T. Holtum JAM. Carbohydrate Partitioning in the Leaves of Bromeliaceae Performing C_3 Photosynthesis or Crassulacean Acid Metabolism. Australian Journal of Plant Physiology (1998). , 25(3), 371-376.

[51]   Kenyon, W. H, Severson, R. F, & Black, C. C. Maintenance Carbon Cycle in Crassulacean Acid Metabolism Plant Leaves: Source and Compartmentation of Carbon for Nocturnal Malate Synthesis. Plant Physiology (1985). , 77(1), 183-189.

[52]   Chen, L. S, & Nose, A. Day-Night Changes of Energy-Rich Compounds in Crassulacean Acid Metabolism (CAM) Species Utilizing Hexose and Starch. Annals of Botany (2004). , 94(3), 449-455.

[53]   Leary, O, Park, B, & Plaxton, J. WC. The Remarkable Diversity of Plant PEPC (Phosphoenolpyruvate Carboxylase): Recent Insights into the Physiological Functions and Post-Translational Controls of Non-Photosynthetic PEPCs. Biochemical Journal (2011). , 436(1), 15-34.

[54]   Gehrig, H. H, Heute, V, & Kluge, M. Toward a Better Knowledge of the Molecular Evolution of Phosphoenolpyruvate Carboxylase by Comparison of Partial cDNA Sequences. Journal of Molecular Evolution (1998). , 46(1), 107-114.

[55]   Nimmo, H. G. The Regulation of Phosphoenolpyruvate Carboxylase in CAM Plants. Trends in Plant Science (2000). , 5(2), 75-80.

[56]   Fontaine, V, Hartwell, J, Jenkins, G. I, & Nimmo, H. G. *Arabidopsis thaliana* Contains two Phosphoenolpyruvate Carboxylase Kinase Genes with Different Expression Patterns. Plant, Cell & Environment (2002). , 25(1), 115-122.

[57]   Gousset-dupont, A, Lebouteillei, B, Monreal, J, Echevarria, C, Pierre, J. N, Hodges, M, & Vidal, J. Metabolite and Post-Translational Control of Phosphoenolpyruvate Carboxylase from Leaves and Mesophyll Cell Protoplasts of *Arabidopsis thaliana*. Plant Science (2005). , 169(6), 1096-1101.

[58]   Li, B, Zhang, X-Q, & Chollet, R. Phosphoenolpyruvate Carboxylase Kinase in Tobacco Leaves Is Activated by Light in a Similar but Not Identical Way as in Maize. Plant Physiology (1996). , 111(2), 497-506.

[59]   Shenton, M, Fontaine, V, Hartwell, J, Marsh, J. T, Jenkins, G. I, & Nimmo, H. G. Distinct Patterns of Control and Expression amongst Members of PEP Caboxylase Kinase Gene Family in C_4 Plants. The Plant Journal (2006). , 48(1), 45-53.

[60]   Hartwell, J, Gill, A, Nimmo, G. A, Wilkins, M. B, Jenkins, G. I, & Nimmo, H. G. Phosphoenolpyruvate Carboxylase Kinase Is a Novel Protein Kinase Regulated at the Level of Expression. The Plant Journal (1999). , 20(3), 333-342.

[61] Hartwell, J, Nimmo, G. A, Wilkins, M. B, Jenkins, G. I, & Nimmo, H. G. Probing the Circadian Control of Phosphoenolpyruvate Carboxylase Kinase Expression in *Kalanchoë fedtschenkoi*. Functional Plant Biology (2002). , 29(6), 663-668.

[62] Nimmo, H. G. Control of the Phosphorylation of Phosphoenolpyruvate Carboxylase in Higher Plants. Archives of Biochemistry and Biophysics (2003). , 414(2), 189-196.

[63] Taybi, T, Patil, S, Chollet, R, & Cushman, J. C. A Minimal Serine/Threonine Protein Kinase Circadianly Regulates Phosphoenolpyruvate Carboxylase Activity in Crassulacean Acid Metabolism-Induced Leaves of the Common Ice Plant. Plant Physiology (2000). , 123(4), 1471-1481.

[64] Borland, A. M, Hartwell, J, Jenkins, G. I, Wilkins, M. B, & Nimmo, H. G. Metabolite Control Overrides Circadian Regulation of Phosphoenolpyruvate Carboxylase Kinase and CO_2 Fixation in Crassulacean Acid Metabolism. Plant Physiology (1999). , 121(3), 889-896.

[65] Davies, B. N, & Griffiths, H. Competing Carboxylases: Circadian and Metabolic Regulation of Rubisco in C_3 and CAM *Mesembryanthemum crystallinum* L. Plant, Cell & Environment (2012). , 35(7), 1211-1220.

[66] Raven, J. A, & Spicer, R. A. The Evolution of Crassulacean Acid Metabolism. In: Winter K., Smith JAC. (eds.) Crassulacean Acid Metabolism: Biochemistry, Ecophysiology and Evolution. Berlin: Springer-Verlag Berlin Heilderbeg; (1996). , 360-385.

[67] West-eberhard, M. J. Smith JAC., Winter K. Photosynthesis, Reorganized. Science (2011). , 332(6027), 311-312.

[68] Crayn, M. C, & Winter, K. Smith JAC. Multiple Origins of Crassulacean Acid Metabolism and the Epiphytic Habit in the Neotropical Family Bromeliaceae. Proceedings of the National Academy of Sciences of the United States of America (2004). , 101(10), 3703-3708.

[69] Ehleringer, J. R, Sage, R. F, Flanagan, L. B, & Pearcy, R. W. Climate Change and the Evolution of C_4 Photosynthesis. Trends in Ecology & Evolution (1991). , 6(3), 95-99.

[70] Winter, K. Smith JAC. Crassulacean Acid Metabolism. Current Status and Perspectives. In: Winter K., Smith JAC. (eds.) Crassulacean Acid Metabolism. Biochemistry, Ecophysiology and Evolution. Berlin: Springer-Verlag Berlin Heilderbeg; (1996). , 389-426.

[71] Ehleringer, J. R, & Monson, J. K. Evolutionary and Ecological Aspects of Photosynthetic Pathway Variation. Annual Review of Ecology and Systematics (1993). , 24-411.

[72] Green, W. A. The function of the Aerenchyma in Arborescent Lycopsids: Evidence of an Unfamiliar Metabolic Strategy. Proceedings of the Royal Society (2010). , 277(1), 2257-2267.

[73] Tidwell, W. D, & Parker, L. R. *Protoyucca Shadishii* gen. et sp. nov., an Arborescent Monocotyledon with Secondary Growth from Middle Miocene of Northwestern Nevada, U.S.A. Review of Palaeobotany and palinology (1990). , 62-79.

[74] Troughton, J. H, Wells, P. V, & Mooney, H. A. Photosynthetic Mechanisms and Paleoecology from Carbon Isotope Ratios in Ancient Specimens of C_4 and CAM Plants. Science (1974). , 185(4151), 610-612.

[75] Baresch, A. Smith JAC., Winter K., Valerio AL., Jaramillo C. *Karatophyllum bromelioides* L.D. Gómez Revisited: A Probable Fossil CAM Bromeliad. American Journal of Botany (2011). , 98(11), 1905-1908.

[76] Pierce, S, Winter, K, & Griffiths, H. Carbon Isotope Ratio and the Extent of Daily CAM Use by Bromeliaceae. New Phytologist (2002). , 156(1), 4-6.

[77] APGIIIAn Update of the Angiosperm Phylogeny Group Classification for the Orders and Families of Flowering Plants: APG III. Botanical Journal of the Linnean Society (2009). , 161(2), 105-121.

[78] Benzing, D. H. Basic Structure, Function, Ecology and Evolution. In: Benzing DH. (ed.) Bromeliaceae: Profile of an Adaptative Radiation. Cambridge: Cambridge University Press; (2000). , 19-70.

[79] Givnish, T. J, Millam, K. C, Berry, P. E, & Sytsma, K. J. Phylogeny, Adaptive Radiation, and Historical Biogeography of Bromeliaceae Inferred from ndhF Sequence Data. Aliso (2007). , 23(5), 3-26.

[80] Quezada, I. M, & Gianoli, E. Crassulacean Acid Metabolism Photosynthesis in Bromeliaceae: An Evolutionary Key Innovation. Biological Journal of the Linnean Society (2011). , 104(2), 480-486.

[81] Pittendrigh, C. S. The Bromeliad-Anopheles-Malaria Complex in Trinidad. I-The Bromeliad Flora. Evolution (1948). , 2(1), 58-89.

[82] Fontoura, T, & Reinert, F. Habitat Utilization and CAM Occurrence among Epiphytic Bromeliads in a Dry Forest from Southeastern Brazil. Revista Brasileira de Botânica (2009). , 32(3), 521-530.

[83] Popp, M, Janett, H-P, Lüttge, U, & Medina, E. Metabolite Gradients and Carbohydrate Translocation in Rosette Leaves of CAM and C_3 Bromeliads. New Phytologist (2003). , 157(3), 649-656.

[84] Skillman, J. B, Garcia, M, & Winter, K. Whole-Plant Consequences of Crassulacean Acid Metabolism for a Tropical Forest Understory Plant. Ecology (1999). , 80(5), 1584-1593.

[85] Keller, P, & Lüttge, U. Photosynthetic Light-Use by three Bromeliads Originating from Shaded Sites (*Ananas ananassoides*, *Ananas comosus* cv. Panare) and Exposed

Sites (*Pitcairnia pruinosa*) in the Medium Orinoco Basin, Venezuela. Biologia Plantarum (2005). , 49(1), 73-79.

[86] Reinert, F. Russo CAM., Salles LO. The Evolution of CAM in the Subfamily Pitcairnioideae (Bromeliaceae). Biological Journal of the Linnean Society (2003). , 80-261.

[87] Medina, E, Olivares, E, & Diaz, M. Water Stress and Light Intensity Effects on Growth and Nocturnal Acid Accumulation in a Terrestrial CAM Bromeliad (*Bromelia humilis* Jacq.) under Natural Conditions. Oecologia (1986). , 70-441.

[88] Fetene, M. Lee HSJ., Lüttge U. Photosynthetic Acclimation in a Terrestrial CAM Bromeliad, *Bromelia humilis* Jacq. New Phytologist (1990). , 114(3), 399-406.

[89] Maxwell, K. Resistance is Useful: Diurnal Pattern of Photosynthesis in C_3 and Crassulacean Acid Metabolism Epiphytic Bromeliads. Functional Plant Biology (2002). , 29-679.

[90] Freschi, L, Takahashi, C. A, Cambui, C. A, Semprebom, T. R, Cruz, A. B, Mioto, P. T, Versieux, L. M, Calvente, A, & Latansio-aidar, S. Aidar, MMP., Mercier H. Specific Leaf Areas of the Tank Bromeliad *Guzmania Monostachia* Perform Distinct Functions in Response to Water Shortage. Journal of Plant Physiology (2010). , 167(7), 526-533.

[91] Freschi, L, & Rodrigues, M. A. Tiné MAS., Mercier H. Correlation Between Citric Acid and Nitrate Metabolisms During CAM Cycle in the Atmospheric Bromeliad *Tillandsia pohliana*. Journal of Plant Physiology (2010). , 167(18), 1577-1583.

[92] Herrera, A, Martin, C. E, Tezara, W, Ballestrini, C, & Medina, E. Induction by Drought of Crassulacean Acid Metabolism in the Terrestrial Bromeliad, *Puya floccosa*. Photosynthetica (2010). , 48(3), 383-388.

[93] Loeschen, V. S, Martin, C. E, Smith, M, & Eder, S. L. Leaf Anatomy and CO_2 Recycling During Crassulacean Acid Metabolism in twelve Epiphytic Species of *Tillandsia* (Bromeliaceae). International Journal of Plant Sciences (1993). , 154(1), 100-106.

[94] Nowak, E. J, & Martin, C. E. Physiological and Anatomical Responses to Water Deficits in the CAM Epiphyte *Tillandsia ionantha* (Bromeliaceae). International Journal of Plant Sciences (1997). , 158(6), 818-826.

[95] Martin, C. E. Adams III WW. Crassulacean Acid Metabolism, CO_2-Recycling and Tissue Desiccation in the Mexican Epiphyte *Tillandsia schiedeana* Steud (Bromeliaceae) Photosynthesis Research (1987). , 11(3), 237-244.

[96] Haslam, R, Borland, A. M, Maxwell, K, & Griffiths, H. Physiological Responses of the CAM Epiphyte *Tillandsia usneoides* L. (Bromeliaceae) to Variations in Light and Water Supply. Journal of Plant Physiology (2003). , 160(6), 627-634.

[97] Stiles, K. C, & Martin, C. E. Effects of Drought Stress on CO_2 Exchange and Water Relations in the CAM Epiphyte *Tillandsia utriculata* (Bromeliaceae). Journal of Plant Physiology (1996). , 149(6), 721-728.

[98] Lee HSJLüttge U., Medina E., Smith JAC., Cram WJ., Diaz M., Griffiths H., Popp M., Schäffer C., Stimmel KH., Thonke B. Ecophysiology of Xerophytic and Halophytic Vegetation of a Coastal Alluvial Plain in Northern Venezuela. New Phytologist (1989). , 111(2), 253-271.

[99] Freschi, L, Rodrigues, M. A, Domingues, D. S, Purgatto, E, Sluys, M-A, Magalhaes, J. R, Kaiser, W. M, & Mercier, H. Nitric Oxide Mediates the Hormonal Control of Crassulacean Acid Metabolism Expression in Young Pineapple Plants. Plant Physiology 2010; (1971). , 154(4), 1971-1985.

[100] Gentry, A. H, & Dodson, C. H. Contribution of Non-Trees to Species Richness of Tropical Rain Forest. Biotropica (1987). , 19(2), 149-156.

[101] Romero, Q, Nomura, F, & Gonçalvez, A. Z. Dias NYN., Mercier H., Conforto EC., Rossa-Feres DC. Nitrogen Fluxes from Treefrogs to Tank Epiphytic Bromeliads: An Isotopic and Physiological Approach. Oecologia (2010). , 162(4), 941-949.

[102] Kerbauy, G. B, Takahashi, C. A, Lopez, A. M, Matsumura, A. T, Hamachi, L, Félix, L. M, Pereira, P. N, Freschi, L, & Mercier, H. Crassulacean Acid Metabolism in Epiphytic Orchids: Current Knowledge, Future Perspectives. In: Najafpour MM. (ed.) Applied Photosynthesis. InTech, (2012). Available from <http://www.intechopen.com/books/applied-photosynthesis/crassulacean-acid-metabolism-in-epiphytic-orchids-current-knowledge-future-perspectives>. , 81-104.

[103] Martin, C. E. Physiological Ecology of the Bromeliaceae. Botanical Review (1994). , 60(1), 1-82.

[104] Richardson, B. A. The Bromeliad Microcosm and the Assessment of Faunal Diversity in a Neotropical Forest. Biotropica (1999). , 31(2), 321-336.

[105] Ceusters, J, Borland, A. M, Londers, E, Verdoodt, V, Godts, C, & De Proft, M. P. Diel Shifts in Carboxylation Pathway and Metabolite Dynamics in the CAM Bromeliad Aechmea 'Maya' in Response to Elevated CO_2. Annals of Botany (2008). , 102(1), 389-397.

[106] Takahashi, C. A, & Mercier, H. Nitrogen Metabolism in Leaves of a Tank Epiphytic Bromeliad: Characterization of a Spatial and Functional Division. Journal of Plant Physiology (2011). , 168(11), 1208-1216.

[107] Ceusters, J, Borland, A. M, Godts, C, Londers, E, Croonenborghs, S, Van Goethem, D, & De Proft, M. P. Crassulacean Acid Metabolism under Severe Light Limitation: A Matter of Plasticity in the Shadows?. Journal of Experimental Botany (2010). , 6(1), 283-291.

[108] Ceusters, J, Borland, A. M, Ceusters, N, Verdoodt, V, Godts, C, & De Proft, M. P. Seasonal Influences on Carbohydrate Metabolism in the CAM Bromeliad Aechmea 'Maya': Consequences for Carbohydrate Partitioning and Growth. Annals of Botany (2010). , 105(1), 301-309.

[109] Maxwell, K, Griffiths, H, & Borland, A. M. Broadmeadow MSJ., McDavid CR. Photo-
       inhibitory Responses of the Epiphytic Bromeliad *Guzmania monostachia* During the
       Dry Season in Trinidad Maintain Photochemical Integrity under Adverse Conditions.
       Plant, Cell & Environment (1992). , 15(1), 37-47.

[110] Maxwell, K, Griffiths, H, & Young, A. J. Photosynthetic Acclimation to Light Regime
       and Water Stress by the C$_3$- CAM Epiphyte *Guzmania Monostachia*: Gas-Exchange
       Characteristics, Photochemical Efficiency and the Xanthophyll Cycle. Functional
       Ecology (1994). , 8(6), 746-754.

[111] Maxwell, K, Griffiths, H, Borland, A. M, & Young, A. J. Broadmeadow MSJ., Ford-
       ham MC. Short-term Photosynthetic Responses of the C$_3$-CAM Epiphyte *Guzmania
       monostachia* var. *monostachia* to Tropical Seasonal Transitions under Field Conditions.
       Australian Journal of Plant Physiology (1995). , 22(5), 771-781.

[112] Olivares, E, & Medina, E. Carbon Dioxide Exchange, Soluble Carbohydrates and
       Acid Accumulation in a Fructan Accumulating Plant: *Fourcroya humboldtiana* Treal.
       Journal of Experimental Botany (1990). , 41(226), 579-585.

[113] Benzing, D. H, & Renfrow, A. The Biology of the Epiphytic Bromeliad *Tillandsia circi-
       nata* Schlecht. I. The Nutrient Status of Populations in South Florida. American Jour-
       nal of Botany (1971). , 58(9), 867-873.

[114] Lüttge, U, & Stimmel, K-H. Smith JAC., Griffiths H. Comparative Ecophysiology of
       CAM and C$_3$ Bromeliads. II: Field Measurements of Gas Exchange of CAM Bromeli-
       ads in the Humid Tropics. Plant, Cell & Environment (1986). , 9(5), 377-383.

[115] Lüttge, U. Photosynthetic Flexibility and Ecophysiological Plasticity : Questions and
       Lessons from *Clusia*, the only CAM Tree, in the Neotropics. New Phytologist (2006). ,
       171(1), 7-25.

[116] Simpson, J, & Hernandez, A. M. Juarez MJA., Sandoval SD., Villarreal AS., Romero
       CC. Genomic Resources and Transcriptome Mining in *Agave tequilana*. Global
       Change Biology Bioenergy (2011). , 3(1), 25-36.

[117] Chambers, D. Holtum JAM. Feasibility of Agave as a Feedstock for Biofuel Produc-
       tion in Australia. In: Holtum JAM. (ed) Rural Industries Research and Development
       Corporation. Kingston: RIRDC; (2010).

[118] Escamilla-treviño, L. Potential of Plants from the Genus *Agave* as Bioenergy Crops.
       Bioenergy Research (2011). , 5(1), 1-19.

[119] Garcia, M. A. Distribution of Agave (Agavaceae) in Mexico. Cactus and Succulent
       Journal (2002). , 74(1), 177-187.

[120] Garcia, M. A. Los Agaves de Mexico. Ciencias, Universidad Autónoma de México
       (2007). , 87-14.

[121] Good-avila, S, Souza, V, Gaut, B, & Eguiarte, L. Timing and Rate of Specialization in Agave (Agavaceae). Proceedings of the National Academy of Sciences of the United States of America (2006). , 103(24), 9124-9129.

[122] Arakaki, M, Christin, P-A, Nyffeler, R, Lendel, A, Eggli, U, Ogburn, M, Spriggs, E, Moore, M, & Edwards, E. Contemporaneous and Recent Radiations of the World's Major Succulent Plant Lineages. Proceedings of the national academy of sciences of the United States of America (2011). , 108(20), 8379-8384.

[123] Szarek, S. R, & Ting, I. P. Occurrence of Crassulacean Acid Metabolism among Plants. Photosynthetica (1977). , 11-330.

[124] Nobel, P. S. Environmental Biology of Agaves and Cacti. Cambridge: Cambridge University Press; (1988).

[125] Kemp, P. R, & Gardetto, P. E. Photosynthetic Pathway Types of Evergreen Rosette Plants (Liliaceae) of the Chihuahuan Desert. Oecologia (1982). , 55(2), 149-156.

[126] Huxman, T. E, Hamerlynck, E. P, Loik, M. E, & Smith, S. D. Gas Exchange and Chlorophyll Fluorescence Responses of three South-Western *Yucca* Species to Elevated CO_2 and High Temperature. Plant, Cell & Environment (1998). , 21(12), 1275-1283.

[127] Eickmeier, W. G, & Adams, M. S. Gas Exchange in *Agave lecheguilla* Torr. (Agavaceae) and its Ecological Implications. The Southwestern Naturalist (1978). , 23(3), 473-485.

[128] Nobel, P. S. Water Relations and Photosynthesis of a Desert CAM plant, *Agave deserti*. Plant Physiology (1976). , 58(4), 447-602.

[129] Nobel, P. S, & Hartsock, T. L. Resistance Analysis of Nocturnal Carbon Dioxide Uptake by a Crassulacean Acid Metabolism Succulent, *Agave deserti*. Plant Physiology (1978). , 61(4), 510-514.

[130] Nobel, P. S, & Hartsock, T. L. Shifts in the Optimal Temperature for Nocturnal CO_2 Uptake Caused by Changes in Growth Temperature for Cacti and Agaves. Physiologia Plantarum (1981). , 53(4), 523-527.

[131] Woodhouse, R. M, Williams, J. G, & Nobel, P. S. Leaf Orientation, Radiation Interception, and Nocturnal Acidity Increases by the CAM Plant *Agave deserti* (Agavaceae). American Journal of Botany (1980). , 67(8), 1179-1185.

[132] Hartsock, T. L, & Nobel, P. S. Watering Converts CAM Plant to Daytime CO_2 Uptake. Nature (1976). , 262 (5569), 574-576.

[133] Pimienta-barrios, E, Robles, M. C, Nobel, P. S, & Net, C. O. Uptake for *Agave tequilana* in a Warm and Temperate Environment. Biotropica (2001). , 33(2), 312-318.

[134] Pimienta-barrios, E, Zañudo, H. J, & Garcia, G. J. Fotosíntesis Estacional en Plantas Jóvenes de Agave Azul. Agrociencias (2006). , 40(6), 699-709.

[135]   Nobel, P. S, Castañeda, G, North, G, Pimienta-barrios, E, & Ruiz-corral, J. A. Temper-
        atures Influences on Leaf CO_2 Exchanges, Cell Viability and Cultivation Range for
        Agave tequilana. Journal of Arid Environments (1998)., 39(1), 1-19.

[136]   Lujan, R, Lledías, F, Martínez, L, Barreto, R, Cassab, G, & Nieto-sotelo, J. Small Heat-
        Shock Proteins and Leaf Cooling Capacity Account for the Unusual Heat Tolerance
        of the Central Spike Leaves in *Agave tequilana* var. Weber. Plant, Cell & Environment
        (2009)., 32(12), 1791-1803.

[137]   Nobel, P. S. Water Relations and Carbon Dioxide Uptake of *Agave deserti:* Special
        Adaptations to Desert Climates. Desert Plants (1985)., 7(1), 51-56.

[138]   Nobel, P. S. PAR, Water and Temperature Limitations on the Productivity of Culti-
        vated *Agave fourcroydes* (henequen). Journal of Applied Ecology (1985)., 22(1),
        157-173.

[139]   French, A. D. Chemical and Physical-Properties of Fructans. Journal of Plant Physiol-
        ogy (1989)., 134(2), 125-136.

[140]   Vijin, I, & Smeekens, S. Fructan: More than a Reserve Carbohydrate?. Plant Physiolo-
        gy (1999)., 120(2), 351-360.

[141]   Risema, T, & Smeekens, S. Fructans: Beneficial for Plants and Humans. Current
        Opinion in Plant Biology (2003)., 6(3), 223-230.

[142]   Lopez, M. G, Mancilla-margalli, N. A, & Mendoza-diaz, G. Molecular Structures of
        Fructans from *Agave tequilana* Weber var. azul. Journal of Agricultural and Food
        Chemistry (2003)., 51(27), 7835-7840.

[143]   Raveh, E, Wang, N, & Nobel, P. S. Gas Exchange and Metabolite Fluctuations in
        Green and Yellow Bands of Variegated Leaves of the Monocotyledonous CAM Spe-
        cies *Agave americana*. Physiologia Plantarum (1998)., 103(1), 99-106.

[144]   Christopher, J. T. Holtum JAM. Patterns of Carbon Partitioning in Leaves of Crassu-
        lacean Acid Metabolism Species During Deacidification. Plant Physiology (1996).,
        112(1), 393-399.

[145]   Ting, I. P. CO_2 and Crassulacean Acid Metabolism Plants: A Review. In: Tolbert NE.,
        Preiss J. (eds.) Regulation of Atmospheric CO_2 and O_2 by Photosynthetic Carbon Me-
        tabolism. New York: Oxford University Press; (1994)., 176-183.

[146]   Winter, K, Engelbrecht, B, & Short-term, C. O. Responses of Light and Dark CO_2 Fixa-
        tion in the Crassulacean Acid Metabolism Plant *Kalanchoë pinnata*. Journal of Plant
        Physiology (1994).

[147]   Holtum JAHO'Leary MH., Osmond CB. Effect of Varying CO_2 Partial Pressure on
        Photosynthesis and on Carbon Isotope Composition of Carbon-4 Of Malate from the
        Crassulacean Acid Metabolism Plant *Kalanchoë daigremontiana* Hamet et Perr. Plant
        Physiology (1983)., 71(3), 602-609.

[148] Gouk, S. S. Yong JWH., Hew CS. Effects of Super-Elevated CO_2 on the Growth and Carboxylating Enzymes in an Epiphytic CAM Orchid Plantlet. Journal of Plant Physiology (1997). , 151(2), 129-136.

[149] Zhu, J, Bartholomew, D. P, & Goldstein, G. Effect of Elevated Carbon Dioxide on the Growth and Physiological Responses of Pineapple, a Species with Crassulacean Acid Metabolism. Journal of the American Society of Hoticultural Science (1997). , 122-233.

[150] Zhu, J, Bartholomew, D. P, & Goldstein, G. Effects of Temperature, CO_2 and Water Stress on Leaf Gas Exchange and Biomass Accumulation of Pineapple. Acta Horticulturae (1997). , 425-297.

[151] Zhu, J, Bartholomew, D. P, & Goldstein, G. Gas Exchange and Carbon Isotope Composition of *Ananas comosus* in Response to Elevated CO_2 and Temperature. Plant, Cell & Environment (1999). , 22(7), 999-1007.

[152] Drennan, P. M, & Nobel, P. S. Responses of CAM Species to Increase Atmospheric CO_2 Concentrations. Plant, Cell & Environment (2000). , 23(8), 767-781.

[153] Li, C. R, Gan, I. J, Xia, K, Zhou, X, & Hew, C. S. Responses of Carboxylating Enzymes, Sucrose Metabolizing Enzymes and Plant Hormones in a Tropical Epiphytic CAM Orchid to CO_2 Enrichment. Plant, Cell & Environment (2002). , 25(3), 369-377.

[154] Huerta, A, & Ting, I. P. Effects of Various Levels of CO_2 on the Induction of Crassulacean Acid Metabolism in *Potulacaria afra* (L.) Jacq. Plant Physiology (1988). , 88(1), 183-188.

[155] Winter, K, Zotz, G, Baur, B, & Dietz, K-J. Light and Dark CO_2 Fixation in *Clusia uvitana* and the Effects of Plant Water Status and CO_2 Availability. Oecologia (1992). , 91(1), 47-51.

[156] Grahan, E. A, & Nobel, P. S. Long-Term Effects of a Doubled Atmospheric CO_2 Concentration on the CAM Species *Agave deserti*. Journal of Experimental Botany (1996). , 47(1), 61-69.

[157] Nowak, E. J, & Martin, C. E. Effect of Elevated CO_2 on Nocturnal Malate Accumulation in the CAM Species *Tillandsia ionantha* and *Crassula arborescens*. Photosynthetica (1995). , 31-441.

[158] Ziska, L. H, Hogan, K. P, Smith, A. P, & Drake, B. G. Growth and Photosynthetic Response of Nine Tropical Species with Long-Term Exposure to Doubled Carbon Dioxide. Oecologia (1991). , 86(3), 383-389.

[159] Croonenborghs, S, Ceusters, J, Londers, E, & De Proft, M. P. Effects of Elevated CO_2 on Growth and Morphological Characteristics of Ornamental Bromeliads. Scientia Horticulturae (2009). , 121(2), 192-198.

[160] Nobel, P. S. Responses of Some North American CAM Plants to Freezing Temperatures and Doubled CO_2 Concentrations: Implications of Global Climate Change for
Extending Cultivation. Journal of Arid Environments (1996). , 34(2), 187-196.

[161] Nobel, P. S, & Hartsock, T. L. Leaf and Stem CO_2 Uptake in the three Subfamilies of
the Cactaceae. Plant Physiology (1986). , 80(4), 913-917.

[162] Szarek, S. R, Holthe, P. A, & Ting, I. P. Minor Physiological Response to Elevated CO_2
by the CAM Plant *Agave vilmoriniana.* Plant Physiology (1987). , 83(4), 938-940.

[163] Idso, B. S, Kimball, B. A, Anderson, M. G, & Szarek, S. R. Growth Response of a Succulent Plant, *Agave vilmoriniana,* to Elevated CO_2. Plant Physiology (1986). , 80(3),
796-797.

[164] Garcia-moya, E, Romero-manzanares, A, & Nobel, P. S. Highlights for Agave Productivity. Global Change Biology Bioenergy (2011). , 3(1), 4-14.

Stress Effects

Influence of Biotic and Abiotic Stress Factors on Physiological Traits of Sugarcane Varieties

Leandro Galon, Germani Concenço,
Evander A. Ferreira, Ignacio Aspiazu,
Alexandre F. da Silva, Clevison L. Giacobbo and
André Andres

Additional information is available at the end of the chapter

1. Introduction

The application of knowledge with strong physiological basis of crop yield, allied to genetic and environmental factors, is essential in developing proper practices for crop management aiming high yields. Several aspects determine the performance of a particular crop plant in a given environment, such as temperature, water availability, incidence of pests, plant genetics and management applied. Although it is virtually impossible to control all these factors, plant behavior can be assessed when submitted to different levels of these factors to understand how the responses of the plant to that given stress are formed [1; 2].

Sugarcane is the most widely planted crop in Brazil for bioenergy, being grown in about 8.1 million hectares in the year 2011 [3]. The high yields associated to the suitability for fabrication of ethanol, as well as easiness for bulk processing in high-capacity industrial facilities, catches the attention from businessmen and researchers. Furthermore, this crop is currently considered the best option for biofuel generation from the economical, energetic and environmental points of view.

There is still a big gap between physiological, high-specialized studies and application of these results for practical everyday crop management. Crop scientists, usually do not use physiological parameters in association to the directly measured variables as tools for supporting their findings. Basic research materials, which support applied studies [1; 4; 2; 5], propose changes to this scenario.

Among the factors which limit sugarcane yield, the interference caused by weed species can be highlighted. Weeds compete with sugarcane for resources as water, CO_2, nutrients, physical space and light, being responsible for considerable losses in crop yield and quality, also reducing the useful life of the sugarcane plantation under high infestation levels. The useful life of a sugarcane plantation is defined by the number of times the crop is able to sprout and form new canes, which grow from the stubble, left behind from previous harvest (called *ratoon crop*). Weed control is mandatory in sugarcane plantations, being the chemical control the most widely used method, due to its high efficiency, easiness of use and low cost compared to other control methods [6]. When plants are subjected to strong competition in the plant community, physiological traits of growth and development are usually changed. This results in differences in the use of environmental resources, especially water, which directly affects the availability of CO_2 in leaf mesophyll as well as leaf temperature, therefore, the photosynthetic efficiency. Water competition can also affect the absorption of soil nutrients by sugarcane plants.

If from one side weeds are the main biotic factor that causes reduction in sugarcane yield, herbicides used for weeds management are usually considered the main abiotic factor impacting yields. Several herbicides are usually applied at high doses in sugarcane, often causing high toxicity levels to the crop [7]. Herbicidal toxicity to sugarcane should not be evaluated based only in visual symptoms, once there are herbicides known to reduce crop yields with almost no external symptoms. There are some herbicides, however, known to cause severe external toxicity symptoms, which disappear after some days with little to no impact on final crop yield.

It is known that sugarcane varieties behave differently to some herbicides, and as consequence it is common to observe different levels of intoxication by given herbicides [7]. Thus, there is the need to classify varieties in terms of susceptibility to the most widely used herbicides for weed control in sugarcane fields. The application of herbicides will, as mandatory, cause some level of harm to crops – from almost no harm to near total plant death. These impacts are sometimes not easily visible externally at the plant, but the physiological parameters would be imbalanced resulting in lower plant performance. Because of that, more susceptible parameters like the ones associated to the photosynthesis and water use of plants are essential tools for monitoring herbicide safety for crops.

The data presented in this chapter includes the state-of-the-art about sugarcane susceptibility to herbicides, plus the most up-to-date knowledge about this subject generated by a group of researchers which are reference in herbicide physiology, from different Brazilian institutions. Parameters associated to the photosynthetic efficiency, as CO_2 concentration in the environment and in the leaf mesophyll, leaf temperature, and parameters associated to the dynamics of use of water - including Water Use Efficiency - impact the photosynthesis rate, thus affecting yield.

2. Chemical weed control

The wide acceptance of chemical weed control with herbicides can be attributed to: (1) less demand of human labor; (2) efficient even under rainy seasons; (3) efficient in controlling weeds at the crop row with no damage to crop root system; (4) essential tool for no-till planting systems; (5) efficient in controlling vegetatively propagated weed species; and (6) allows free decision about planting system (in rows, sowing) and crop row spacing [8].

It is important to consider, however, that an herbicide is a chemical molecule that should be correctly managed to avoid human intoxication as well as environmental contamination [8]. The knowledge in plant physiology, chemical herbicide groups and technology of pesticide application is essential for the success of the chemical weed control [9]. Surely there are risks involved at this method, but if they are known they can be avoided and controlled.

Chemical weed control should be applied as an auxiliary method. Efforts should be focused on the cultural method of weed management once it allows the best conditions for the development of crops while at the same time creating barriers for the proper development of seedlings of weed species [8; 9].

3. Traits related to the photosynthesis and water use efficiency

The photosynthesis rate surely is one of the main processes responsible for high crop yields, but the liquid photosynthesis is a result of interaction among several processes, and each one of these processes, alone or in sets, may limit plant gain in terms of photoassimilates [9]. The genetic variation among species and even among biotypes of the same species may shift the enzymatic mechanism and make a given species more capable than other in extracting or using efficiently a given environmental resource, aiming to maximize its photosynthetic rate. Until recently, it was widely accepted that light affected indirectly the stomatal opening through the CO_2 assimilation dependent on light – i.e., light increased the photosynthesis rate, which would reduce the internal CO_2 concentration in the leaf and as a consequence the stomata would open [5; 10]. More accurate studies, however, concluded that stomatal response is less connected to the internal CO_2 concentration of the leaf than anticipated; most of the response to light in stomatal opening is direct, not mediated by CO_2 [10].

Distinct light regimes, both in terms of quantity and composition, influence almost all physiological processes like photosynthesis and respiration rates, affecting also variables like plant height, fresh and dry mass and water content of the plant [11]. Water content, on the other hand, shifts both the stem length and leaf area of the plant, in a way to adapt the plant to the amount or quality of light intercepted [12]. Interspecific and intraspecific plant competition affects the amount and the quality of the final product, as well as its efficiency in utilization of environmental resources [13; 14]. This is noted when assessing physiological traits associated to photosynthesis, such as concentration of internal and external gases [15], light composition and intensity [16] and mass accumulation by plants under different conditions.

Although gas exchange capability by stomata is considered a main limitation for photosynthetic CO_2 assimilation [17], it is unlikely that gas exchange will limit the photosynthesis rate when interacting with other factors. However, photosynthetic rate is directly related to the photosynthetically active radiation (composition of light), to water availability and gas exchange. Plants have specific needs for light, predominantly in bands of red and blue [23]. When plants do not receive these wavelengths in a satisfactory manner, they need to adapt themselves in order to survive [25]. When under competition for light, the red and far-red ratio affected by shading is also important [16] and influences the photosynthetic efficiency [18].

4. Physiological parameters

Table 1 presents the main physiological parameters evaluated by the equipment called Infra Red Gas Analyzer (IRGA). Details on the principles of measurements by this equipment, as well as cares to be taken to avoid reading errors, can be found in [19], [20] and in the technical manual of the equipment. Please note that the parameters available vary among equipments from different manufacturers, as well as among models of the same manufacturer. Some of them (ΔC, Tleaf, ΔT and WUE) are usually not automatically supplied, but may be easily calculated based on parameters supplied by most of the equipments. The measuring units of the parameters also vary, and the most common units were adopted at Table 1. Water use efficiency (WUE) has several interpretations and may be presented in several different units, for distinct purposes. For a more comprehensive overview of this parameter, please consult [21].

Parameter	Usual Unit	Name and Description
A	$\mu mol\ m^{-2}\ s^{-1}$	*Photosynthesis rate* – Rate of incorporation of carbon molecules from the air into biomass. Supplied by equipment.
E	$mol\ H_2O\ m^{-2}\ s^{-1}$	*Transpiration* – Rate of water loss through stomata. Supplied by equipment.
Gs	$mol\ m^{-1}\ s^{-1}$	*Stomatal conductance* - Rate of passage of either water vapor or carbon dioxide through the stomata. Supplied by equipment.
Ci	$\mu mol\ mol^{-1}$	*Internal CO_2 concentration* – Concentration of CO_2 in the leaf mesophyll . Supplied by equipment.
E_{an}	mBar	*Vapor pressure at sub-stomatal chamber* – Water pressure at the sub-stomatal zone within the leaf. Not supplied by some equipments.
ΔC	$\mu mol\ mol^{-1}$	*Carbon gradient* – Gradient of CO_2 between the interior and the exterior of the leaf. Usually not supplied by equipment. May be calculated by the difference between the CO_2 of reference (supplied by equipment) and its concentration at the mesophyll (Ci – supplied by equipment).
T_{leaf}	°C	*Leaf temperature* – Supplied by equipment in °C or °F.
ΔT	Δ °C	*Temperature gradient* – Not supplied by equipment. May be calculated by the difference between the temperature of the leaf (supplied by equipment) and the environmental temperature (supplied by equipment).
WUE	$\mu mol\ CO_2\ mol\ H_2O^{-1}$	*Water use efficiency* – Describes the relation between the rate of incorporation of CO_2 into biomass and the amount of water lost at the

Parameter	Usual Unit	Name and Description
		same time interval. Usually not supplied. May be calculated by dividing photosynthesis (A) by transpiration (E). May be presented in several distinct units.
A/Gs	Curve*	*Intrinsic water use efficiency* – Not widely used, but describes a relation between the actual photosynthesis rate and the stomatal conductance. The original units of each parameter is maintained.
A/Ci	Curve*	*Photosynthesis / CO$_2$ relation* – Describes the curve of photosynthesis rate as the concentration of CO$_2$ within the leaf is increased. The original units of each parameter is maintained.

Table 1. Physiological parameters usually available when using an Infra Red Gas Analyzer (IRGA). Parameters available vary among manufacturers as well as among models of the same manufacturer. The equipment supplies some parameters while others have to be calculated. * These data should be represented as a graph showing the relation between variables as the concentration of one of them is increased; thus, the original units are maintained.

5. Photosynthesis

The photosynthesis (A) and thus the respiration, depend upon a constant flux of CO$_2$ and O$_2$ in and out of the cell; this free flux is a function of the concentration of CO$_2$ (Ci) and O$_2$ at the intercellular spaces, which depend on the stomatal opening, major controller of the gas flux through stomata [22; 23]. This is mainly controlled by the turgescence both of the guard cells (which control stomatal opening) as well as by the epidermic cells at the stomata [24]. A low water potential will promote reduction in stomatal opening and reduce the leaf conductance, inhibiting photosynthesis and also the respiration [25], and increasing the gradient of CO$_2$ concentration between the leaf mesophyll and the exterior of the leaf (ΔC).

6. Water use efficiency

When plants are studied in terms of the efficiency they present when using water, the parameters stomatal conductance of water vapor (Gs), vapor pressure at the sub-stomatal chamber (E$_{an}$), transpiration rate (E) and water use efficiency (WUE) should be considered. The WUE is obtained by the relation between CO$_2$ incorporated in the plant and the amount of water lost by transpiration during the same period [2]. The more efficient water use is directly related to the photosynthetic efficiency as well as the dynamics of stomatal opening, because while the plant absorbs CO$_2$ for the photosynthesis, it also loses water to the atmosphere by transpiration, in rates that depend on the potential gradient between the interior and the exterior of the leaf [9]. Water exchange also allows the plant to keep adequate temperature levels, which can be evaluated by the leaf temperature (T$_{leaf}$), as well as by the difference between the leaf temperature and the temperature of the air surrounding the leaf (ΔT).

7. Interference of *Brachiaria brizantha* in the morphology and physiology of sugarcane

The study was installed at the experimental station of Horta Nova, owned by the Federal University of Viçosa (UFV), Viçosa-MG, Brazil, in an Acrisol. The planting of sugarcane was carried out under conventional system of 1-year sugarcane, after the operations of plowing and harrowing, in rows spaced in 1.4m. Planting density was 18 buds m^{-1} with fertilization performed at planting according to the results of physicochemical analysis of the soil and following the recommendations for the crop, using 500 kg ha^{-1} NPK 08-28-16, supplemented by topdressing of 160 kg ha^{-1} KCl.

The experimental unit measured 8.4m wide by 5.0m long (42 m^2 per plot). Treatments consisted of 12 populations of *B. brizantha*, planted among three varieties of sugarcane, at the following densities : 0, 1, 3, 7, 15, 32, 40, 32, 64, 92, 88, and 112 plants m^{-2}; 0, 1, 4, 14, 10, 18, 28, 30, 36, 54, 52 and 72 plants m^{-2}; 0, 1, 3, 6, 14, 20, 24, 26, 26, 32, 46 and 56 plants m^{-2}, for the varieties RB72454, RB867515 and SP801816, respectively.

The populations of *B. brizantha* were obtained by sowing 10 kg ha^{-1} of seeds, 10 days before the emergence of sugarcane; densities for competition were established when plants were at the stage of two leaves to one tiller through thinning, carried out with application of the herbicide MSMA (Volcane® - 2 L ha^{-1}). Random plants were covered with plastic cups before application, in order to escape from the herbicide, resulting in the desired densities. Weeds other than *B. brizantha* were controlled with 2,4-D or through manual hoeing, especially the new emergences of *B. brizantha*. Herbicides were applied with a CO_2-propelled backpack sprayer, coupled to a 2m wide bar with four spray nozzles Teejet TT 110.02, calibrated to spray 150 L ha^{-1}.

The evaluations of the final populations of *B. brizantha* in each treatment were accomplished 90 days after crop emergence (DAE), with scores in two areas of 0.25 m^2 (0.5 x 0.5 m) in each experimental plot. At 120 DAE, the stalks diameter, the number of stalks, the plant height and number of leaves per plant were measured. After these determinations, sugarcane plants were sectioned at the soil level, and stalks were separated from leaves. The leaves were placed in polystyrene boxes containing ice, to prevent dehydration, in order to determine the leaf area (LA) in electronic equipment (Licor, model LI-3100C). After LA determination, stalks were added to the corresponding leaves, and the material was placed inside ovens at 65 °C with continuous air circulation, for determination of shoot dry mass.

The yield of stalks was estimated by counting stalks present in the four central rows of 0.5m, disregarding borders on each side and front portion. Later, 30 stalks were collected at random then weighed. With the average weight of stalks and number of stalks per unit area, sugarcane yield was estimated in t ha^{-1}.

At the same time of the morphological evaluations, the physiological parameters of the treatments were evaluated by using an Infra Red Gas Analyzer (IRGA), model LCA Pro⁺. The sub-stomatal CO_2 concentration (Ci - $\mu mol\ mol^{-1}$), photosynthesis rate (A - $\mu mol\ m^{-2}\ s^{-1}$), a stomatal conductance (GS – $mol\ m^{-1}\ s^{-1}$), transpiration rate (E – $mol\ H_2O\ m^{-2}\ s^{-1}$) water use efficiency (EUA – $mol\ CO_2\ mol\ H_2O^{-1}$) were obtained. To avoid excessive interference of the environmental parameters on the readings, each block of the experiment was evaluated in a different day, between 07:00 and 10:00 am.

Data were tested for homoscedasticity and then submitted to analysis of variance. Subsequently, analyses of linear and nonlinear regression were performed to evaluate the effects of populations of *B. brizantha* on morphological traits of each sugarcane variety. The choice of the most suitable model for each variable was based on the statistical significance (F-test), fitting correlation coefficient (R^2) and the biological significance of the model.

The leaf area (LA) of sugarcane was reduced as the population of *B. brizantha* was increased, for varieties RB72454 and SP80-1816 (Figure 1A). Variety RB867515 was more tolerant to competition with the weed, for the same variable, because constant values of LA were observed with the increasing in the population of the competitor. Under competition, plants tend to have lower leaf area, due to the allocation of a higher proportion of dry mass in stalks, making them longer and raising the leaves in search for light. In addition, under this situation leaves usually have smaller dimensions. Similar results were found by [26] who observed that the increase in the population of ryegrass under competition with wheat, caused a decrease in leaf area but did not alter the thickness of the leaves. In some cases, under competition leaves can also become thinner, changing the leaf area while keeping constant weight. For varieties in which leaf area was not affected, it may be necessary to determine whether there were changes in leaf thickness in order to elucidate the mechanism of tolerance.

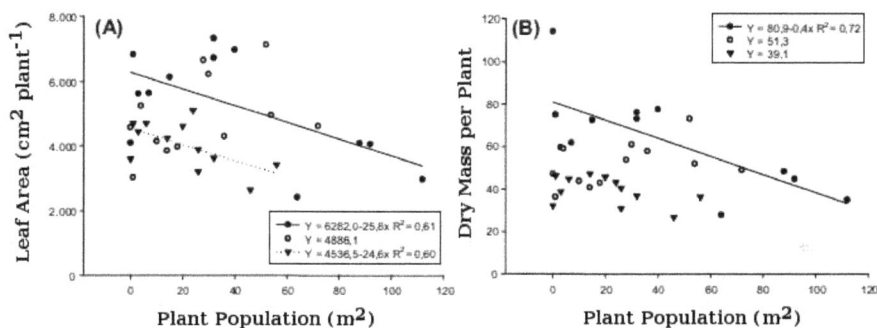

Figure 1. Leaf area (A) and shoot dry mass (B) of sugarcane varieties (•) RB72454, (o) RB758515, and (▼) SP801816 under competition with populations of *Brachiaria brizantha*. DFT/UFV, Viçosa-MG-Brazil, 2008/09.

The results show that there was less accumulation of dry mass (DM) in the variety RB72454 as a function of increase in population of *B. brizantha*. Since the DM of varieties RB867515 and SP80-1816 did not change with the increasing in weed population, it may be hypothesized they are less affected by competition in the early stages of development (Figure 1B). High populations of weeds extract more resources from the environment and therefore interfere with the growth and accumulation of assimilates in crop plants, which is reflected in slower development of LA and the accumulation of plant DM [27].

With the increase in the population of *B. brizantha*, it was observed increase in stalk diameter, except for the variety RB72454 (Figure 2A). Diameter increase is generally attributed to a smaller number of tillers. The stalk diameter is directly related to the availability of resources for the crop; in case of lower resource availability, some tillers may be aborted with a possible increasing in the thickness of the remaining ones [28]. In a global context, plants are highly responsive to stress, whether natural or man-imposed, by changing their morphology towards a necessary adaptation for their survival [29].

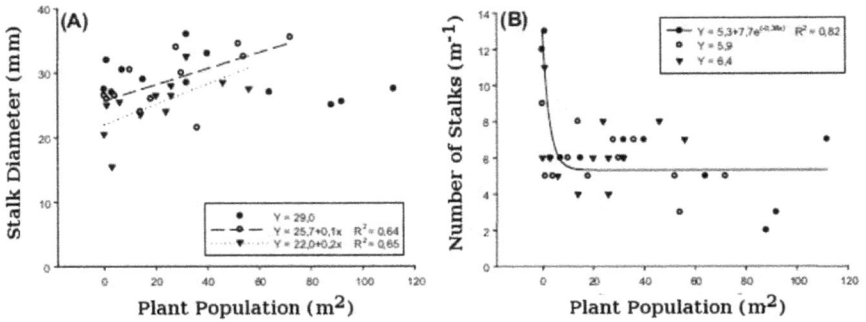

Figure 2. Stalk diameter (A) and number of stalks (B) of the sugarcane varieties (•) RB72454, (o) RB758515, and (▼) SP801816 under competition with populations of *Brachiaria brizantha*. DFT/UFV, Viçosa-MG-Brazil, 2008/09.

It was also observed a reduction in the number of stalks per meter, with increases in the population of the weed for RB72454, while the other varieties were not affected (Figure 2B). This demonstrates that RB72454 is more susceptible to competition compared to the other varieties. The number of stalks of RB72454 was reduced until the population of 10 plants m^{-2} of *B. brizantha* (Figure 2B).

Differences in competitive ability of crops versus weeds were observed by several researchers: sugarcane with signalgrass [30], soybean with oil radish [34], rice with barnyardgrass [31], rice with red rice [32], and sorghum with Johnsongrass [33]. Temperature, humidity, soil fertility and light are the major determinants of plant growth, each of these elements may be more or less limiting in specific situations [29].

With increases in the population of *B. brizantha*, there was an increase in plant height of varieties RB867515 and SP80-1816 (Figure 3A). The competition mainly for light, leads to the tendency of plants to invest more photosynthates in height, increasing the interception and shadowing other competing plants [35]. Competition for light between plant communities begins very early, affecting early apical dominance [36].

Some authors hypothesized that ecotypes (or varieties) of the same species might have differential capacity to adapt to distinct light intensities [37; 38]. Researches emphasize that all individuals of the same species are technically able to adjust their physiology to the light

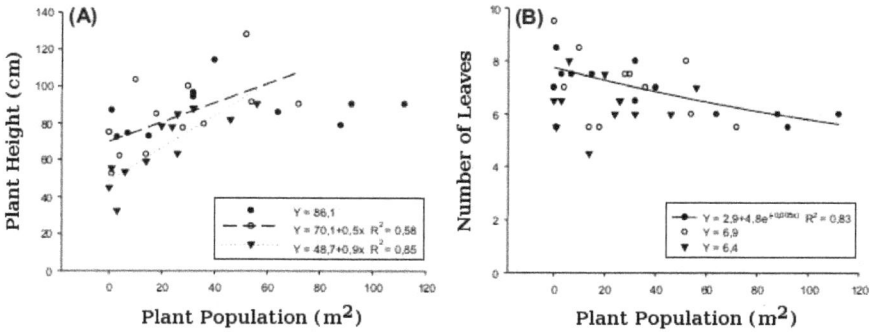

Figure 3. Plant height (A) and number of leaves per plant (B) of the sugarcane varieties (•) RB72454, (o) RB758515, and (▼) SP801816 under competition with populations of *Brachiaria brizantha*. DFT/UFV, Viçosa-MG-Brazil, 2008/09.

intensity to which they are exposed, i.e., an individual which grew originally in the sun can adapt itself to a shaded environment and vice versa [39; 40]. Thus, it is implied that the varieties of sugarcane that develop simultaneously and more rapidly in height and leaf area compete more effectively for light.

The number of leaves per plant of the variety RB72454 was negatively influenced by increasing the plant population of *B. brizantha* (Figure 3B), and under these circumstances the variety RB72454 again proved to be less competitive than the others when in the presence of the competitor.

The stalk yield was reduced with the increase in the population of *B. brizantha*, stabilizing losses in populations of about 40 plants m^{-2} of the competitor, for all varieties (Figure 4). It should be noted that increased competition between sugarcane and *B. brizantha* caused reduction also in crop tillering and, consequently, fewer stems were obtained per area. According to [41] and [8], a fast and uniform initial growth is essential for obtaining a good stand, enabling the rapid closure of the canopy, which leads to better utilization of light energy and more effective suppression of weeds. Beyond 40 plants m^{-2} of the competitor the intraspecific competition between plants of *B. brizantha* could have been too strong, thus reducing the potential for competition of individual plants.

The substomatal CO_2 concentration (Ci) decreased with increasing plant population of *B. brizantha* for RB72454 (Figure 5). Ci is mainly influenced by stomatal conductance and photosynthetic activity, being considered a physiological variable influenced by many environmental factors such as water availability, light and others [9]. The variety RB72454 presented higher Ci when compared to SP80-1816 [42]. Differences in Ci among rice varieties were observed [43], and only one cultivar was influenced by the competition in this variable. Changes in the values of Gs (Figure 6) for RB72454 help explaining the lower Ci found for the same variety. With lower Gs there is a lower influx of CO_2 to the inner space of the leaf, thus, reducing the concentration of substrate (C) leading to limitations in the photosynthetic activity.

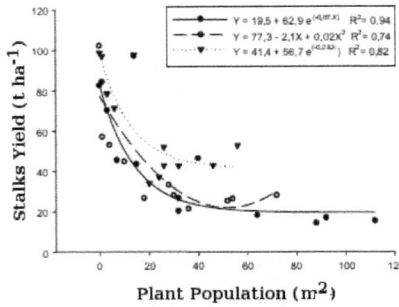

Figure 4. Stalks yield (t ha⁻¹) of the sugarcane varieties (•) RB72454, (o) RB758515, and (▼) SP801816 under competition with populations of Brachiaria brizantha. DFT/UFV, Viçosa-MG-Brazil, 2008/09.

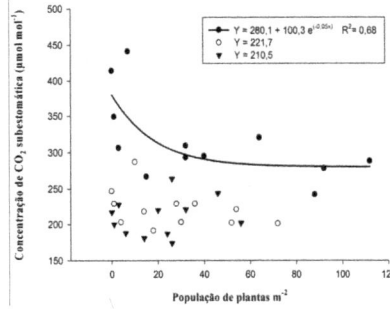

Figure 5. Substomatal CO_2 concentration (Ci - μmol mol⁻¹) of sugarcane varieties (●) RB72454, (o) RB857515 and (▼) SP8018-16 under competition with populations of *Brachiria brizantha* DFT/UFV, Viçosa-MG, Brazil, 2008/09.Figure

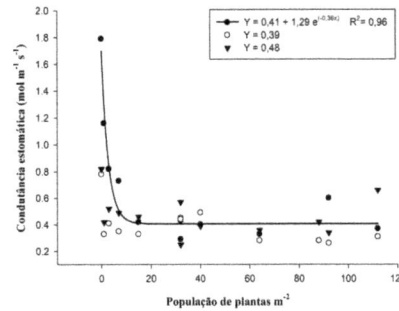

Figure 6. Stomatal conductante (Gs - mol m⁻² s⁻¹) of sugarcane varieties (●) RB72454, (o) RB857515 and (▼) SP80-1816 under competition with populations of *Brachiria brizantha* DFT/UFV, Viçosa-MG, Brazil, 2008/09.

There was a reduction in stomatal conductance (Gs) of the sugarcane with increasing plant population of B. brizantha (up to 16 plants m^{-2}) for the variety RB72454; the others presented stable Gs (Figure 6). Corroborating these results, there were no differences between the values of Gs among varieties in a study conducted by [42] under field conditions. Differential responses of reduced Gs in plants subjected to competition were also observed for the rice crop, where only one variety showed reduction in this variable [43].

The transpiration rate (E) for RB867515 and RB72454 was reduced as the density of plants of B. brizantha was increased; for the former, this reduction was observed until 40 plants m^{-2} of the competitor, being stabilized under higher competition intensities (Figure 7). The reduction of E is directly related to water availability to plants [9], thus as the plant density was increased, the competition for water in the soil was also more serious.

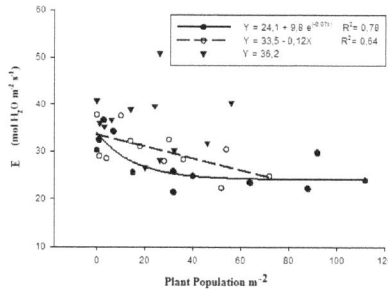

Figure 7. Transpiration rate (E - mol H$_2$O m^{-2} s^{-1}) of sugarcane varieties (●) RB72454, (○) RB857515 and (▼) SP80-1816 under competition with populations of Brachiria brizantha DFT/UFV, Viçosa-MG, Brazil, 2008/09.

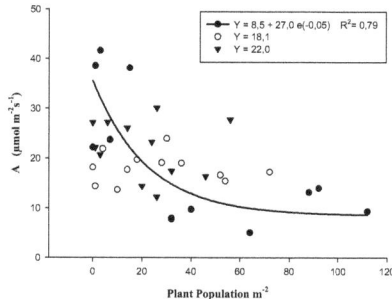

Figure 8. Photosynthesis rate (A - µmol m^{-2} s^{-1}) of sugarcane varieties (●) RB72454, (○) RB857515 and (▼) SP80-1816 under competition with populations of Brachiria brizantha DFT/UFV, Viçosa-MG, Brazil, 2008/09.

A reduction in the photosynthesis rate (A) for RB72454 was observed, with the increase in the density of the competitor, being the same not observed for the other varieties (Figure 8). Photosynthesis rate is limited, in this case, by the reduced availability of CO_2 in the inner space of the leaf due to the lower influx of CO_2, as observed for Ci and Gs. However, some plants are more efficient in the stomatal adjustment, so that even with a small reduction in water availability there is not necessarily a reduction in the photosynthesis rate [9].

Some environmental factors impact the stomatal regulation, such as water stress, because highly negative water potentials induce stomatal closure with a consequent reduction in stomatal conductance [9], reducing the influx of CO_2 to the leaves and limiting the photosynthesis by deficiency of substrate. The nutritional deficiency may also lead to deficiency, for example, of proteins, enzymes and nutrients, such as potassium, which acts directly on the stomatal adjustment, and such deficiency could limit the photosynthesis rate [14]. From the results it is unlikely that the photosynthetic activity of plants was limited by the interference of radiation, or by the smaller change in the proportion of red and far-red light, because in this case, there would most likely be no changes in Ci and Gs [9].

The water use efficiency (WUE) is characterized as the amount of water lost by transpiration while a certain amount of CO_2 is captured for dry mass accumulation. Thus, plants, which present high photosynthesis rate while losing small amounts of water through transpiration, are considered more efficient in the use of water [9]. As the density of B. brizantha was increased, there was a reduction in WUE of the variety RB72454, with higher effect in the lower populations of the competitor (Figure 9). This may be an indicator that this variety, under competition with Brachiaria, is able to keep the photosynthesis rate under conditions of lower water availability. The other varieties presented a reduction in A compatible with the reductions observed in E, which resulted in stable WUE. Some researchers have observed that differences occur between the WUE in varieties of rice [43] and sugarcane [42] when subjected to both intra- and inter-specific competitions.

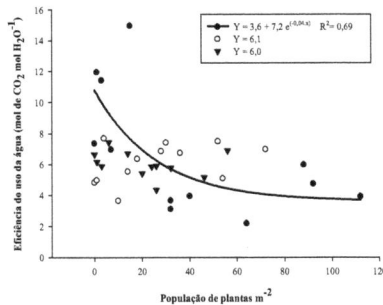

Figure 9. Water use efficiency (WUE - mol CO_2 mol H_2O^{-1}) of sugarcane varieties (●) RB72454, (○) RB857515 and (▼) SP80-1816 under competition with populations of Brachiaria brizantha DFT/UFV, Viçosa-MG, Brazil, 2008/09.

Based on the results, the infestation of *B. brizantha* interferes in the morphological components of sugarcane varieties, by reducing leaf area, shoot dry mass, diameter and number of stalks, plant height and number of leaves. The average yields of stalks for the three varieties studied showed a reduction of about 60 t ha^{-1} when competing with the higher levels of infestation; the variety RB72454 was the most affected by competition with *B. brizantha*.

In relation to the physiological components, the variety RB72454 suffered the greatest effect by the competition for environmental resources, particularly for light and water, as the density of *B. brizantha* was increased. The photosynthesis rate and water use efficiency from sugarcane are limited as the competition with *B. brizantha* is increased, due to factors that lead to lower influx of CO_2 into the leaves of the crop, and the consequent higher transpiration. The varieties of sugarcane present differential susceptibility to the competition imposed by *B. brizantha*, and the variety RB72454 is highlighted as the most sensitive to competition among the tested.

8. Changes in physiological traits of sugarcane varieties under herbicide application

The experiment was installed in a randomized block design, with three replications in a factorial scheme 3 x 8. Factor A consisted of sugarcane varieties (RB867515, RB855156 and SP80-1816) and factor B by the herbicides tembotrione (Soberan® 200 mL ha^{1} + Aureo® 1,0 L ha^{-1}); MSMA (Volcane® 3,0 L ha^{-1}); diuron + hexazinone (Velpar-K GRDA® 2,0 kg ha^{-1}); sulfentrazone (Solara® 1,2 L ha^{-1}); trifloxysulfuron-sodium (Envoke® 30,0 g ha^{-1}+ Extravon® 0,2% $^v/_v$); tebuthiuron (Combine 500 SC® 2,0 L ha^{-1}); clomazone (Gamit® 3,0 L ha^{-1}), plus a check treatment with no herbicide.

The experimental units consisted of plastic pots containing 12 dm^3 of substrate - Red Latosol, previously limed and fertilized according to the analysis: pH water 4.3; OM = 2.5 dag kg^1; P = 1.5 mg dm^3; K = 40 mg dm^3; Al^{3+} = 0.5 cmol c dm^3 ; Ca^{2+} = 1.3 cmol c dm^3; Mg^{2+} = 0.2 cmol c dm^3; CTC(t) = 2.1 cmol c dm^3; CTC(T) = 6.39 cmol c dm^3; H+Al = 4.79 cmol c dm^3; SB = 1.6 cmol c dm^3; V = 25%; and clay = 38%. After filling out pot with the substrate, two buds of sugarcane were planted in each experimental unit. Ten days after emergence (DAE) only one plant was left per pot. Herbicide application was accomplished 50 DAE of sugarcane, at the stage of four to six fully expanded leaves. For the application, it was used a backpack sprayer, propelled by CO_2 and connected to a 2m wide bar with four nozzles Teejet TT 110.02, calibrated to spray 150 L ha^1. Forty five days after application of the herbicides, physiological evaluations were done in the middle third of the first fully expanded leaf of the main tiller of the sugarcane plants. For that, an Infra Red Gas Analyzer (IRGA LCA Pro⁺) was used, in an open greenhouse, allowing free movement of the air. The photosynthesis rate (mol m^2 s^1), stomatal conductance of water vapor (Gs - mol m^1 s^1), transpiration rate (E - mol H_2O m^2 s^1) and the water use efficiency (WUE - mol CO_2 mol H_2O^1) were obtained. The evaluations were conducted between 7:00 and 10:00 am. All data was analyzed by the F-test, and when significant, means were grouped by the test of Scott-knot at 5% probability.

There was interaction between herbicides and varieties for the photosynthesis rate, stomatal conductance, transpiration rate and the water use efficiency. The variety RB867515, when sprayed with sulfentrazone, tebuthiuron or clomazone, showed lower transpiration rate (E) compared with the check with no herbicide (Table 2). It is noteworthy that these herbicides act directly or not in the photosynthetic apparatus of plants that are applied. Sulfentrazone inhibits the enzyme protoporphyrinogen oxidase (PROTOX), acting indirectly on chlorophyll synthesis in sensitive plants. Tebuthiuron is an inhibitor of photosystem II acting on D1 protein, and clomazone inhibits the synthesis of carotenoids, whose function is the protection of chlorophyll against excess of light [8].

	Sugarcane Variety		
Herbicide	RB867515	RB855156	SP80-1816
Tembotrione	2.66 Aa	2.57 Aa	2.62 Aa
MSMA	2.76 Aa	2.56 Aa	2.06 Bb
Diuron+hexazinone	2.87 Aa	2.54 Aa	3.05 Aa
Sulfentrazone	2.08 Bb	2.17 Bb	2.81 Aa
Trifloxysulfurom-sodium	2.70 Aa	2.24 Bb	1.82 Bb
Tebuthiuron	2.24 Bb	2.03 Bb	2.83 Aa
Clomazone	2.35 Ab	2.07 Ab	2.77 Aa
Check	2.85 Aa	2.87 Aa	2.72 Aa
CV (%)	14.20		

Table 2. Transpiration rate (E - mol H_2O m^{-2} s^{-1}) of sugarcane varieties under competition with populations of *Brachiria brizantha* DFT/UFV, Viçosa-MG, Brazil, 2008/09.

Means followed by the same letter, uppercase in the rows and lowercase in the columns, are included in the same group by the test of Scott-Knot at 5% probability.

The application of sulfentrazone, trifloxysulfuron-sodium, clomazone and tebuthiuron on the variety RB855156 caused reduction in E. For variety SP80-1816, however, only trifloxysulfuron-sodium and MSMA had negative impact (Table 2). Trifloxysulfuron-sodium is an inhibitor of the enzyme acetolactate synthase (ALS), with indirect effect on phososynthesis because it acts on the production of the branched-chain amino acids valine, leucine and isoleucine, responsible for the production of proteins in plants.

With regard to the water use efficiency (WUE), no differences were observed inside variety or herbicide (Table 3). The water use efficiency is characterized as the amount of water transpired by a plant to produce a certain amount of dry mass [8]. Thus, more efficient crops in the water use can produce bigger amounts of dry mass per gram of water transpired. The most efficient use of water is directly related to stomatal opening, therefore, while the plant absorbs CO_2 for

Herbicide	Sugarcane Variety		
	RB867515	RB855156	SP80-1816
Tembotrione	8.06 Aa	9.50 Aa	6.40 Aa
MSMA	6.36 Aa	5.65 Aa	6.29 Aa
Diuron+hexazinone	7.37 Aa	8.43 Aa	6.95 Aa
Sulfentrazone	5.10 Aa	6.88 Aa	8.69 Aa
Trifloxysulfurom-sodium	7.53 Aa	7.04 Aa	7.49 Aa
Tebuthiuron	7.88 Aa	7.95 Aa	6.66 Aa
Clomazone	7.28 Aa	6.14 Aa	6.64 Aa
Check	10.12 Aa	8.41 Aa	9.56 Aa
CV (%)	22.62		

Table 3. Water use efficiency (WUE - mol CO_2 mol H_2O^{-1}) of sugarcane varieties under competition with populations of *Brachiria brizantha* DFT/UFV, Viçosa-MG, Brazil, 2008/09.

photosynthesis, water is lost with variable intensity, following a gradient of water potentials from the leaf surface to its surrounding air [44].

Means followed by the same letter, uppercase in the rows and lowercase in the columns, are included in the same group by the test of Skott-Knott at 5% probability.

With regard to the photosynthesis rate (A), it was observed that only sulfentrazone affected the variety RB867515, while the remaining treatments did not differ from the check (Table 4). Another research [45] evaluated varieties of sugarcane and found that RB867515 was the least affected by the application of commercially formulated mixture ametryn + trifloxysulfuron-sodium in greenhouse.

The influx of CO_2 in the leaf may be compromised due to the action of herbicides which inhibit the enzyme PROTOX, because there is formation of nitric oxide by the reactive oxygen species (ROS). This oxide stimulates the synthesis and activity of abscisic acid (ABA) - hormone that acts regulating stomatal closure [2]. Stomatal closure can also occur by the direct action of ROS, favoring the accumulation of calcium in the cytosol [46], or due to peroxidation of membranes of cells adjacent to stomata.

For the variety RB855156, MSMA, sulfentrazone, trifloxysulfuron–sodium, tebuthiuron and clomazone caused a reduction in A (Table 3). For SP80-1816, diuron + hexazinone and sulfentrazone did not affect this variable. It is noteworthy that hexazinone + diuron in mixture, even both being photosystem II inhibiting herbicides, did not cause reductions in A for the three varieties compared to control (Table 3). [42], working with six sugarcane varieties, found that under application of ametryn, RB855113 presented A similar to the control treatment, and the other varieties showed a decrease in this parameter under application of ametryn.

Means followed by the same letter, uppercase in the rows and lowercase in the columns, are included in the same group by the test of Scott-Knot at 5% probability.

	Sugarcane Variety		
Herbicide	**RB867515**	**RB855156**	**SP80-1816**
Tembotrione	22.10 Aa	24.48 Aa	16.45 Ab
MSMA	17.54 Aa	14.51 Ab	13.75 Ab
Diuron+hexazinone	21.15 Aa	21.17 Aa	21.22 Aa
Sulfentrazone	10.56 Ba	15.59 Bb	24.94 Aa
Trifloxysulfurom-sodium	20.53 Aa	15.85 Ab	13.90 Ab
Tebuthiuron	17.86 Aa	16.25 Ab	18.64 Ab
Clomazone	16.89 Aa	12.96 Ab	18.45 Ab
Check	28.44 Aa	24.07 Aa	26.03 Aa
CV (%)		27.78	

Table 4. Photosynthesis rate (A - μmol m^{-2} s^{-1}) of sugarcane varieties under competition with populations of *Brachiria brizantha* DFT/UFV, Viçosa-MG, Brazil, 2008/09.

When evaluating varieties under each herbicide, there was no correlation between them for all the products tested, except for sulfentrazone applied over SP80-1816, where there was higher A compared to the other varieties under the same herbicide (Table 4). [42] observed that the photosynthetic rate in sugarcane plants treated with trifloxysulfuron-sodium was similar to the respective control without herbicide application for all varieties evaluated in this study. Most of the herbicides used in this work, except MSMA and trifloxysulfuron-sodium, affect directly the photosynthesis rate because they have as a site of action either the chloroplast, inhibiting the transport of electrons, or the synthesis of chlorophyll, also interfering in the synthesis of pigments responsible for protecting the photosynthetic apparatus.

In RB867515 and SP80-1816, herbicides did not cause differences for stomatal conductance (Gs), but for RB855156, Gs decreased in the presence of all herbicides (Table 5). When evaluating the Gs of varieties within each herbicide, there was no difference among them. However, in the absence of herbicide, RB855156 showed greater Gs than the other two varieties (Table 5). Reduction of stomatal conductance was reported in soybean and *Portulaca oleracea* six hours after application of lactofen, a PROTOX inhibitor [47]. This product may promote stomatal closure due to oxidation processes and the concentration of nitric oxide, as already mentioned, which promotes formation of ABA, which in turn is a hormone that regulates the stomatal closure [2].

Means followed by the same letter, uppercase in the rows and lowercase in the columns, are included in the same group by the test of Scott-Knot at 5% probability.

According to the results, the herbicides tested affected differently the physiological traits of the three sugarcane varieties evaluated. Overall, the variety RB867515 showed the lowest variation in the photosynthesis rate in the presence of the herbicides, compared to the check.

	Sugarcane Variety		
Herbicide	RB867515	RB855156	SP80-1816
Tembotrione	0.67 Aa	0.46 Ab	0.48 Aa
MSMA	0.60 Aa	0.46 Ab	0.28 Aa
Diuron+hexazinone	0.74 Aa	0.44 Ab	0.70 Aa
Sulfentrazone	0.27 Aa	0.34 Ab	0.59 Aa
Trifloxysulfurom-sodium	0.52 Aa	0.31 Ab	0.20 Aa
Tebuthiuron	0.33 Aa	0.28 Ab	0.56 Aa
Clomazone	0.36 Aa	0.27 Ab	0.57 Aa
Check	0.61 Ba	1.00 Aa	0.47 Ba
CV (%)		46.9	

Table 5. Stomatal conductance (Gs - mol m^{-1} s^{-1}) of sugarcane varieties under competition with populations of *Brachiria brizantha* DFT/UFV, Viçosa-MG, Brazil, 2008/09.

The varieties SP80-1816 and RB855156 presented lower photosynthesis rates under application of most of the herbicides evaluated. It was also observed a decrease in stomatal conductance of the variety RB855156 after application of all herbicides.

9. Influence of herbicides on the photosynthetic activity of sugarcane genotypes

According to [48], the application of late post emergence herbicides in sugarcane fields may result in high toxicity to the crop, limiting yields. These authors attribute this to physiological changes in sugarcane plants, which would result in negative effects, also on the quality of the harvest. [42] studied the following herbicides, applied over several sugarcane varieties: ametryn - 2000 g ha^{-1}; trifloxysulfuron-sodium – 22.5 g ha^{-1}; and a commercial mixture containing ametryn + trifloxysulfuron-sodium at 1463 + 37.0 g ha^{-1}, respectively. Treatments were compared against a check with no herbicide. Results are summarized in Table 6.

In general terms, the CO_2 consumed by photosynthesis (ΔC) was smaller in treatments including the herbicide ametryn. There were also remarkable differences among varieties. The ΔC is directly related to the photosynthesis rate of the plant by the time of the evaluation. In this situation, it is possible to observe that the variety RB72454 and SP80-1816 were less susceptible to ametryn than the other genotypes.

The concentration of CO_2 within the leaf (Ci) was affected by the herbicide treatments, being also observed once more differences among genotypes. As expected, this parameter presented, in general terms, opposite behavior in comparison to ΔC. The application of the herbicide

Treatment	Sugarcane Genotype					
	RB72454	RB835486	RB855113	RB867515	RB947520	SP801816
	Shoot Dry Mass (g plant[-1])					
TC	A 5.77 a	A 5.46 a	A 4.93 a	A 5.02 a	A 5.02 a	B 2.46 a
HA	B 2.83 b	B 3.69 ab	B 2.89 b	A 5.97 a	B 3.38 ab	B 3.63 a
HB	A 5.21 ab	B 2.81 b	B 3.81 ab	B 2.99 b	B 2.03 b	B 2.30 a
HC	A 2.90 b	A 1.99 b	A 2.46 ab	A 2.09 b	A 1.66 b	A 1.30 a
	Consumed CO_2 — ΔC (μmol mol^{-1})					
TC	A 124 ab	AB 149 a	B 120 a	AB 139 a	A 177 a	B 117 a
HA	A 109 b	B 75 c	AB 105 b	AB 91 b	AB 102 b	A 110 a
HB	AB 119 ab	B 106 b	AB 114 ab	AB 118 ab	A 144 ab	B 108 a
HC	A 131 a	A 107 b	AB 112 ab	A 95 b	A 118 b	A 111 a
	Internal CO_2 Concentration — Ci (μmol mol^{-1})					
TC	AB 102 a	A 177 a	AB 104 b	AB 123 b	B 68 c	AB 120 b
HA	A 165 a	A 136 ab	A 169 a	A 178 a	A 134 a	A 157 a
HB	AB 126 ab	AB 127 ab	AB 128 ab	AB 114 b	B 88 b	A 137 ab
HC	AB 146 ab	B 87,7 b	AB 125 ab	A 179 a	AB 145 a	A 158 a
	Photosynthesis Rate — A (μmol m^{-2} s^{-1})					
TC	AB 45.1 a	AB 51.2 a	B 41.3 a	AB 47.9 a	A 60.7 a	AB 48.0 a
HA	A 37.5 b	B 25.8 b	A 36.1 a	B 28.9 b	A 37.0 b	A 37.3 b
HB	B 41.1 ab	B 36.6 ab	B 38.9 a	B 40.3 ab	A 49.5 ab	B 38.8 b
HC	A 42.1 ab	A 36.4 ab	A 38.4 a	B 32.8 b	A 40.5 b	A 40.1 b

Table 6. Physiological variables evaluated in sugarcane genotypes as a function of herbicide treatment. **TC**: control with no herbicide; **HA**: ametryn at 2000 g a.i. ha^{-1}; **HB**: trifloxysulfuron-sodium at 22.5 g a.i. ha^{-1}; **HC**: ametryn + trifloxysulfuron-sodium at 1463 + 37.0 g a.i. ha^{-1}. Means followed by the same letter at te column, inside each variable, are not different by the DMRT test at 5% probability.

ametryn, a photosynthesis II (PSII) inhibitor, resulted in higher concentrations of CO_2 within the leaf, once the photosynthesis of the genotypes under application of this herbicide was more severely affected. The CO_2 concentration within the leaf was about 50% higher in treatments involving ametryn than to the check with no herbicide. Trifloxysulfuron-sodium also caused changes in Ci, but not at the same magnitude of ametryn.

In general terms, the photosynthesis rate (A) observed at the treatment with trifloxysulfuron alone was similar to the control with no herbicide. On the same way, treatments involving the PSII inhibitor presented photosynthesis rate inferior to rates observed at the control treatment. When considering the treatment containing ametryn + trifloxysulfuron, it was possible to highlight the variety RB947520. The authors report that, even the damages caused by ametryn being more easily identified by parameters associated to the photosynthesis, variations due to the application of trifloxysulfuron-sodium were also detectable by changes in these parameters by using an Infra Red Gas Analyzer (IRGA). In other words, herbicide damage on crops can be effectively quantified by evaluating direct and indirect damage to the photosynthetic route. Furthermore, the accumulation of dry mass did not correlate directly with most of the studied physiological parameters, because plant growth is a result of biomass accumulation since the emergence until the moment of the evaluation. In this way, the authors remark the importance of evaluating both types of variables, physiological and biomass/growth-related, before concluding about the efficacy or impact of a given herbicide treatment. In addition, the authors remark the existence of differences among varieties in terms of susceptibility to herbicides, which were effectively identified by physiological parameters.

10. Conclusions

Both the abiotic and biotic factors, represented in this study respectively by the environmental conditions and the presence of a competitor, present impact on the sugarcane crop. This impact is differential on distinct varieties, and some should be preferred over others in situations where the types of stress differ. Sugarcane varieties to be planted should be chosen, besides commercial traits, based on their ability to avoid or to overcome the negative impact of the stress to which they are submitted. This will help reaching high yields under field conditions.

Brachiaria brizantha interferes both in the morphological and physiological components of sugarcane varieties. The variety RB72454 was the most affected by competition with *B. brizantha*, both in the morphological and physiological components. Thus, in areas with high infestation of weeds, there is an indication that the variety RB72454 should be avoided; if this variety is going to be used due to other positive traits, high levels of weed control are demanded in such fields.

Under application of herbicides, the variety RB867515 showed to be less sensitive to most of the herbicides tested, compared to SP80-1816 and RB855156. Herbicide damage to sugarcane can be effectively quantified by using an IRGA. Furthermore, the accumulation of dry mass usually did not correlate directly with most of the studied physiological parameters.

It is highlighted the importance of evaluating both types of variables, physiological and biomass/growth-related, before concluding about the efficacy or impact of a given herbicide treatment. It is also remarkable the existence of differences among sugarcane varieties in terms of susceptibility to herbicides, which were effectively identified by physiological parameters.

Author details

Leandro Galon[1], Germani Concenço[2], Evander A. Ferreira[3], Ignacio Aspiazu[4],
Alexandre F. da Silva[5], Clevison L. Giacobbo[1] and André Andres[6]

1 Federal University of the Southern Border, Brazil

2 Embrapa Western Region Agriculture, Brazil

3 Federal University of Jequitinhonha e Mucuri Valleys) ANDA, Brazil

4 State University of Montes Claros, Brazil

5 Embrapa Maize and Sorghum, Brazil

6 Embrapa Temperate Climate, Brazil

References

[1] Radosevich, S.R., Holt, J.S. & Ghersa, C.M. (2007). *Ecology of Weeds and Invasive Plants: Relationship to Agriculture and Natural Resource Management* (3rd Ed.), John Wiley & Sons, ISBN 978-047-1767-79-4, Hoboken, USA.

[2] Gurevitch, J., Scheiner, S.M. & Fox, G.A. (2009). *Ecologia Vegetal* (2nd Ed.), Artmed, ISBN 978-853-6319-18-6, Porto Alegre, Brazil.

[3] CONAB – Companhia Nacional do Abastecimento. Safra 2011. Available at http://www.conab.gov.br. Access in June 134, 2012.

[4] Larcher, W. *Ecofisiologia vegetal*. São Carlos: RiMa, 2004. 531p.

[5] Aliyev, J.A. (2010). Photosynthesis, photorespiration and productivity of wheat and soybean genotypes. *Proceedings of ANAS (Biological Sciences)*, Vol.65, No.5/6, pp.7-48.

[6] GALON, L. et al. (2009). Influência de herbicidas na qualidade da matéria-prima de genótipos de cana-de-açúcar. *Planta Daninha*, v. 27, n. 3, p. 555-562, ISSN 0100-8358.2009.

[7] FERREIRA, E.A., SILVA, A.F., SILVA, A.A., SILVA, D.V., GALON, L., FRANÇA, A.C., & SANTOS, J.B. (2009). Toxidade de herbicidas a genótipos de cana-de-açucar. *Trópica*, v.6, n.1, on-line, ISSN 1982-4831.

[8] Silva, A.A., Ferreira, F.A., Ferreira, L.R. & Santos, J.B. (2007). Biologia de plantas daninhas, In: *Tópicos em Manejo de Plantas Daninhas*, Silva, A.A. & Silva, J.F., pp.17-61, Universidade Federal de Viçosa, ISBN 978-857-2692-75-5, Viçosa, Brazil.

[9] Floss, E.L. (2008). *Fisiologia das Plantas Cultivadas* (4th Ed.), Universidade de Passo Fundo, ISBN 978-857-5156-41-4, Passo Fundo, Brazil.

[10] Sharkey, T.D. & Raschke, K. (1981). Effect of light quality on stimatal opening in leaves of *Xanthium strumarium*. *Plant Physiology*, Vol.68, No.5, (November 1981), pp. 1170-1174, ISSN 0032-0889.

[11] Pystina, N.V. & Danilov, R.A. (2001). Influence of light regimes on respiration, activity of alternative respiratory pathway and carbohydrates content in mature leaves of *Ajuga reptans* L. *Revista Brasileira de Fisiologia Vegetal*, Vol.13, No.3, (December 2001), pp. 285-292, ISSN 0103-3131.

[12] Aspiazú, I., Concenço, G., Galon, L., Ferreira, E.A. & Silva, A.F. (2008). Relação colmos/folhas de biótipos de capim-arroz em condição de competição. *Revista Trópica*, Vol.2, No.1, pp.22-30, ISSN 1982-4831.

[13] Vanderzee, D. & Kennedy, R.A. (1983). Development of photosynthetic activity following anaerobic germination in rice-mimic grass (*Echinochloa crus-galli* var. oryzicola). *Plant Physiology*, Vol.73, No.2, (October 1983), pp.332-339, ISSN 0032-0889.

[14] Melo, P.T.B.S., Schuch, L.O.B., Assis, F. & Concenço, G. (2006). Comportamento de populações de arroz irrigado em função das proporções de plantas originadas de sementes de alta e baixa qualidade fisiológica. *Revista Brasileira de Sementes*, Vol.12, No. 1, pp.37-43, ISSN 0101-3122.

[15] Kirschbaum, M.U.F. & Pearcy, R.W. (1988). Gas exchange analysis of the relative importance of stomatal and biochemical factors in photosynthetic induction in *Alocasia macrorrhiza*. *Plant Physiology*, Vol.86, No.3, (March 1988), pp.782-785, ISSN 0032-0889.

[16] Merotto Jr., A., Fischer, A.J. & Vidal, R.A. (2009). Perspectives for using light quality knowledge as an advanced ecophysiological weed management tool. *Planta Daninha*, Vol.27, No.2, (April/June 2009), pp.407-419, ISSN 0100-8358.

[17] Hutmacher, R.B. & Krieg, D.R. (1983). Photosynthetic rate control in cotton. *Plant Physiology*, Vol.73, No.3, (November 1983), pp.658-661, ISSN 0032-0889.

[18] Da Matta, F.M., Loos, R.A., Rodrigues, R. & Barros, R. (2001). Actual and potential photosynthetic rates of tropical crop species. *Revista Brasileira de Fisiologia Vegetal*, Vol.13, No.1, (Abril 2001), pp.24-32, ISSN 0103-3131.

[19] Dutton, R.G., Jiao, J., Tsujita, J. & Grodzinski, B. (1988). Whole plant CO2 exchange measurements for non destructive estimation of growth. *Plant Physiology*, Vol.86, No. 2, (February 1988), pp.355-358, ISSN 0032-0889.

[20] Long, S.P. & Bernacchi, C.J. (2003). Gas exchange measurements, what can they tell us about the underlying limitations to photosynthesis? Procedures and sources of error. *Journal of Experimental Botany*, Vol.54, No.392, (November 2003), pp.2393-2401, ISSN 0022-0957.

[21] Tambuci, E.A., Bort, J. & Araus, J.L. (2011). Water use efficiency in C3 cereals under mediterranean conditions: a review of some physiological aspects. *Options Méditerranéennes*, Series B, No.57, pp.189-203, ISSN 1016-1228.

[22] Taylor Jr., G.E. & Gunderson, C.A. (1986). The response of foliar gas exchange to exogenously applied ethylene. *Plant Physiology*, Vol.82, No.3, (November 1986), pp. 653-657, ISSN 0032-0889.

[23] Messinger, S.M., Buckley, T.N. & Mott, K.A. (2006). Evidence for involvement of photosynthetic processes in the stomatal response to CO2. *Plant Physiology*, Vol.140, No. 2, (February 2006), pp.771-778, ISSN 0032-0889.

[24] Humble, G.D. & Hsiao, T.C. (1970). Light-dependent influx and efflux of potassium of guard cells during stomatal opening and closing. *Plant Physiology*, Vol.46, No.3, (September 1970), pp.483-487, ISSN 0032-0889.

[25] Attridge, T.H. (1990). The natural environment, In: *Light and Plant Responses*, Attridge, T.H. (Ed.), pp.1-5, Edward Arnold, ISBN 978-052-1427-48-7, London, England.

[26] Fereira, E.A. et al. (2008). Potencial competitivo de biótipos de azevém (*Lolium multiflorum*). *Planta Daninha*, Vol.26, No.2, (June 2008), pp.261-269, ISSN 0100-8358.

[27] Rizzardi, M.A. et al. (2001). Competição por recursos do solo entre ervas daninhas e culturas. *Ciencia Rural*, Vol.31, No.4, (December 2001), pp.707-714, ISSN 0103-8478.

[28] Pedrosa, R.M.B. (2005). Avaliação dos parâmetros dos colmos da cana-de-açúcar, segunda folha, submetida a níveis de irrigação e adubação. *R. Biol. Ci. Terra*, Vol.5, No. 1, (March 2005), pp.1-5, ISSN 1519-5228.

[29] Santos, B.R. et al. (2006). Estresse ambiental e produtividade agrícola, In: *Fisiologia e produção vegetal*, Paiva, R., Oliveira, L.M. (Eds.), PP.71-91, Universidade Federal de Lavras, ISBN 85-87693-30-5, Lavras, Brazil.

[30] Kuva, M.A. et al. (2003). Período de interferência de plantas daninhas na cultura da cana-de-açúcar. III - capim-braquiária (*Brachiaria decumbens*) e capim-colonião (*Panicum maximum*). *Planta Daninha*, Vol.21, No.1, (March 2003), ISSN 0100-8358.

[31] Galon, L., & Agostinetto, D. (2009). Comparison of empirical models for predicting yield loss of irrigated rice (*Oryza sativa*) mixed with *Echinochloa* spp. *Crop Protection*, Vol.28, No.10, (October 2009), pp.825-830, ISSN 0261-2194.

[32] Pantone, D.J. & Baker, J.B. (1991). Reciprocal yield analysis of red rice (*Oryza sativa*) competition in cultivated rice. *Weed Science*, Vol.39, No.1, (January 1991), pp.42-47, ISSN 1939-747X.

[33] Hoffman, M.L. & Buhler, D.D. (2002). Utilizing *Sorghum* as a functional model of crop weed competition. I. Establishing a competitive hierarchy. *Weed Science*, Vol.50, No.4, (April 2002), pp.466-472, ISSN 1939-747X.

[34] Bianchi, M.A., Fleck, N.G. & Lamego, F.P. (2006). Proporção entre plantas de soja e plantas competidoras e as relações de interferência mútua. *Ciencia Rural*, Vol.36, No. 5, (May 2006), pp.1380-1387, ISSN 0103-8478.

[35] BALLARÉ Ballaré, C.L., Scopel, A.L. & Sánchez, R.A. (1990). Far-red radiation reflected from adjacent leaves an early signal of competition in plant canopies. *Science*, Vol. 247, No.4940, (January 1990), pp. 329-332, ISSN 0036-8075.

[36] Almeida, M.L. & Mundstock, C.M. (2001). A qualidade da luz afeta o afilhamento em plantas de trigo, quando cultivadas sob competição. *Ciencia Rural*, Vol.31, No.3, (September 2001), pp.401-408, ISSN 0103-8478.

[37] Bjorkman, O. & Holmgren, P. (1966). Photosynthetic adaptation to light intensity in plants native to shaded and exposed habitats. *Plant Physiology*, Vol.19, No.3, (September 1966), pp.854-859, ISSN 0032-0889.

[38] Bjorkman, O. (1968). Further studies on differentiation of photosynthetic properties in sun and shade ecotypes of *Solidago virgaurea*. *Plant Physiology*, Vol.21, No.1, pp. 84-99, ISSN 0032-0889.

[39] Walker, G.K., Blackshaw, R.E. & Dekker, J. (1988). Leaf area and competition for light between plant species using direct sunlight transmission. *Weed Technology*, Vol.2, No. 2, (March 1988), pp.159-165, ISSN 0890-037X.

[40] Sims, D.A. & Kelley, S. (1998). Somatic and genetic factors in sun and shade population differentiation in *Plantago lanceolata* and *Anthoxanthum odoratum*. *New Phytologist*, Vol.140, No.1, (February 1998), pp.75-84, ISSN 1469-8137.

[41] Wiedenfeld, B. (2003). Enhanced sugarcane establishment using plant growth regulators. *Journal of the American Society of Sugarcane Technology*, Vol.23, No.1, (February 2003), pp.48-61.

[42] Galon, L., Ferreira, F.A., Silva, A.A., Concenço, G., Ferreira, E.A., Barbosa, M.H.P., Silva, A.F., Aspiazú. I., França, A.C. & Tironi, S.P. (2010). Influência de herbicidas na atividade fotossintética de genótipos de cana-de-açúcar. *Planta Daninha*, Vol.28, No.3, (June 2010), pp.591-597, ISSN 0100-8358.

[43] Concenço, G. et al. (2009). Uso da água por plantas híbridas ou convencionais de arroz irrigado. *Planta Daninha*, Vol.27, No.3, (September 2009), pp.447-453, ISSN 0100-8358.

[44] Concenço, G. et al. (2007) Uso da água em biótipos de azevém (*Lolium multiflorum*) em condição de competição. *Planta Daninha*, Vol.25, No.3, (September 2007), pp. 449-455, ISSN 0100-8358.

[45] Ferreira, E.A. et al. (2005). Sensibilidade de cultivares de cana-de-açúcar à mistura trifloxysulfuron-sodium + ametryn. *Planta Daninha*, Vol.23, No.1, (March 2005), pp. 93-99, ISSN 0100-8358.

[46] Taiz, L. & Zeiger, E. (2009). *Fisiologia Vegetal*. 4.ed. Porto Alegre: Artmed, 848p. ISBN 9788536316147.

[47] Wichert, R.A. & Talbert, R.E. (1993). Soybean [*Glycine max* (L.)] response to lactofen. *Weed Science*, Vol.41, No.1, (February 1993), pp.23-27, ISSN 1939-747X.

[48] Azania, C.A.M. & Azania, A.A.P.M. (2005). Cana: limpa e lucrativa. *Caderno Técnico Cultivar Grandes Culturas*, No.79, pp.3-10, not indexed.

Chapter 7

Temperature-Dependent Photoregulation in Oceanic Picophytoplankton During Excessive Irradiance Exposure

Gemma Kulk, Pablo de Vries, Willem H. van de Poll,
Ronald J. W. Visser and Anita G. J. Buma

Additional information is available at the end of the chapter

1. Introduction

The phytoplankton community of open oligotrophic oceans is dominated by prokaryotic *Prochlorococcus* spp., *Synechococcus* spp., and eukaryotic pico- and nanophytoplankton [1-3]. The competitive success of these phytoplankton species depends on different factors, including the response to the (dynamic) irradiance conditions encountered in the water column. With the occurrence of different ecotypes, picophytoplankton species such as *Prochlorococcus* spp., *Synechococcus* spp., and *Ostreococcus* spp. can grow over a broad range of irradiance conditions [4-7]. For example, the low light adapted ecotypes of *Prochlorococcus* are well adapted to the irradiance intensity and spectral composition of the deep chlorophyll maximum with high chlorophyll *b/a* ratios and low optimal growth irradiances [4,5,8]. In contrast, the high light adapted ecotypes of *Prochlorococcus* spp. can competitively grow in the (upper) mixed layer with low chlorophyll *b/a* ratios and higher optimal growth irradiances [4,5,8]. Similar differences in pigmentation, absorption, and photosynthetic characteristics have been found in ecotypes of marine *Synechococcus* spp. [9-11] and *Ostreococcus* spp. [7,12,13]. In addition to the genetically defined (photo)physiology of the different ecotypes, the photoacclimation potential of specific (pico)phytoplankton species may play an important role in the response to (dynamic) irradiance conditions [11].

Phytoplankton irradiance exposure is strongly influenced by physical processes in the ocean [14]. During stratification, phytoplankton can be trapped in a shallow upper mixed layer, thereby enhancing exposure to photosynthetically active radiation (PAR, 400-700 nm) and ultraviolet radiation (UVR, 280-400 nm), or can experience limiting irradiance conditions at the deep chlorophyll maximum. In seasonally stratified regions, the period of stratification is interchanged with periods of deep convective mixing that can reach below the euphotic zone. This causes a strong reduction in the daily experienced irradiance, with occasional interruptions of excessive irradiance exposure. Consequently, phytoplankton irradiance exposure in open ocean ecosystems can vary by several orders of magnitude on a time scale ranging from seconds to days. Moreover, short wavelength solar radiation (UVB, 280-315 nm) can penetrate to significant depths in clear oligotrophic waters [15,16].

High irradiance exposure may have considerable effects on photosynthesis and viability in oceanic picophytoplankton species such as *Prochlorococcus* spp., *Synechococcus* spp., and *Ostreococcus* spp. [17-19]. When residing near the surface, picophytoplankton can experience irradiance intensities that exceed photosynthetic requirements. Exposure to excessive PAR and UVR causes photoinhibition, a process in which an over-reduction of the photosynthetic electron transport chain reduces photosynthetic efficiency by a decrease in functional photosystem II (PSII) reaction centers [20]. Moreover, prolonged exposure to excessive irradiance can lead to the uncontrolled formation of reactive oxygen species and viability loss [21,22]. To prevent photoinhibition and viability loss during excessive irradiance exposure, phytoplankton regulate light harvesting and other photosynthetically important processes. In prokaryotic species, the utilization of light harvesting energy can be regulated by state transitions, in which the light harvesting antenna of the phycobillisome (PBS) is redistributed between the reaction centers of photosystem I (PSI) and PSII [23,24]. In addition, light harvesting energy can be regulated by the thermal dissipation of excess energy. This photoprotective process can occur within seconds after irradiance changes in both prokaryotic and eukaryotic phytoplankton species, but the underlying mechanisms are considerably different. In eukaryotic species, the thermal dissipation of excess energy involves the xanthophyll pigment cycle. Epoxidized xanthophyll cycle pigments assist in light harvesting, whereas de-epoxidized equivalents dissipate excess energy in the form of heat [25]. In PBS containing cyanobacteria, the thermal dissipation of excess energy involves the orange carotenoid protein [24,26]. In *Prochlorococcus* spp., these proteins are not observed and the underlying mechanism remains unknown [24,27]. In addition to the regulation of light harvesting, photoinhibition and viability loss may be avoided by the increase of photochemical quenching by enhancing alternative electron transport and (non-)enzymatic scavenging of reactive oxygen species [28,29]. Simultaneously, phytoplankton can counteract the effects of photoinhibition by photorepair, a process in which damaged D1 proteins are removed from PSII and replaced by newly synthesized D1 proteins [20].

Although it has previously been reported that temperature may have a positive effect on the survival of picophytoplankton under high irradiance conditions [30], no direct assessment of the temperature-dependency of photoregulation during high PAR and UVR exposure is available for this specific phytoplankton group. A recent study showed that both prokaryotic and eukaryotic picophytoplankton may be less susceptible to the negative effects of high irradiance intensities at elevated temperatures [31]. In the prokaryotic species *Prochlorococcus* spp. (eMED4 and eMIT9313) and the eukaryotic species *Ostreococcus* sp. (clade B) and

Pelagomonas calceolata, acclimation to elevated temperatures enhanced photoacclimation to higher irradiance intensities and reduced photoinhibition [31]. This has also been found in larger phytoplankton species, such as the diatom species *Chaetoceros gracilis, Thalassiosira pseudonana,* and *Thalassiosira weissflogii* [32-34]. In cyanobacteria and eukaryotic nanophytoplankton, reduced levels of photoinhibition at elevated temperatures may be associated with enhanced rates of state transitions [24], enhanced enzymatic conversions of the xanthophyll pigment cycle [35], enhanced D1 repair [36], and the potential enhancement of Rubisco activity [34]. However, the potential role of these photoregulating mechanisms at elevated temperatures remains unknown in oceanic picophytoplankton.

In the present study, a comparative analysis of the high irradiance sensitivity of oceanic picophytoplankton was performed to study the combined effect of elevated temperatures and irradiance levels near the surface of open oligotrophic oceans. To this end, two prokaryotic and two eukaryotic strains were acclimated to 16 °C, 20 °C, and 24 °C, after which they were exposed to a single high PAR dose, with and without UVR. The response to and the recovery after high irradiance exposure was assessed by analysis of PSII fluorescence and pigmentation in order to investigate immediate photoinhibition and photoprotective processes. The results are discussed in the context of differences between oceanic picophytoplankton species and are used to unravel the importance of photoinhibition in structuring the phytoplankton community in open oligotrophic oceans.

2. Method

2.1. Culture conditions

Cultures were obtained from the Roscoff Culture Collection (RCC) and the Provasoli-Guillard National Center for Marine Algae and Microbiota (NCMA). The strains were all isolated from oligotrophic regions and are representative for low light (LL) and high light (HL) adapted species in open ocean ecosystems. *Prochlorococcus marinus* strain CCMP2389 (ecotype MED4, HL) and *Prochlorococcus* sp. strain RCC407 (ecotype MIT9313, LL) were cultured in K/10-Cu medium based on natural oceanic seawater as described by [37]. *Ostreococcus* sp. strain RCC410 (clade B, LL) and *Pelagomonas calceolata* strain RCC879 (LL) were cultured in K medium as described by [38]. Cultures were maintained in 100 ml glass Erlenmeyer flasks at 9 µmol photons m^{-2} s^{-1} (*Prochlorococcus* sp. and *P. calceolata*) and 68 µmol photons m^{-2} s^{-1} (*P. marinus* and *Ostreococcus* sp.) in a diurnal cycle of 12:12 h light:dark at 20 °C.

2.2. Experimental design

Cultures of *P. marinus, Prochlorococcus* sp., *Ostreococcus* sp., and *P. calceolata* were transferred to 500 ml glass Erlenmeyer flasks and incubated in triplicate at 16 °C, 20 °C, and 24 °C. Experiments were carried out in a temperature controlled U-shaped lamp setup as described by [39]. The temperature in the setup was maintained at 16 °C, 20 °C, and 24 °C by a thermostat (RK 8 KS, edition 2000, Lauda Dr. R. Wobser & Co.) and deviated less than ± 0.5 °C. During the experiments, 50 µmol photons m^{-2} s^{-1} PAR (Biolux and Skywhite lamps,

Osram) was provided as a square wave function with a 12:12 h light:dark cycle (monitored with a QSL-100, Biospherical Instruments). Prior to the experiments, the picophytoplankton strains were kept in exponential growth phase and acclimated to the experimental irradiance and temperature conditions for at least three weeks. In mid-exponential growth phase, the response to high photosynthetically active radiation (PAR, 400-700 nm), with and without ultraviolet radiation (UVR, 290-400 nm), was assessed at growth temperature by pigment and PSII chlorophyll fluorescence analysis. To this end, 200 ml of each replicate culture was exposed to high PAR and PAR+UVR for 10 min in a temperature controlled (RTE-211, Neslab Instruments Inc.) irradiance set-up at 16 °C, 20 °C, or 24 °C. The irradiance set-up provided ± 500 μmol photons m^{-2} s^{-1} by a 250 W MHN-TD lamp (Philips) and two 20 W TL/12 UVB fluorescent lamps (Philips), in which the PAR and PAR +UVR conditions were obtained by using the long pass filters GG395 and WG305 (Schott AG, Mainz), respectively (Table 1). Prior to exposure (t = 0), samples for the analysis of pigmentation and the maximum quantum yield of PSII (F_v/F_m) were collected and measured as described below. After exposure, treated culture samples were transferred to dim light conditions at growth temperature (16 °C, 20 °C, or 24 °C). Subsequently, samples for the analysis of pigmentation were taken at t = 10, 20, and 40 min and recovery of the quantum yield of PSII (Φ_{PSII}) was determined at t = 10, 15, 20, 25, 30, 35, 40, 60, 80, and 100 min for both PAR and PAR+UVR treated cultures. Culturing of *Prochlorococcus* sp. at 16 °C and 50 μmol photons m^{-2} s^{-1} was attempted several times, but this condition exceeded the limit for growth of this individual strain. No measurements were performed for *Prochlorococcus* sp. under these conditions.

	PAR	PAR+UVR
PAR	172	157
UVA	8.69	15.3
UVB	0.05	1.79

Table 1. Doses (W m^{-2}) for photosynthetically active radiation (PAR, 400-700 nm) and ultraviolet radiation A (UVA, 315-400 nm) and B (UVB, 290-315 nm) are given for the PAR and PAR+UVR treatments during the experiments. Total irradiance intensity was ± 500 μmol photons m^{-2} s^{-1} in both treatments.

2.3. Photosystem II chlorophyll fluorescence characteristics

PSII fluorescence analyses were performed on a WATER-PAM chlorophyll fluorometer (Waltz GmbH) equipped with a WATER-FT flow-through emitter-detector unit and analyzed using WinControl software (version 2.08, Waltz GmbH) according to [40] and references therein. Prior to exposure to PAR and PAR+UVR (t = 0), 5-15 ml culture samples were dark-adapted for 20 min at 16 °C, 20 °C, or 24 °C. For analysis, the measuring light was turned on and F_0 was recorded as the minimal fluorescence. During a saturating light flash, F_m° was then recorded as the maximum fluorescence in the dark-adapted state. The maximum quantum yield of PSII (F_v/F_m) was calculated as (F_m° - F_0) / F_m°. After exposure (t = 10-100), the quantum yield of PSII (Φ_{PSII}) was determined by measuring F_t as the steady state fluorescence prior to the saturating light flash and F_m' as the maximum fluorescence in the light. Φ_{PSII} was calculated as $(F_m' - F_t) / F_m'$. From the F_v/F_m measurements at $t = 0$ and the Φ_{PSII} measurements at $t = 10$, total non photochemical quenching (NPQ) was calculated as $(F_m^{\circ} - F_m') / F_m'$. Relaxation analysis was performed to estimate the contribution of slowly and rapidly relaxing non photochemical quenching. Relaxation of NPQ on a time scale of minutes is associated with photoprotective processes such as state transitions, relaxation of the xanthophyll pigment cycle or other forms of thermal dissipation [35, 40,41]. Processes that relax over a longer period of time (hours) are referred to as photoinhibition, i.e. damage to the reaction centers of PSII [40,42]. To estimate photoprotection and photoinhibition, the recorded F_m' was corrected for baseline quenching by subtracting F_0 and was log transformed for further analysis. Transformed F_m' values of the final 60 min of the Φ_{PSII} recovery curve were extrapolated to calculate the value of F_m' that would had been attained if only slowly relaxing quenching was present in the light (F_m^r). Slow relaxing non photochemical quenching (NPQ$_S$) was then calculated as $(F_m^{\circ} - F_m^r) / F_m^r$ and fast relaxing non photochemical quenching (NPQ$_F$) as $(F_m^{\circ} / F_m') - (F_m^{\circ} - F_m^r)$. In addition, the contribution of UVR to the decrease in quantum yield of PSII during irradiance exposure was calculated as $(\Phi_{PSII,PAR} - \Phi_{PSII,PAR+UVR}) / \Phi_{PSII,PAR} \cdot 100$ [43].

2.4. Pigment composition

Samples (25-30 ml) for untreated (t = 0), PAR treated (t = 10, 20, 40), and PAR+UVR (t = 10, 20, 40) treated cultures were filtered onto 25 mm GF/F filters (Whatman), snap frozen in liquid nitrogen, and stored at -80 °C until further analysis. Pigments were quantified using High Performance Liquid Chromatography (HPLC) as described by [44]. In short, filters were freeze-dried for 48 h and pigments were extracted in 3 ml 90% acetone (v/v, 48 h, 4 °C). Detection of pigments was carried out using a HPLC (Waters 2695 separation module, 996 photodiode array detector) equipped with a Zorbax Eclipse XDB-C$_8$ 3.5 μm column (Agilent Technologies, Inc.). Peaks were identified by retention time and diode array spectroscopy. Pigments were quantified using standards (DHI LAB products) of chlorophyll a_1, chlorophyll a_2, diadinoxanthin (Dd), diatoxanthin (Dt), violaxanthin (Vio), antheraxanthin (Ant), and zeaxanthin (Zea). From here on, chlorophyll a (Chl-a) will refer to chlorophyll a_2 in P. marinus and Prochlorococcus sp. and to chlorophyll a_1 in Ostreococcus sp. and P. calceolata. The de-epoxidation state (DPS) of the xanthophyll pigment cycle was calculated as (Ant + Zea) / (Vio + Ant + Zea) for Ostreococcus sp. and as Dt / (Dd + Dt) for P. calceolata. In addition to the DPS, the rate of de-epoxidation of the xanthophyll pigment cycle (k_{DPS} in min^{-1}) was estimated as the increase in DPS during exposure to high PAR and PAR+UVR [45].

2.5. Statistical analysis

All measurements were performed for triplicate cultures ($n = 3$) at each temperature. Differences between the three temperature conditions, differences between irradiance treatments, and differences between species were statistically tested by analysis of variance (ANOVA) using STATISTICA software (version 8.0 and 10.0, StatSoft Inc.). Before analysis, data were tested for normality and homogeneity of variances. Differences were considered significant when $p < 0.05$.

3. Results

3.1. Non photochemical quenching and photosystem II recovery

P. marinus, Prochlorococcus sp., *Ostreococcus* sp., and *P. calceolata* all showed non photochemical quenching (NPQ) upon exposure to high photosynthetically active radiation (PAR), with and without ultraviolet radiation (UVR) (Figure 1). The effect of temperature on NPQ was most pronounced in the prokaryotic strains *P. marinus* and *Prochlorococcus* sp. (Figure 1). Although total NPQ did not change with temperature in *P. marinus*, the proportion of slow and fast non photochemical quenching changed significantly. Slow relaxing non photochemical quenching (NPQ$_S$) decreased with increasing temperature ($p < 0.05$, not significant between 20 °C and 24 °C), whereas fast relaxing non photochemical quenching (NPQ$_F$) increased significantly with increasing temperature ($p < 0.05$). In *Prochlorococcus* sp., total NPQ increased from 20 °C to 24 °C (Figure 1). The proportion of NPQ$_S$ and NPQ$_F$ was also affected by temperature in *Prochlorococcus* sp., with a significant increase in NPQ$_F$ with increasing temperature ($p < 0.05$) and unchanged levels of NPQ$_S$. In the eukaryotic species *Ostreococcus* sp., temperature had no effect on NPQ (Figure 1). In *P. calceolata*, total NPQ decreased with increasing temperature ($p < 0.05$, not significant between 20 °C and 24 °C). This was associated with a decrease in NPQ$_S$ with increasing temperature ($p < 0.05$, not significant between 16 °C and 20 °C), whereas NPQ$_F$ remained unaffected by temperature.

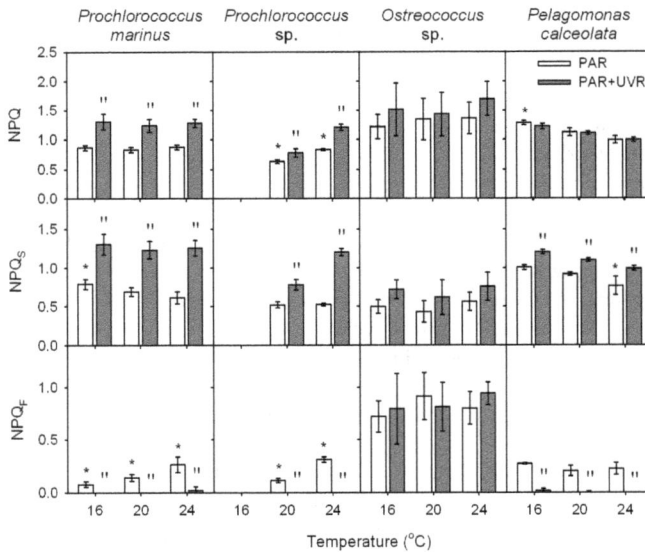

Figure 1. Non photochemical quenching. Mean (± standard deviation, $n = 3$) total non photochemical quenching (NPQ), slow relaxing NPQ (NPQ$_S$), and fast relaxing NPQ (NPQ$_F$) are given for *Prochlorococcus marinus* eMED4, *Pro-*

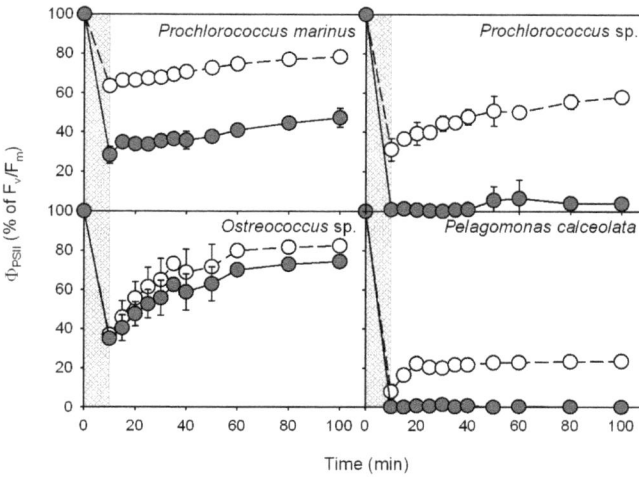

Figure 2. Recovery of PSII after high irradiance exposure. Mean (± standard deviation, $n = 3$) quantum yield of PSII (Φ_{PSII} in % of F_v/F_m) during and after exposure to high irradiance for *Prochlorococcus marinus* eMED4, *Prochlorococcus* sp. eMIT9313, *Ostreococcus* sp. clade B, and *Pelagomonas calceolata* acclimated to 20 °C. The picophytoplankton strains were exposed to high photosynthetically active radiation (PAR, white circles) and high PAR with ultraviolet radiation (PAR+UVR, dark grey circles) for 10 minutes (light grey area).

chlorococcus sp. eMIT9313, *Ostreococcus* sp. clade B, and *Pelagomonas calceolata* at 16 °C, 20 °C and 24 °C. The picophytoplankton strains were exposed to high photosynthetically active radiation (PAR, white bars) and high PAR with ultraviolet radiation (PAR+UVR, grey bars) for 10 minutes. Significant effects ($p < 0.05$) of the growth temperature (*) and the spectral composition of the irradiance treatment (") are indicated.

The spectral composition of the irradiance treatment influenced non photochemical quenching and the recovery of the quantum yield of PSII (Φ_{PSII}) considerably in the prokaryotic strains (Figure 1, Figure 2). In both *P. marinus* and *Prochlorococcus* sp., total NPQ and NPQ_S were significantly higher during exposure to PAR+UVR compared with PAR, whereas NPQ_F decreased significantly during exposure to UVR ($p < 0.05$) (Figure 1). In *Prochlorococcus* sp., this was associated with almost no recovery of Φ_{PSII} after exposure to PAR+UVR (Figure 2). In the eukaryotic species *Ostreococcus* sp., the spectral composition of the irradiance treatment did not have a significant effect on NPQ (Figure 1). However, recovery of Φ_{PSII} in *Ostreococcus* sp. was lower after exposure to PAR+UVR compared with PAR (significant for t = 60-100, $p < 0.05$, Figure 2). In *P. calceolata*, exposure to PAR+UVR significantly increased NPQ_S and decreased NPQ_F ($p < 0.05$), but total NPQ remained unaffected by the spectral composition of the irradiance treatment (Figure 1). *P. calceolata* showed no recovery of Φ_{PSII} after exposure to PAR +UVR (Figure 2).

Comparison of NPQ between the different picophytoplankton strains demonstrated significantly lower total NPQ in the prokaryotic species *P. marinus* and *Prochlorococcus* sp. compared with the eukaryotic species *Ostreococcus* sp. and *P. calceolata* ($p < 0.05$) (Figure 1). In *P. calceolata*, NPQ_S was significantly higher compared with the other species ($p < 0.05$, not significant at 24 °C). *P. marinus* and *Prochlorococcus* sp. showed intermediate levels of NPQ_S, whereas *Ostreococcus* sp. showed significantly lowest NPQ_S ($p < 0.05$, not significant at 24 °C). The relative low levels of NPQ_S in *Ostreococcus* sp. were accompanied by significantly higher NPQ_F compared with the other species ($p < 0.005$). No differences in NPQ_F were found between *P. marinus*, *Prochlorococcus* sp., and *P. calceolata*.

3.2. Inhibition of photosystem II by ultraviolet radiation

The inhibition of Φ_{PSII} due to UVR was affected by temperature in *P. marinus*, *Ostreococcus* sp., and *P. calceolata* (Table 2). In *P. marinus*, UVR inhibition decreased significantly with increasing temperature ($p < 0.01$ for 16 °C compared with 24 °C). In the eukaryotic species *Ostreococcus* sp. (not between 20 °C and 24 °C) and *P. calceolata*, UVR inhibition of Φ_{PSII} also decreased with increasing temperature, but not significantly. In *Prochlorococcus* sp., no effect of temperature was found on the UVR inhibition of Φ_{PSII}. Comparison of the different picophytoplankton strains showed that *Ostreococcus* sp. was least inhibited by UVR ($p < 0.001$) (Figure 2, Table 2). *P. marinus* showed intermediate levels of UVR inhibition, whereas Φ_{PSII} was most inhibited by UVR in *Prochlorococcus* sp. and *P. calceolata* ($p < 0.001$).

	Prochlorococcus marinus	*Prochlorococcus* sp.	*Ostreococcus* sp.	*Pelagomonas* calceolata
16 °C	66.4 ± 3.9[a]	n/a	24.5 ± 18.1	100.0 ± 0.0
20 °C	55.6 ± 4.9	97.9 ± 3.6	5.3 ± 5.4	97.3 ± 4.7
24 °C	49.5 ± 4.8[a]	97.3 ± 3.8	13.0 ± 5.5	77.0 ± 20.7

Table 2. Mean (± standard deviations, $n = 3$) inhibition by ultraviolet radiation (% of photosynthetically active radiation treatment) after 10 min high irradiance exposure in *Prochlorococcus marinus* eMED4, *Prochlorococcus* sp. eMIT9313, *Ostreococcus* sp. clade B, and *Pelagomonas calceolata* acclimated to 16 °C, 20 °C, and 24 °C. *abc* indicate significant effects of the temperature treatment within each species. n/a: data not available, growth was not observed under the used conditions and no additional measurements were performed.

3.3. Photoprotective pigmentation

Temperature acclimation affected the initial photoprotective pigment pool in *P. marinus* (t = 0, Table 3), with higher zeaxanthin per chlorophyll *a* levels at lower temperatures ($p < 0.001$). In *Prochlorococcus* sp., no significant effect of temperature acclimation was observed in the initial zeaxanthin per chlorophyll *a* level. In both prokaryotic strains, exposure to high irradiance did not influence photoprotective pigmentation, as the zeaxanthin per chlorophyll *a* levels remained similar during and after high irradiance exposure (Table 3). In *Ostreococcus* sp., acclimation to higher temperatures increased the initial xanthophyll cycle pigment pool (30-40%), but not significantly (t = 0, Table 3). In response to high irradiance exposure, large fluctuations in the sum of violaxanthin, antheraxanthin, and zeaxanthin per chlorophyll *a* were observed and no significant effect of temperature acclimation on the photoprotective pigment pool was found (Table 3). In *P. calceolata*, the initial photoprotective pigments per chlorophyll *a* ratio was highest at 24 °C (19 %, not significant). Temperature had no effect on the total xanthophyll cycle pigment pool in response to high irradiance in *P. calceolata* as the sum of diadinoxanthin and diatoxanthin per chlorophyll *a* remained unchanged during and after exposure to high irradiance (Table 3). No significant effect of the spectral composition of the irradiance treatment was observed in the photoprotective pigments pools of *P. marinus*, *Prochlorococcus* sp., *Ostreococcus* sp., and *P. calceolata* (Table 3).

	PAR			PAR+UVR		
	16 °C	20 °C	24 °C	16 °C	20 °C	24 °C
Prochlorococcus marinus						
t = 0	0.647±0.060[a]	0.499±0.004[a]	0.431±0.007[a]	0.647±0.060[b]	0.499±0.004[b]	0.431±0.007[b]
t = 10	0.644 ± 0.081	0.488 ± 0.019	0.434 ± 0.017	0.649 ± 0.057	0.488 ± 0.013	0.426 ± 0.017
t = 20	0.657 ± 0.066	0.493 ± 0.006	0.426 ± 0.031	0.655 ± 0.071	0.494 ± 0.002	0.421 ± 0.031
t = 40	0.661 ± 0.067	0.498 ± 0.009	0.424 ± 0.014	0.663 ± 0.053	0.492 ± 0.013	0.437 ± 0.014
Prochlorococcus sp.						
t = 0	n/a	1.062 ± 0.034	1.025 ± 0.023	n/a	1.062 ± 0.034	1.025 ± 0.023
t = 10	n/a	1.206 ± 0.076	0.976 ± 0.009	n/a	1.198 ± 0.039	0.946 ± 0.039
t = 20	n/a	1.209 ± 0.093	0.966 ± 0.036	n/a	1.189 ± 0.059	0.936 ± 0.032
t = 40	n/a	1.226 ± 0.088	0.996 ± 0.041	n/a	1.192 ± 0.044	0.974 ± 0.039
Ostreococcus sp.						
t = 0	0.079 ± 0.030	0.057 ± 0.024	0.061 ± 0.014	0.079 ± 0.030	0.057 ± 0.024	0.061 ± 0.014
t = 10	0.109 ± 0.017	0.062 ± 0.026	0.084 ± 0.032	0.058 ± 0.004	0.053 ± 0.012	0.060 ± 0.013
t = 20	0.091 ± 0.036	0.052 ± 0.020	0.060 ± 0.009	0.109 ± 0.003	0.082 ± 0.002	0.105 ± 0.008
t = 40	0.106 ± 0.005	0.078 ± 0.006	0.074 ± 0.031	0.109 ± 0.003	0.079 ± 0.014	0.077 ± 0.020
Pelagomonas calceolata						
t = 0	0.089 ± 0.008	0.089 ± 0.005	0.106 ± 0.012	0.089 ± 0.008	0.089 ± 0.005	0.106 ± 0.012
t = 10	0.096 ± 0.009	0.095 ± 0.002	0.106 ± 0.013	0.092 ± 0.010	0.094 ± 0.004	0.103 ± 0.014
t = 20	0.093 ± 0.005	0.096 ± 0.007	0.107 ± 0.014	0.080 ± 0.031	0.093 ± 0.005	0.105 ± 0.014
t = 40	0.093 ± 0.009	0.094 ± 0.006	0.109 ± 0.013	0.090 ± 0.008	0.092 ± 0.005	0.104 ± 0.012

Table 3. Mean (± standard deviations, *n* = 3) photoprotective pigments per chlorophyll *a* ratio in *Prochlorococcus marinus* eMED4 (zeaxanthin), *Prochlorococcus* sp. eMIT9313 (zeaxanthin), *Ostreococcus* sp. clade B (violaxanthin, antheraxanthin, and zeaxanthin), and *Pelagomonas calceolata* (diadinoxanthin and diatoxanthin) acclimated to 16 °C, 20 °C, and 24 °C. Pigment ratios were obtained before (t = 0) and after (t = 10, 20, 40) exposure to high photosynthetically active radiation (PAR), with and without ultraviolet radiation (UVR). *abc* indicate significant effects of the temperature treatment within each species. n/a: data not available, growth was not observed under the used conditions and no additional measurements were performed.

3.4. De-epoxidation of the xanthophyll cycle

In both *Ostreococcus* sp. and *P. calceolata*, the de-epoxidation state (DPS) of the xanthophyll pigment cycle increased significantly during exposure to high irradiance ($p < 0.001$) (Figure 3). In both strains, the DPS of the xanthophyll pigment cycle decreased over time, but the DPS did not return to initial values after 30 min of recovery in low light conditions (t= 40, Figure 3). In *Ostreococcus* sp., the de-epoxidation of the xanthophyll pigment cycle mainly included the de-epoxidation of violaxanthin to antheraxanthin, whereas the de-epoxidation of antheraxanthin to zeaxanthin was small. Temperature had an effect on the DPS of the xanthophyll

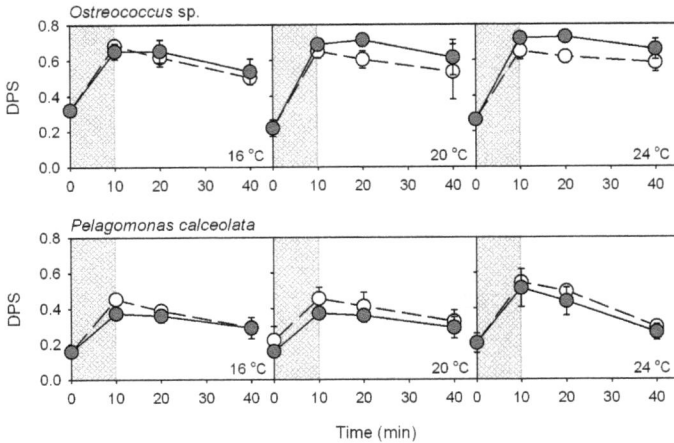

Figure 3. De-epoxidation of the xanthophyll pigment cycle. Mean (± standard deviation, $n = 3$) de-epoxidation state (DPS) of the xanthophyll pigment cycle in *Ostreococcus* sp. clade B and *Pelagomonas calceolata* are given during and after 10 minutes of exposure to high photophotosynthetically active radiation (PAR, white circles) and high PAR with ultraviolet radiation (PAR+UVR, grey circles) at 16 °C, 20 °C, and 24 °C.

pigment cycle in *Ostreococcus* sp. (Figure 3, Table 4), but differences were mostly not significant. The initial DPS of the xanthophyll pigment cycle ($t = 0$) in *Ostreococcus* sp. was 21-47% higher at 16 °C compared with 20 °C and 24 °C. During exposure to high PAR and PAR+UVR, the increase in the DPS was fastest at 20 °C (Table 4), as was the epoxidation of the xanthophyll pigment cycle after exposure to high irradiance (Figure 3). In *P. calceolata*, the initial DPS of the xanthophyll pigment cycle was 22-28% lower at 16 °C compared with the higher temperatures (not significant) (Figure 3). During irradiance exposure, the rate of de-epoxidation of the xanthophyll pigment cycle increased with increasing temperature in *P. calceolata* (not significant) (Figure 3, Table 4). In accordance with the rate of de-epoxidation, the epoxidation of the xanthophyll pigment cycle was fastest at 24 °C ($p < 0.05$).

The effect of the spectral composition of the irradiance treatment on the de-epoxidation of the xanthophyll pigment cycle was most evident in *Ostreococcus* sp. (Figure 3). During irradiance exposure ($t = 0$-10), the DPS in *Ostreococcus* sp. did not differ significantly between the PAR and PAR+UVR treatment (Figure 3, Table 4). However, in the PAR treatment, epoxidation of the xanthophyll pigment cycle started directly after exposure ($t = 10$), whereas the epoxidation was delayed in the PAR+UVR treatment and started after 10 minutes of recovery in low light ($t = 20$). After 30 minutes of recovery ($t = 40$), the DPS in *Ostreococcus* sp. was similar in both PAR and PAR+UVR treatments (Figure 3). In *P. calceolata*, no significant effect of the spectral composition of the irradiance treatment was found, but it seemed that exposure to UVR limited the de-epoxidation of the xanthophyll pigment cycle, especially at lower temperatures (Figure 3).

When the dynamics of the xanthophyll pigment cycle of both species were compared, it was shown that *Ostreococcus* sp. had a significantly higher DPS compared with *P. calceolata* ($p <$ 0.05) (Figure 3). In addition, the increase in de-epoxidation of the xanthophyll pigment cycle during high irradiance exposure was faster in *Ostreococcus* sp. ($p < 0.05$) (Table 3), whereas no differences in epoxidation rate were observed between *Ostreococcus* sp. and *P. calceolata*.

		Ostreococcus sp.	*Pelagomonas calceolata*
PAR			
	16 °C	$0.036 \pm 3.75 \times 10^{-3}$	$0.029 \pm 2.76 \times 10^{-4}$
	20 °C	$0.043 \pm 6.01 \times 10^{-3}$	$0.030 \pm 4.53 \times 10^{-3}$
	24 °C	$0.038 \pm 6.33 \times 10^{-3}$	$0.034 \pm 5.39 \times 10^{-3}$
PAR+UVR			
	16 °C	$0.033 \pm 4.79 \times 10^{-3ab}$	$0.021 \pm 6.38 \times 10^{-4}$
	20 °C	$0.047 \pm 4.08 \times 10^{-3a}$	$0.023 \pm 5.33 \times 10^{-3}$
	24 °C	$0.045 \pm 1.12 \times 10^{-3b}$	$0.031 \pm 7.34 \times 10^{-3}$

Table 4. Mean (± standard deviation, $n = 3$) rate of increase in the de-epoxidation state of the xanthophyll pigment cycle (k_{DPS} in min⁻¹) in *Ostreococcus* sp. clade B and *Pelagomonas calceolata* during exposure to high photosynthetically active radiation (PAR) and high PAR with ultraviolet radiation (PAR+UVR) at 16 °C, 20 °C, and 24°C. *abc* indicate significant effects of the temperature treatment within each species.

4. Discussion

Climate change is expected to mediate a rise in seawater temperature by 1.5-4.5 °C over the next century [46]. This rise in seawater temperature will lead to changes in water column stratification in open oligotrophic oceans [47,48]. The subsequent modifications in mixed layer dynamics increase the exposure of phytoplankton to high levels of photosynthetic active radiation (PAR) and ultraviolet radiation (UVR). Because temperature and irradiance conditions play an important role in the success of specific oceanic phytoplankton species [4,49,50], it is important to understand how oceanic phytoplankton will respond to elevated temperatures and whether this will affect their (photo)physiological performance. The present study focused on the temperature-dependence of photoinhibition and photoregulating processes that are essential for survival during high (dynamic) irradiance conditions.

During short periods of high irradiance exposure, both the prokaryotic picophytoplankton strains *P. marinus* and *Prochlorococcus* sp., as the eukaryotic picophytoplankton strains *Ostreococcus* sp. and *P. calceolata* were susceptible to photoinhibition. The response to high irradiances was species specific and appeared to be related to the genetically defined light adaptation of the different strains. In the prokaryotic species, the low light adapted ecotype *Prochlorococcus* sp. (eMIT9313) was highly sensitive to high PAR and UVR, whereas the high light adapted ecotype *P. marinus* (eMED4) showed lower sensitivity. Similar differences in photoinhibition during high irradiance exposure were observed for other low and high light adapted ecotypes of *Prochlorococcus* spp. during exposure to high blue irradiance [18]. The differential response to excessive irradiance intensities found in the present study related well to the occurrence of different *Prochlorococcus* ecotypes in the upper mixed layer (eMED4) and the deep chlorophyll maximum (eMIT9313) [4,49]. In the eukaryotic species, the levels of total non photochemical quenching induced by a tenfold increase in irradiance intensity were similar compared with earlier observations for *Ostreococcus* sp. and *P. calceolata* [12,51]. Although the two eukaryotic species were both isolated at 100 m depth from oceanic regions, *Ostreococcus* sp. showed considerably lower levels of photoinhibition compared with *P. calceolata*, especially during UVR exposure. It therefore seems that *Ostreococcus* sp. clade B is not specifically adapted to low light [7], but rather adapted to open ocean irradiance conditions (also see [50,52]) with a relatively low sensitivity to high irradiance intensities compared with other oceanic picophytoplankton [this study, 11,31]. The low light adapted ecotype *P. calceolata* showed highest levels of photoinhibition during exposure to high PAR compared with the other species. However, photoinhibition increased dramatically in the prokaryotic strains during exposure to UVR. This confirms the relative sensitivity of *Prochlorococcus* spp. to high levels of UVR, as has been observed in oligotrophic waters [53,54].

Temperature acclimation influenced photoinhibition and related processes during high irradiance exposure in *P. marinus*, *Prochlorococcus* sp., *Ostreococcus* sp., and *P. calceolata*. The effect was not uniform among the different strains, but temperature acclimation influenced the response to high irradiance exposure by changes in the relative contribution of photoinhibition and photoprotective mechanisms to non photochemical quenching in all strains. This general response corresponds well with the observation that both prokaryotic and eukaryotic picophytoplankton may benefit from high irradiance intensities at elevated temperatures by alterations in photophysiology and electron transport [31]. In addition, elevated temperatures had a beneficial effect on the response to high irradiance intensities by partially counteracting the UVR-induced photoinhibition in *P. marinus*, *Ostreococcus* sp., and *P. calceolata*. This was earlier observed in several diatom species and related to an increase in Rubisco activity and gene expression in *Thalassiosira weissflogii* [34], an increase in repair rates in *T. pseudonana* [32], and an increase in photoprotection by the dissipation of excess energy in *T. weissflogii* and *C. gracillis* [33]. In this study, fast relaxing non photochemical processes, i.e. photoprotection, and the influence of temperature acclimation on these processes was further investigated in the response to excessive irradiance intensities in oceanic picophytoplankton.

Both low and high light adapted *Prochlorococcus* strains were capable of producing fast relaxing non photochemical quenching (NPQ_F). Interestingly, the level of NPQ_F in the low light adapted strain *Prochlorococcus* sp. (eMIT9313/clade LLIV) was considerably higher compared with that of another low light adapted strain of *Prochlorococcus* (strain SS120/clade LLII) [27]. It therefore seems that some low light adapted ecotypes of *Prochlorococcus* are capable of inducing high levels of NPQ_F comparable to that of high light adapted ecotypes (this study), but others are not [27]. This might possibly be related to the differential occurrence of *pcb* genes encoding the major chlorophyll binding and light harvesting antenna proteins in both low and high light adapted ecotypes of *Prochlorococcus* [27,55]. Although the precise underlying mechanism remains unknown, the process of NPQ_F in *P. marinus* and *Prochlorococcus* sp. was sensitive to changes in temperature. It is therefore likely that the underlying mechanisms of NPQ_F in *Prochlorococcus* spp. involves an enzymatic reaction or changes due to the improved fluidity of the thylakoid membrane at elevated temperatures [56,57]. This contrasts to earlier observations of NPQ_F in phycobillisome containing cyanobacteria [58] (for a review see [24]), which supports the notion that the underlying mechanisms are different between *Prochlorococcus* spp. and other prokaryotic species [24]. It was further shown in the present study that the mechanism of photoprotection in *P. marinus* and *Prochlorococcus* sp. was highly sensitive to UVR, possibly related to increased oxidative stress on the thylakoid membrane [59]. Fast relaxing non photochemical quenching was not related to changes in pigmentation during high irradiance exposure in *P. marinus* and *Prochlorococcus* sp. The xanthophyll pigment zeaxanthin is not regulated by an epoxydation/de-epoxidation cycle in prokaryotic species and its function is often debated [60,61]. However, the photoprotective role of zeaxanthin is not excluded, since the concentration of zeaxanthin increases relative to chlorophyll *a* in high light acclimated cells [8,11,61] and zeaxanthin is found in high concentrations in the field [62,63]. The presence of zeaxanthin might have overestimated the calculation of photoinhibition by slowly relaxing non photochemical quenching in the light-harvesting antenna of PSII (F_0 quenching) [40,64]. This was however, not observed in *P. marinus* and *Prochlorococcus* sp. (data not shown), suggesting that slowly relaxing non photochemical quenching related to damage to the reaction center of PSII in these strains.

In the eukaryotic picophytoplankton species *Ostreococcus* sp. and *P. calceolata*, fast relaxing non photochemical quenching coincided with the de-epoxidation of the xanthophyll pigment cycle. The rate of de-epoxidation of the xanthophyll pigment cycle in *Ostreococcus* sp. and *P. calceolata* was within the range reported for other eukaryotic pico- and nanophytoplankton [45], as was the relative increase in the de-epoxidation state of the xanthophyll pigment cycle upon high irradiance exposure [12,19,45,51]. For *Ostreococcus* sp. clade B it was previously shown that both the xanthophyll pigment cycle [19] and alternative electron transport [13] play an important role in the response to high irradiance, whereas photorepair is relatively slow compared with other *Ostreococcus* ecotypes [19]. This study showed that the photoprotective processes were also effective during UVR exposure, since *Ostreococcus* sp. was the only strain used in this study that showed substantial NPQ_F during UVR exposure. The influence of temperature acclimation was also most pronounced during UVR exposure, especially on the xanthophyll pigment cycle. Different effects may add to the high levels of fast relaxing non photochemical quenching observed in *Ostreococcus* sp. The xanthophyll cycle pigments may have an additional photoprotective function in *Ostreococcus* sp., including the stabilization of the thylakoid membrane by antheraxanthin and zeaxanthin, providing protection against reactive oxygen species under conditions of a highly reduced electron transport chain (for a review see [65]). In addition, the de-epoxidation of the xanthophyll pigment cycle and the consequent non photochemical quenching in *Ostreococcus* sp. may be promoted by an increase in the trans-membrane proton gradient due to the presence of chlororespiratory electron flow [13,65]. In *P. calceolata*, the rate of de-epoxidation and the relative de-epoxidation of the xanthophyll pigment cycle increased at elevated temperature, but this was not associated with an increase in fast relaxing non photochemical quenching. It is possible that the membrane stability necessary for the dissipation of excess energy trough the xanthophyll pigment cycle was affected by oxidative stress [66,67]. This could also explain the diminished fast relaxing non photochemical levels during UVR exposure in this species. Because *P. calceolata* is a low light adapted ecotype, this species might possibly use additional photoprotective mechanisms, such as the chlororespiratory electron flow observed in *Ostreococcus* sp., to a lesser extent.

This study showed that oceanic picophytoplankton were susceptible to photoinhibition during short periods of high irradiance. The genetically defined light adaptation of *P. marinus*, *Prochlorococcus* sp., *Ostreococcus* sp., and *P. calceolata* played an important role in their PAR and UVR sensitivity, likely related to the presence of different (combinations of) photoprotective mechanisms. Temperature acclimation influenced the response to excessive irradiance exposure by changes in the relative contribution of photoinhibition and photoprotective mechanisms to non photochemical quenching. These changes were found to be species specific. Acclimation to elevated temperatures increased the dissipation of excess energy in both *P. marinus* and *Prochlorococcus* sp., indicating a strong dependence on temperature of this photoprotective mechanism. In combination with decreased photoinhibition during both PAR and UVR exposure at elevated temperature, the high light adapted ecotype *P. marinus* may benefit considerably from elevated temperatures in response to high irradiance intensities encountered in the upper mixed layer of open oligotrophic oceans. Considering exposure to UVR, the effect of elevated temperature was most pronounced in the eukaryotic strain *Ostreococcus* sp., indicating that this species can effectively regulate light harvesting in relatively warm, UVR rich waters near the surface of the open oligotrophic ocean. Even though *Prochlorococcus* sp. and *P. calceolata* are unlikely to experience high irradiance intensities in the deep chlorophyll maximum, photoinhibition in these low light adapted ecotypes is highly relevant, since damage to PSII can occur at relatively low irradiance intensities [18,31,68]. At elevated temperatures, the prokaryotic strain *Prochlorococcus* sp. benefitted by increasing dissipation of excess energy, whereas the eukaryotic strain *P. calceolata* was less susceptible to photoinhibition. Overall, the differential response to high irradiance may have considerably effect on phytoplankton species distribution and community composition in the open oligotrophic oceans, with some ecotypes and/or species being more susceptible to photoinhibition than others. Photoinhibition and/or photoprotective processes may be positively affected by the rise in seawater temperature associated with climate change, but species specific differences in (photo)physiology remain important in the performance of oceanic picophytoplankton.

Acknowledgements

Remote access to the Roscoff Culture Collection of strains RCC407 and RCC879 was facilitated by ASSEMBLE grant number 227799 (GK). This work was supported by the Netherlands

Organization for Scientific Research (NWO), grant numbers 817.01.009 (GK) and 839.08.422 (WHP).

Author details

Gemma Kulk[1*], Pablo de Vries[1], Willem H. van de Poll[2], Ronald J. W. Visser[1] and Anita G. J. Buma[1]

*Address all correspondence to: g.kulk@rug.nl

1 Department of Ocean Ecosystems, Energy and Sustainability Research Institute Groningen, University of Groningen, Groningen, The Netherlands

2 Department of Biological Oceanography, Royal Netherlands Institute for Sea Research, Den Burg, The Netherlands

References

[1] Li WKW. Primary production of prochlorophytes, cyanobacteria, and eukaryotic ultraphytoplankton - measurements from flow cytometric sorting. Limnology and Oceanography 1994;39 169-175.

[2] DuRand MD, Olson RJ, Chisholm SW. Phytoplankton population dynamics at the Bermuda Atlantic Time-series station in the Sargasso Sea. Deep-Sea Research Part II 2001;48 1983-2003.

[3] Worden AZ, Nolan JK, Palenik B. Assessing the dynamics and ecology of marine picophytoplankton: the importance of the eukaryotic component. Limnology and Oceanography 2004;49 168-179.

[4] Moore LR, Rocap G, Chisholm SW. Physiology and molecular phylogeny of coexisting *Prochlorococcus* ecotypes. Nature 1998;393 464-467.

[5] Moore LR, Chisholm SW. Photophysiology of the marine cyanobacterium *Prochlorococcus*: ecotypic differences among cultured isolates. Limnology and Oceanography 1999;44 628-638.

[6] Fuller NJ, Marie D, Partensky F, Vaulot D, Post AF, Scanlan DJ. Clade-specific 16S ribosomal DNA oligonucleotides reveal the predominance of a single marine *Synechococcus* clade throughout a stratified water column in the Red Sea. Applied and Environmental Microbiology 2003;69 2430-2443.

[7] Rodriguez F, Derelle E, Guillou L, Le Gall F, Vaulot D, Moreau H. Ecotype diversity in the marine picoeukaryote *Ostreococcus* (Chlorophyta, Prasinophyceae). Environmental Microbiology 2005;7 853-859.

[8] Moore LR, Goericke R, Chisholm SW. Comparative physiology of *Synechococcus* and *Prochlorococcus*: influence of light and temperature on growth, pigments, fluorescence and absorptive properties. Marine Ecology Progress Series 1995;116, 259-275.

[9] Barlow RG, Alberti RS. Photosynthetic characteristics of phycoerythrin-containing marine *Synechococcus* spp. I. Response to growth photon flux density. Marine Biology 1985;86 63-74.

[10] Six C, Thomas J-C, Garczarek L, Ostrowski M, Dufresne A, Blot N, Scanland DJ, Partensky F. Diversity and evolution of phycobilisomes in marine *Synechococcus* spp.: a comparative genomics study. Genome Biology 2007b;8 R259.

[11] Kulk G, Van de Poll WH, Visser RJW, Buma AGJ. Distinct differences in photoacclimation potential between prokaryotic and eukaryotic oceanic phytoplankton. Journal of Experimental Marine Biology and Ecology 2011;398 63-72.

[12] Six C, Finkel ZV, Rodriguez F, Marie D, Partensky F, Campbell DA. Contrasting photoacclimation costs in ecotypes of the marine eukaryotic picoplankter *Ostreococcus*. Limnology and Oceanography 2008;53 255-265.

[13] Cardol P, Bailleul B, Rappaport F, Derelle E, Béal D, Breyton C, Bailey S, Wollman FA, Grossman AR, Moreau H, Finazzi G. Original adaptation of photosynthesis in the green alga *Ostreococcus*. Proceedings of the National Academy of Sciences of the United States of America 2008;105 7881-7886.

[14] Kirk TO, editor. Light and Photosynthesis in Aquatic Ecosystems, 3rd ed. Cambridge: Cambridge University Press; 2010.

[15] Boelen P, Obernosterer I, Vink AA, Buma AGJ. Attenuation of biologically effective UV radiation in tropical Atlantic waters measured with a biochemical DNA dosimeter. Photochemistry and Photobiology 1999;69 34-40.

[16] Dishon G, Dubinsky Z, Caras T, Rahav E, Bar-Zeev E, Tzubery Y, Iluz D. Optical habitats of ultraphytoplankton groups in the Gulf of Eilat (Aqaba), Northern Red Sea. International Journal of Remote Sensing 2012;33 2683-2705

[17] Agustí S, Llabrés M. Solar radiation-induced mortality of marine pico-phytoplankton in the oligotrophic ocean. Photochemistry and Photobiology 2007;83 793-801.

[18] Six C, Finkel ZV, Irwin AJ, Campbell DA. Light variability illuminates niche-partitioning among marine picocyanobacteria. PLoS One 2007a;12 1-6

[19] Six C, Sherrard R, Lionard M, Roy S, Campbell DA. Photosystem II and pigment dynamics among ecotypes the green alga *Ostreococcus*. Plant Physiology 2009;151 379-390.

[20] Aro E-M, Virgin I, Andersson B. Photoinhibition of photosystem II. Inactivation, protein damage and turnover. Biochimica et Biophysica Acta 1993;1143 113-134.

[21] Gechev TS, Van Breusegem F, Stone JM, Denev I, Laloit C. Reactive oxygen species as signals that modulate plant stress responses and programmed cell death. BioEssays 2006;28 1091-1101.

[22] Van de Poll WH, Alderkamp A-C, Janknegt PJ, Roggeveld J, Buma AGJ. Photoacclimation modulates excessive photosynthetically active and ultraviolet radiation effects in a temperate and an Antarctic marine diatom. Limnology and Oceanography 2006;51 1329-1248.

[23] Campbell D, Hurry V, Clarke AK, Gustafsson P, Öquist G. Chlorophyll fluorescence analysis of cyanobacterial photosynthesis and acclimation. Microbiology and Molecular Biology Reviews 1998;62 667-683.

[24] Bailey S, Grossman AR. Photoprotection in cyanobacteria: regulation of light harvesting. Photochemistry and Photobiology 2008;84 1410-1420.

[25] Olaizola M, La Roche J, Kolber Z, Falkowski PG. Non-photochemical fluorescence quenching and the diadinoxanthin cycle in a marine diatom. Photosynthesis Research 1994;392 585-589.

[26] Wilson A, Ajlani G, Verbavatz J-M, Vass I, Kerfeld CA, Kirilovsky D. A soluble carotenoid protein involved in phycobilisome-related energy dissipation in cyanobacteria. Plant Cell 2006;18 992-1007.

[27] Bailey S, Mann NH, Robinson C, Scanlan DJ. The occurrence of rapidly reversible non-photochemical quenching of chlorophyll a fluorescence in cyanobacteria. FEBS Letters 2005;79 275-280.

[28] Häder D-P, Kumar HD, Smith RC, Worrest RC. Effects of solar UVR radiation on aquatic ecosystems and interactions with climate change. Photochemical and Photobiological Sciences 2007;6 267-285.

[29] Raven JA. The cost of photoinhibition. Physiologia Plantarum 2011;142 87-104.

[30] Alonso-Laita P, Agustí S. Contrasting patterns of phytoplankton viability in the subtropical NE Atlantic Ocean. Aquatic Microbial Ecology 2006;43 67-78.

[31] Kulk G, De Vries P, Van de Poll WH, Visser RJW, Buma AGJ. Temperature-dependent growth and photophysiology of prokaryotic and eukaryotic oceanic picophytoplankton. Marine Ecology Progress Series 2012;466 43-55.

[32] Sobrino C, Neale PJ. Short-term and long-term effects of temperature on photosynthesis in the diatom *Thalassiosira pseudonana* under UVR exposures. Journal of Phycology 2007;43 426-436.

[33] Halac SR, Villafañe VE, Helbling EW. Temperature benefits the photosynthetic per-
      formance of the diatoms *Chaetoceros gracilis* and *Thalassiosira weissflogii* when exposed
      to UVR. J. Photochemistry and Photobiology B 2010;101 196-205.

[34] Helbling EW, Buma AGJ, Boelen P, Van der Strate HJ, Giordanino MVF, Villafañe
      VE. Increase in Rubisco activity and gene expression due to elevated temperature
      partially counteracts ultraviolet radiation-induced photoinhibition in the marine dia-
      tom *Thalassiosira weissflogii*. Limnology and Oceanography 2011;56 1330-1342.

[35] Demmig-Adams B, Adams WW. Photoprotection and other responses of plants to
      high light stress. Annual Review of Plant Physiology and Plant Molecular Biology
      1992;43 599-626.

[36] Bouchard JN, Roy S, Campbell DA. UVB Effects on the photosystem II-D1 protein of
      phytoplankton and natural phytoplankton communities. Photochemistry and Photo-
      biology 2006;82: 936-951.

[37] Chisholm SW. What limits phytoplankton growth. Oceanus 1992; 35 36-46.

[38] Keller MD, Selvin RC, Claus W, Guillard RRL. Media for the culture of oceanic ultra-
      phytoplankton. Journal of Phycology 1987;23 633-638.

[39] Van de Poll WH, Visser RJW, Buma AGJ. Acclimation to a dynamic irradiance re-
      gime changes excessive irradiance sensitivity of *Emiliania huxleyi* and *Thalassiosira
      weissflogii*. Limnology and Oceanography 2007;52 1430-1438.

[40] Maxwell K, Johnson GN. Chlorophyll fluorescence – a practical guide. Journal of Ex-
      perimental Botany 2000;51 659-668.

[41] Walters RG, Horton P. Resolution of components of non-photochemical chlorophyll
      fluorescence quenching in barley leaves. Photosynthesis Research 1991;27 121-133.

[42] Osmond CB. What is photoinhibition? Some insights from comparisons of sun and
      shade plants. In Baker NR, Bowyer JR. (eds) Photoinhibition of photosynthesis: from
      molecular mechanisms to the field. Oxford: Bios Scientific Publishers; 1994. p1-24.

[43] Villafañe V, Marcoval MA, Helbling EW. Photosynthesis versus irradiance character-
      istics in phytoplankton assemblages off Patagonia (Argentina); temporal variability
      and solar UVR effects. Marine Ecology Progress Series 2004;284 23-34.

[44] Hooker SB, Van Heukelem L, Thomas CS, Claustre H, Ras J, Schlüter L, Clementson
      L, Van der Linde D, Eker-Develi E, Berthon J-F, Barlow R, Sessions H, Ismail H, Perl
      J. The third SeaWiFS HPLC Analysis Round-Robin Experiment (SeaHARRE-3).
      NASA Tech. Memo 2009-215849, NASA Goddard space flight center, Greenbelt,
      Maryland 20771; 2009.

[45] Dimier C, Giovanni S, Ferdinando T, Brunet C. Comparative ecophysiology of the
      xanthophyll cycle in six marine phytoplankton species. Protist 2009a;160 397-411.

[46] Houghton JT, Meir Filho LG, Callander BA, Harris N, Kattenberg A, Maskell K, editors. Climate change 1995: the science of climate change. Cambridge: Cambridge University Press; 1995.

[47] Behrenfeld MJ, O'Malley RT, Siegel DA, McClain CR, Sarmiento JL, Feldman GC, Milligan AJ, Falkowski PG, Letelier RM, Boss ES. Climate-driven trends in contemporary ocean productivity. Nature 2006;444 752-755.

[48] Polovina JJ, Howell EA, Abecassis M. Ocean's least productive waters are expanding. Geophysical Research Letters. 2008;35 L03618.

[49] Johnson ZI, Zinser ER, Coe A, McNulty NP, Malcolm E, Woodward S, Chisholm SW. Niche partitioning among *Prochlorococcus* ecotypes along ocean scale environmental gradients. Science 2006;311 1737-1740.

[50] Demir-Hilton E, Sudek S, Cuvelier ML, Gentemann CL, Zehr JP, Worden AZ. Global distribution patterns of distinct clades of photosynthetic picoeukaryote *Ostreococcus*. The International Society for Microbial Ecology Journal 2011;5 1095-1107.

[51] Dimier C, Brunet C, Geider R, Raven J. Growth and photoregulation dynamics of the picoeukaryote *Pelagomonas calceolata* in fluctuating light. Limnology and Oceanography 2009b;54 823-836.

[52] Worden AZ. Picoeukaryote diversity in coastal waters of Pacific Ocean. Aquatic Microbial Ecology 2006;43 165-175.

[53] Llabrés M, Agustí S. Picophytoplankton cell death induced by UV radiation: Evidence for oceanic Atlantic communities. Limnology and Oceanography 2006;51: 21-29.

[54] Sommaruga R, Hofer JS, Alonso-Sáez L, Gasol JM. Differential sunlight sensitivity of picophytoplankton from surface Mediterranean coastal water. Applied and Environmental Microbiology 2005;71 2154-2157.

[55] Bibby TS, Mary I, Nield J, Partensky F, Barber J. Low-light-adapted *Prochlorococcus* species possess specific antennae for each photosystem. Nature 2003;424 1051-1054.

[56] Geider RJ. Light and temperature dependence of the carbon to chlorophyll a ratio in microalgae and cyanobacteria: implications for physiology and growth of phytoplankton. New Phytologist 1987;106 1-34.

[57] Davison IR. Environmental effects on algal photosynthesis: temperature. Journal of Phycology 1991;27 2-8.

[58] El Bissati K, Delphin E, Murata N, Etienne AL, Kirilovsky D. Photosystem II fluorescence quenching in the cyanobacterium *Synechocystis* PCC 6803: involvement of two different mechanisms. Biochimica et Biophysica Acta: Bioenergetics 2000;1457 229-242.

[59] Lesser MP. Oxidative stress in marine environments: biochemistry and physiological ecology. Annual Reviews of Physiology 2006;68 253-278.

[60] Siefermann-Harms D. Carotenoids in photosynthesis 1. Location in photosynthetic membranes and light-harvesting function. Biochimica et Biophysica Acta 1985;811 325-355

[61] Partensky F, Hoepffner N, Li WKW, Ulloa O, Vaulot D. Photoacclimation of *Prochlorococcus* sp. (Prochlorophyta) strains isolated from the North Atlantic and the Mediterranean Sea. Plant Physiology 1993;101 285-296.

[62] Letelier RM, Bidigare RR, Hebel DV, Ondrusek M, Winn CD, Karl DM. Temporal variability of phytoplankton community structure based on pigment analysis. Limnology and Oceanography 1993;38 1420-1437.

[63] Claustre H. The trophic status of various provinces as revealed by phytoplankton pigment signatures. Limnology and Oceanography 1994;39 1206-1210.

[64] Horton P, Ruban AV, Walters RG. Regulation of light harvesting in green plants. Annual Review of Plant Physiology and Plant Molecular Biology 1996;47 655-684.

[65] Goss R, Jakob T. Regulation and function of xanthophyll cycle-dependent photoprotection in algae. Photosynthesis Research 2010;106 103-122.

[66] Van de Poll WH, Buma AGJ. Does ultraviolet radiation affect the xanthophyll cycle in marine phytoplankton? Photochemical and Photobiological Sciences 2009;8: 1295-1301.

[67] Rijstenbil JW. UV- and salinity-induced oxidative effects in the marine diatom *Cylindrotheca closterium* during simulated emersion. Marine Biology 2005;147 1063-1073.

[68] Mackey KRM, Payton A, Grossman AR, Bailey S. A photosynthetic strategy for coping in a high-light, low-nutrient environment. Limnology and Oceanography 2008;53 900-913.

Chapter 8

Effect of Abiotic Stress on Photosystem I-Related Gene Transcription in Photosynthetic Organisms

Dilek Unal Ozakca

Additional information is available at the end of the chapter

1. Introduction

Photosystem I (PSI) from cyanobacteria and chloroplasts is a multisubunit membrane-protein complex that catalyses electron transfer from reduced plastocyanin in the thylakoid lumen to oxidized ferredoxin in the chloroplast stroma or cyanobacterial cytoplasm [1-4]. PSI is responsible for NADP$^+$ reduction and cyclic photophosphorylation and consists of at least 8 polypeptides. Its major components are the P700 chlorophyll a A1 and A2 apoproteins whose molecular weights vary between 60 and 70 kd, depending on the species [5]. In this chapter, we discuss recent progress on several topics related to the functions of the PSI complex, like the protein composition of the complex in the plant and algae, the structure and organization of the PSI subunits and the regulation of photosystem I-related gene under abiotic stress conditions. Furthermore, PSI seems to be well protected from photoinhibition in vivo in many plant and algae species and many environmental conditions. The physiology and molecular mechanism during short term adaptation to changes under oxidative stress is discussed in functional and structural terms. Finally, such characteristics of PSI photoinhibition with special emphasis on the relationship between two photosystems as well as the protective mechanism of PSI in vivo is reviewed with respect to function of the thylakoid membrane.

2. Structure of photosystem I

PSI catalyzes the light-driven electron transfer from the soluble electron carrier plastocyanin on the luminal side of thylakoid membrane, to ferredoxin on the stromal side of thylakoid membrane. In plants, the PSI complex consists of at least 19 protein subunits, approximately 175 chlorophyll molecules, 2 phylloquinones and 3 Fe_4S_4 clusters [6]. The crystal structure of

PSI from *Synechococcus elongatus* has also been recently determined at a resolution of less than 4 Å which is sufficient for distinguishing 31 transmembrane and 14 parallel helices and the three ironsulfur clusters [7].

The PSI complex of most plants and algae consists of 13 subunits: at least five chloroplast-encoded subunits (PsaA, PsaB, PsaC, PsaI and PsaJ) and eight nucleusencoded subunits (PsaD, PsaE, PsaF, PsaG, PsaH, PsaK, PsaL, PsaN) and numerous redox cofactors and antenna chlorophylls [8]. An additional subunit, PsaM, has only been found in cyanobacteria and in the chloroplast genomes of some lower plants and algae. The PSI subunits PsaG, PsaH and PsaN are only found in eukaryotic photosynthetic organisms and are missing in cyanobacteria[8,9,10].

2.1. Subunits of PSI

2.1.1. PsaA and PsaB

The major subunits of photosystem I, *PsaA* and *PsaB*, show strong sequence homology [10, 11] and it was suggested that they have been evolved via gene duplication [12]. They are from the central heterodimer holding the reaction center p700 and components of the electron transport chain (ETC), A_o (a CHLa), A_1 (phylloquinon) and F_x (a (4Fe-4S) cluster). In addition, the heterodimer coordinates 80 chlorophylls that function as the instric light-harvesting antenna [13]

Previous studies showed that mutants deficient in PsaB are unable to synthesize both PsaB and PsaA whereas mutants affected primarily in PsaA synthesis are still able to produce PsaB [2,3,11]. Based on these results it was proposed that PsaB is an anchor protein during PSI assembly, which needs to be synthesized and integrated into the thylakoid membrane before the other PSI subunits are synthesized [12]. In its absence these polypeptides are no longer synthesized and/or are rapidly degraded. Elucidating how PsaB is translated and inserted into the thylakoid membrane is thus important for understanding the initial steps of PSI assembly.

2.1.2. PsaC, PsaD and PsaE

The subunits PsaC, PsaD and PsaE do not contain transmembrane α-helices [3,2,14]. They are located on the stromal side of the complex, forming the stromal hump. They are in close contact to the stromal loop regions of PsaA and PsaB. Subunit PsaC carries the two terminal FeS clusters FA and FB, and is located in the central part of the stromal hump. PsaD forms the part of this hump, which is closest to the trimeric axis [15,2,3]. The C-terminal part of PsaD forms a 'clamp' surrounding PsaC. PsaE is located on the side of the hump, which is distal from the trimer axis[16,17].

The clusters of PsaC are characterised by their distinct electron paramagnetic resonance (EPR) spectra [18,19]. PsaC is likely to posses a pseudo-C_2 symmetry axis that is oriented perpendicular to distance vector connecting the two iron-sulfur clusters, F_A and F_B. The role of subunit PsaC in coordination of the two terminal FeS clusters was suggested from the conserved sequence motif CXXCXXCXXXCP which is found twice in the gene of PsaC [3,20]. A homology of subunit PsaC to bacterial ferredoxins, also containing two [4Fe-4S] clusters, was proposed from sequence similarity [21]. The structures of PsaC and these ferredoxins, such as that from *Peptostreptococcus asacharolyticus* [3,21,22], are indeed similar, with pronounced pseudo-two-fold symmetry, concerning previous models which related the structure of subunit PsaC to this ferredoxin [3,2,21].

PsaD is a peripheral subunit of photosystem I (PSI1), an integral protein complex in the thylakoid membrane of oxygenic photosynthetic organisms. Biochemical experiments [20, 21] and analyses of the primary structure of PsaD suggest that it does not posses a transmembranal segment and that it faces the stromal side of the thylakoid membrane. The PsaD is a polypeptide of 139-144 amino acids in cyanobacteria, but has an N-terminal extension of several residues in higher plants, yielding a total length of 158-162 residues [20]. Topological studies [3, 2, 19, 22, 23] and data from an X-ray structure of PSI at 4 Å [24, 25]show that PsaD probably contains an R-helix and is in contact principally with PsaC and PsaE, and also with PsaH and PsaL [25, 26, 27]. The three-dimensional structure of the higher-plant PSI as determined by electron crystallography has been recently reported [27], confirming that the stromal ridge of higher-plant PSI can also be interpreted as being due to the PsaC, -D, and -E subunits. The N-terminal part of the PsaD subunit can be accessed by the proteases, and its C-terminal region is exposed to solvent [23, 14]. Comparison of the amino acid sequences of PsaD from several species shows that the C-terminal part is highly conserved, especially in a region containing many basic residues[23].

In spinach, PsaD is synthesized in the cytoplasm as a precursor of 23.2 kDa [2,23, 21] that is processed to produce the mature 17.9-kDa PasD. In vitro assembly assays indicated that both forms of the protein, pre-PsaD and PsaD, can assemble into the thylakoid membranes, specifically into the PSI complex [2, 14, 22, 24, 25, 26]

PsaD is known to interact strongly with ferredoxin. Chemical cross-linking of PSI and ferredoxin consistently yield a product consisting of PSI-D and ferredoxin [14, 23, 25], and recently the interaction has been shown even with isolated PSI-D and ferredoxin [47]. These observations clearly point to an important function of PSI-D in docking of ferredoxin in both eukaryotes and cyanobacteria. The position of ferredoxin in these crosslinked complexes was also identified by electron microscopy [28]. The same docking site was also found for flavodoxin [29] and is in agreement with a docking site proposed from the structural model of PS I at 6 Å resolution. Subunit PsaD is essential for electron transfer to ferredoxin [28].

PsaE is like PsaD a hydrophilic subunit exposed to the stroma. PsaE is encoded in the nucleus and the mature protein is about 11 kDa. Just like PsaD, the mature PsaE in plants has an extended N-terminal region. The extension is variable from 30-40 amino acid residues. As was the case for PsaD, there is no extension in the chloroplast encoded PsaE in *Odontella* [30], *Porphyra* [31], and *Cyanidium*, nor in the cyanelle-encoded PsaE in *Cyanophora*. *Chlamydomonas* is peculiar in having PsaE with a short extension of only about ten residues compared to the cyanobacterial PsaE [22].

The structure of subunit PsaE (8 kDa) in solution was determined by 1H and 15N-NMR [32, 33]. The loop connecting L-strands 3 and 4 was found to be flexible in the NMR structure[33].

The structure of PsaE in the PSI complex is very similar to the solution structure, with some remarkable deviations in the loop region E-L3L4, which corresponds to the CD loop in the NMR structure. The twist of this loop reported at 4 Å [34] is fully confirmed in the structural model at 2.5 Å resolution. This loop is involved in interactions with PsaA, PsaB and PsaC, suggesting a change of the loop conformation during assembly of the photosystem I complex.

Different functions have been reported for PsaE. PsaE in barley has also been found to be associated with ferredoxin NADP oxidoreductase (FNR) [35]. In cyanobacteria, FNR has a domain linking it to the phycobilisomes [36]. However, recent observations have shown that in spite of this domain, FNR does appear to interact with PsaE [37].

2.1.3. Other subunits of PSI

Six small intrinsic membrane protein components of photosystem I have been identified from the gene sequence in *S. elongatus* [28]: the subunits PsaF (15 kDa), PsaI (4.3 kDa), PsaJ (4.4 kDa), PsaK (8.5 kDa), PsaL (16.6 kDa) and PsaM (3.4 kDa). In the 2.5 Å resolution structure [38, 39, 40], a 12th subunit of PS I, PsaX, which contains one transmembrane K-helix, was identified. All of the small membrane integral subunits are located peripherally to the subunits PsaA and PsaB. The main function of the small subunits is the stabilisation of the antenna system and the quaternary structure of photosystem I. The central Mg^{2+} ions of 10 antenna Chla molecules are axially liganded by amino acid side chains or via water molecules by PsaJ, PsaK, PsaL, PsaM and PsaX. Furthermore, subunits PsaF, PsaI, PsaJ, PsaL, PsaM and to a lesser extent PsaK are in numerous hydrophobic contacts with the carotenoids[40].

The small subunits can also be divided into two groups according to their location in the complex: PsaL, PsaI and PsaM are located in the region where the adjacent monomers face each other in the trimeric PS I complex, whereas PsaF, PsaJ, PsaK and PsaX are located at the detergent exposed surface of photosystem[40].

PsaF binds the luminal electron donor, plastocyanin [41, 42, 43, 44], and It is essential for providing excitation energy transfer from LHCI to the core complex. Early work showed that PsaF (then called subunit III) was required for electron transfer from Pc to P700 [45, 46]. Subsequently, it was demonstrated that Pc cross-linked to PSI is capable of fast electron transfer to P700 and the cross-linking partner was identified as PsaF.

PsaG and PsaK are two small membrane intrinsic proteins of approximately 10-11 kDa each with two transmembrane α-helices connected by a stromal-exposed loop [47, 48] and they show a 30 % sequence homology in *Arabidopsis* [49]. PsaG is unique to plants and green algae. Within the PSI complex PsaK is bound to PsaA and PsaG is bound to PsaB at a roughly symmetry-related position [49, 50]. PsaK is involved in Lhca3/Lhca2 binding as revealed by knock-out and gene knock-down studies [51, 52, 53].

PsaH is a 10 kDa protein with one predicted transmembrane helix [54]. The subunit can be chemically cross-linked to PsaD, PsaI and PsaL [50, 54, 55, 56]. Thus, PsaH must be located near the region that constitutes the domain of interaction between monomers in *Synechococcus* PSI [57]. PSI-H has only been found in plants and green algae. The orientation of PSI-H is not known but based on the positive-inside-rule, the N terminal region is predicted to be in the stroma [58]. Thus, PsaH appears to have about 6 kDa of N terminal region on the stromal side of the membrane and only about 2 kDa facing the lumen. PSI-H is encoded in the nuclear genome. Apparently, PsaH was a late addition to PSI since it has not been found outside Chlorophyta [54].

PsaN is a small extrinsic subunit of about 10 kDa [54]. PsaN is synthesized with a presequence directing it to the lumen and is the only subunit located exclusively on the lumenal side of PSI [59]. The Psa-G, -H, and -N fulfill functions in PSI that are unique to eukaryotic PSI. PsaH has been shown to be involved in state 1–state 2 transitions probably in the interaction with LHCII [54], PsaN is involved in interaction with plastocyanin [41], and PsaG is involved in the stabilization of LHCI and regulation of PSI activity [53]. It is therefore likely that PsaO plays a role in the interaction between the PSI core and other complexes in the thylakoid membrane such as LHCI or LHCII [50, 60]. Alternatively PsaO is involved in the regulation or fine tuning of PSI activity. Together with PsaO, the subunits Psa-G, -H, and -N are unique to higher plants and algae. Structure of plant PSI at 4.4 Å, the structure and position of the Psa-G and -H subunits within the PSI complex are revealed [49]. However, Psa-N and -O are either not resolved at the current resolution or are lost from the complex during preparation and their structure and exact position in the PSI complex are therefore not known.

PsaJ is a hydrophobic subunit of 4-5 kDa. The protein is chloroplast encoded as is the case also for PsaI, which has a similar size and hydrophobicity [54]. PsaJ is located near PsaF as evidenced by cross-linking [56]. The protein has been thought to be membrane spanning, however, the structural model of cyanobacterial PSI suggest that PsaJ may form an unusual bend helix in the plane of the membrane [60]. In the unicellular green alga *Chlamydomonas reinhardtii*, PsaJ is required for the stabilization of the PC-binding site [40]. In the absence of PsaJ, a large fraction of photo-oxidized P700 (chl-a dimer of the PSI reaction centre) is not efficiently reduced by PC or cyt (cytochrome) c6, although the PsaF subunit, which forms the actual binding site for both mobile redox carriers, is still present. This has suggested a role of PsaJ in adjusting the conformation of the PC-binding site. These physiological data are circumstantially supported by structural data [49] and cross-linking studies [15] revealing a localization of the J-subunit adjacent to PsaF

Psa-K from spinach may be tightly associated with the PSIA/B heterodimer [61, 62]. However, PsaK from spinach, pea, and barley was depleted from the PSI core by methods used for separation of LHCI from PSI [65, 64]. Treatment of thylakoids with proteases resulted in degradation of PsaK, indicating that part of the PSI-K polypeptide is exposed on the stromal side of the thylakoid membrane. It has therefore been proposed that the membrane-spanning PsaK subunit is located near the rim of the PSI complex between the PSI and LHCI and is thus easily lost upon detergent treatment [51].

There is significant sequence similarity between PsaG and PsaK from eukaryotes [64]. A computer comparison of PsaG and PsaK from *Arabidopsis* displays approximately 30% amino acid identity. In fact, the cyanobacterial PsaK is equally similar to plant PSI-G and PsaK. However, there is no evidence that cyanobacterial PSI contains more than one copy of PsaK. In the genome sequence of *Synechocystis* PCC6803 two open reading frames have been assigned as potential *psaK* genes[51, 65]. The deduced primary sequences of the two open reading frames show only 42% overall identity. The role of one of the *psaK* genes, which encoded a peptide with an amino terminus corresponding to that of the PsaK peptide purified from PSI, has recently been analyzed, and it was shown that the gene product was dispensable for growth, photosynthesis, and the formation of PSI trimers in *Synechocystis* [65]. The role of the other potential *psaK* gene remains unknown. However, this gene does not encode a protein with higher similarity to PsaG.

2.2. Light harvesting complex I

X-ray crystallography of the PSI core from cyanobacteria [22, 25, 50, 55] as well as modeling studies indicates strong interpigment interactions and unique protein environment as a source for the low energy shifts in absorption of PSI. Biochemical and spectroscopic studies of Light Harvesting Complex I (LHCI) suggest that in the PSI-LHCI super complexes the peripheral antenna and the PSI core antenna have structurally and spectrally distinct pools of red pigments [22]. As in the PSI core antenna, excitonically coupled dimers or trimers of Chl *a* or Chl *b* in the LHCI were suggested to form a pool of red pigments in the LHCI [2, 66, 67]. Recent experiments with antisense inhibited *Arabidopsis* plants *in vivo* [68], and reconstitution of the polypeptides *in vitro* suggests that the presence of low energy pigments is a feature of all four Lhca polypeptides (Lhca1, Lhca2, Lhca3, and Lhca4). Lhca1 seems to possess the less red-shifted spectral forms (684 nm) [69, 70], which is in agreement with its close relatedness to a minor light harvesting polypeptide of PSII, CP29. On the contrary, the Lhca4 binds the "reddest" pigments [71]. The presence of the strong fluorescence at 730 nm in Lhca4 even at room temperature indicates a specific molecular organization of the pigments that efficiently localizes the excitation and could dissipate the excess excitation energy as a nonphotochemical sink when the LHCI antenna is energetically uncoupled from the PSI. Despite a consensus that all the chlorophyll *a/b* binding proteins (cab) have common evolutionary origin, similar structure and protein sequences [67, 72], there are local structural differences that determine the assembly of Lhca polypeptides around PSI as dimers (Lhca2/Lhca2, Lhca3/Lhca3, and Lhca1/Lhca4) in contrast to the trimer-forming Lhcb polypeptides in LHCII.

3. Photosytem I antenna

3.1. Core antenna

The antenna of PSI consists of two structurally and functionally parts; the core antenna and preripheral antenna.

The core antenna of PSI contains in total appromeximately 100 chla and 15 β-carotene of which the majority is bound to the PsaA/PsaB dimmer [49]. The chlorophylls have their Q_4 absorption maxima around 680 nm. A comparison of absorbtion spectra of PSI, LHCI complexes from wild type A, thaliana and from a mutant lacking the PsaL and PsaH subunits revealed that the about five chlorophylls that are bound to these subunits absorb preferentially at 638 and 667 nm [73].

3.2. Prephral antenna

The prepheral antenna of PSI consists of nuclear encoded chlorophyll binding proteins (Lhca) which are transported into the chloroplast and form a light-harvesting complex (LHCI) which increases the light-harvesting capacity of PSI [74].

The protein contacts between the core complex and LHCI appear to be relatively weak, which explains the biochemical sensitivity of the PSI-LHCI supercomplex to detergent attack. It is clear, however, that each of the four light-harvesting proteins fits its specific binding site, because the interface of the core complex formed by subunits PsaG, PsaB, PsaF, PsaJ, PsaA, and PsaK is asymmetric [4, 49]. Lhca1 antenna protein is bound to the core through PsaB and PsaG. Previous studies showed that plants in which *psaG* gene expression was suppressed by antisense technology or eliminated by transposon tagging exhibited compromised stability of the antenna and weaker binding of LHCI to the core. Thus, it was proposed that PsaG somehow stabilizes the peripheral antenna. In addition, Lhca3 appears to interact with PsaA. However, the position of newly traced subunit PsaK suggests that it also binds Lhca3. PsaK [39] was proposed to stabilize LHCI organization, and PSI lacking PsaK, isolated from *Arabidopsis* plants, showed 30–40% less Lhca3, whereas associations with Lhca1 and Lhca4 were unaffected [75].

4. Role of the PSI subunits specific to plants and algae

The extrinsic protein, PsaD, has two reported functions in the PS I complex of cyanobacteria, algae and higher plants. The first function, deduced from *in vitro* reconstitution experiments, is to stabilize PsaC on the PS I reaction center [2, 23]. The second function, inferred from cross-linking studies, is to serve as a "docking" protein to facilitate interaction of soluble ferredoxin with the PS I complex [2]. Recent cross-linking experiments have shown that Lys^{106} of PsaD from *Synechocystis* sp. PCC 6803 can be cross-linked to Glu^{93} in ferredoxin [76]. Therefore these two residues come in physical proximity with each other during at least one stage of electron transfer from PS I to ferredoxin. These results indicate a ferredoxin-docking function of PsaD, but do not illustrate a functional requirement of PsaD for $NADP^+$ photoreduction [77].

While the function of the ten PSI subunits common to plants, algae and cyanobacteria has been studied extensively, the role of the three eukaryotic-specific subunits PsaH, PsaG and PsaN is less well understood. One reason is that these subunits are nucleus-encoded and thus less amenable to genetic manipulation [2, 53]. However it has recently been possible to generate transgenic *Arabidopsis* plants lacking PsaH or PsaN by cosuppression or by using an antisense strategy after transformation of the plants with the corresponding cDNAs in the sense or antisense orientation under the control of a constitutive promoter [41, 53, 78, 79, 80]. PSI complexes lacking the intrinsic PsaH membrane protein contain normal amounts of the other PSI subunits except PsaL which is reduced two-fold [80]. Electron flow through PSI is impaired and the purified PSI complex is highly unstable in the presence of urea. Plants lacking the lumenal PsaN subunit are impaired in the interaction between plastocyanin and photosystem I and the steady-state reduction of $NADP^+$ is decreased two-fold [2, 41].

The analysis of the PsaF-deficient strain and its suppressor reveals that in the presence of a functional antenna, an intact donor side of PSI is required for protection of *C. reinhardtii* cells from photo-oxidative damage in high light [43]. It is therefore possible that the development of light-harvesting systems for PSI and PSII in eukaryotic organisms demanded an improved donor side of PSI, especially with regard to its functional interaction with the electron donors [81]. This may have led to the evolution of the recognition site within the N-terminal part of PsaF that is essential for fast electron transfer between PSI and its electron donors under high light. It is also possible that plastocyanin replaced cytochrome c6 as electron donor for PSI during evolution because it is slightly more efficient than the heme-containing protein in reducing P700$^+$ [2, 15, 81, 82] and should therefore provide a beter protection against photo-oxidative damage.

The function of the PsaF protein (15 kDa) at the lumenal side has been subject to discussion. In intact cells of the green alga Chlamydomonas reinhardtii, PsaF is implicated in the electron transfer from plastocyanin to oxidized P700 by providing a docking site for the electron donor: psaFÿ mutants of this organism had a dramatically reduced electron transfer rate [15, 45, 76]. In contrast, a psaFÿ mutant of the cyanobacterium Synechocystis PCC 6803 exhibited normal electron transfer to P700., implying that PsaF is not essential for the docking of either cytochrome c6 or plastocyanin to PSI [15, 76, 82]. While PSI is extracted as a mixture of trimers and monomers from thylakoid membranes of wild-type cyanobacteria, PSI from mutants that lack the PsaL protein (16 kDa) exists exclusively as a monomer after membrane solubilization [73]. In addition, proteolysis studies have shown PsaL to be located about the 3-fold axis of the trimer, thus holding it together [5, 15]. Little is known about the function of the four other membrane intrinsic subunits (PsaI, -J, -K and -M) that have molecular masses ranging from 3 to 8 kDa [63].

Comparison of deduced primary sequences indicates that the PsaL subunits contain a greater diversity than seen in other subunits [15, 54]. Function of PsaL in the formation of PS I trimers was revealed by the inactivation of the psaL gene in *Synechocystis* sp. PCC 6803 [3, 54, 176]. The requirement of PsaL for the trimer formation was later conformed in a PsaL-deficient mutant from *Synechococcus* sp. PCC 7002 [76, 83]. Recent studies showed that the C-terminus of PsaL is embedded inside the monomeric PS I complex and is involved in trimer formation. Another role of PsaL is in binding to some antenna chlorophyll a molecules.

5. Gene of photosystem I

Photosystem I (PSI) is a multiprotein complex in the thylakoid membrane of chloroplasts, providing an interesting system for studying the nucleo-choloroplast relationship in plants.. The core subunits of the PSI reaction centre are still encoded by the chloroplast genome, whereas the genes for the peripheral subunits are located in the nucleus in green algae and land plants. In this study, we dissected the promoter architecture of a nuclear-encoded PSI gene in tobacco, and investigated whether the characteristics found in this promoter are shared by those of the other photosynthesis nuclear genes.

Sequencing of these proteins and/or their corresponding genes have registered two genes, *psaA* and *psaB*, for the P700-carrying proteins [84], 6 genes, *psaC* to *psaH*, for the core subunits and at least three types of *cab* genes for the LHCI subunits [87]. In green plants, *psaA*, *psaB* and *psaC* are encoded by chloroplast genome while the others are encoded by nuclear genome. Sequence comparison indicates that the two P700-carrying proteins and most of the core subunits are more or less conserved between cyanobacteria and green plants, although LHCI is absent in cyanobacteria. In addition to these components, Ikeuchi et al. [88] found that three new low-molecular-mass components of 4.8 kDa, 5.2 kDa and 6.8 kDa in cyanobacterial PSI complex and reported their partial amino acid sequences. Interestingly, the sequence of this cyanobacterial 4.1 kDa component corresponded to ORF42/44 of higher plant chloroplast DNA, although its product had not yet been found in plants. Scheller et al. [22] reported the presence of two small proteins below 4 kDa in higher plant PSI complex. These prompted us to search for the product(s) of ORF42/44 in higher plant PSI complex as well. Here we report the presence of three lowmolecular- mass proteins in spinach and pea PSI complexes and their unambiguous identification by N terminal sequencing.

The genes for the two subunits, psaA and psaB, are located adjacent to each other in the large single-copy region of circular plastid genome in higher plants [89]. Gene psaB is followed by rps14 encoding the chloroplast ribosomal protein CS14 [90]. The psaA-psaB-rps14 gene cluster was found to co-transcribe into a 5- to 6-kb polycistronic mRNA in spinach [91], tobacco [90], and rice [89]. Cheng et. al. [90] have performed a detailed transcriptional analysis of the promoter of rice psaA-psaB-rps14 operon with deleted mutants in vitro. They showed that two functional promoters denoted as "-175" and "-129" were revealed.

Some of the known signals which are targeted to these genes are light, plastid signal(s) and hormones. Nuclear encoded genes of PSI are effectively activated by white and red light, while blue light is less effective. Essential promoter units which are responsive to light were identified by deletions and mutational analyses. Another set of signals is produced by the plastids reporting the state of the plastids to the nucleus. When plastids were bleached by the addition of norflurazon, transcription of nuclear PSI genes was decreased [92]. The nature of these plastid-derived signals has still to be elucidated, however it appears that the communication between the two organelles is mediated by more than one signal [93, 94]. The responsive units within the promoter appear to be the same as the light-responsive elements [93]. The third group of signals acting on genes for thylakoid proteins is plant hormones. Kusnetsov et al. [95] suggested that it could be shown that cytokinin stimulated the transcription of *AtpC* and *Cab* genes in the dark. Even the formation of thylakoid membranes could be stimulated by cytokinin. Antagonistic to the cytokinin effect abscisic acid decreased transcription of target genes. In addition to these types of regulation PSI genes were also regulated by daytime. The peaks of transcriptional activities of *PsaE, PsaF, PsaG* and *PsaH* were one hour after the onset of light [92]. The oscillation stopped in continuous light. The regulatory interactions leading to this phenomenon are not known at present. However, it became obvious that all of these signaling systems controlling nuclear gene expression can only operate fully when functional plastids are available [92]. Therefore we examined whether the expression of a photosynthesis promoter-driven reporter gene construct is also expressed in roots and whether elements can be distinguished which respond differently to externally applied signals.

In the spinach *PsaF* promoter a CAAT box is located near the transcription start site (at position -178/-174). A similar motif has been found in the light-reactive promoter unit of the *PsaD* gene in *Nicotiana sylvestris* [96]. A protein factor binding to such a motif has been isolated from *Arabidopsis thaliana* and was shown to be regulated by light, cytokinin, and the stage of the plastid [97]. While the CAAT box region appears to be sufficient for regulated expression of *AtpC*, hardly any expression could be detected for a CAAT box containing *PsaF* fragment lacking the region upstream of -179 [97, 98]. In gel retardation studies binding of a nuclear factor to the adjacent *PsaF* promoter region -220/-179 could be demonstrated [99]. The -220/-179 region was not only shown to be a stimulator of *PsaF* expression, but also enhances expression in response to different stimuli. It includes a GT–1 binding site (–194/–180) [100]. Results of this study show that in transgenic tobacco roots this promoter region is crucial for the activation by cytokinin, mastoparan and Ca^{2+}. This suggests that phosphoinositides, such as inositol-3-phosphate, and a Ca^{2+}-dependent kinase are involved in regulating transcription factor activity operating on this promoter region.

In higher plants, the function of the nuclear-encoded subunits has been elucidated in recent years using RNAi (RNA interference), antisense techniques and insertional mutagenesis in *Arabidopsis thaliana* [101]. This work has revealed non-essential functions of most nuclear-encoded PSI subunits, with the exception of PsaD, which is essential for PSI accumulation and is involved in the formation of the ferredoxinbinding site [101]. All other knockout mutants displayed much weaker phenotypes, often including a reduced PSI content due to destabilization of the complex and sometimes showing altered antenna binding and exciton transfer to the reaction centre [16].

6. Photoinhibition of PSI and response to abiotic stress

Plant chloroplasts include two large pigment-protein complexes, such as photosystem I and photosystem II that are located within thylakoid membranes. The reaction centres of PSI and PSII are formed by chlorophyll a-binding heterodimers, PsaA/PsaB and PsbA/PsbD proteins, respectively. PSI and PSII are both organized into supercomplexes with variable amounts of nuclear-encoded chlorophyll a/b-binding proteins forming light-harvesting antenna complexes around PSI (LHCI) and PSII (LHCII) (Figure 1).

Environmental factors such as temperature, UV-light, irradiance, drought and salinity are known to affect photosynthesis in both cyanobacteria and plants (Figure 1). In cyanobacteria, several studies have been reported on photosynthetic electron transport activities both under salt and high light stress conditions in whole cells as well as thylakoid membranes[102,103].

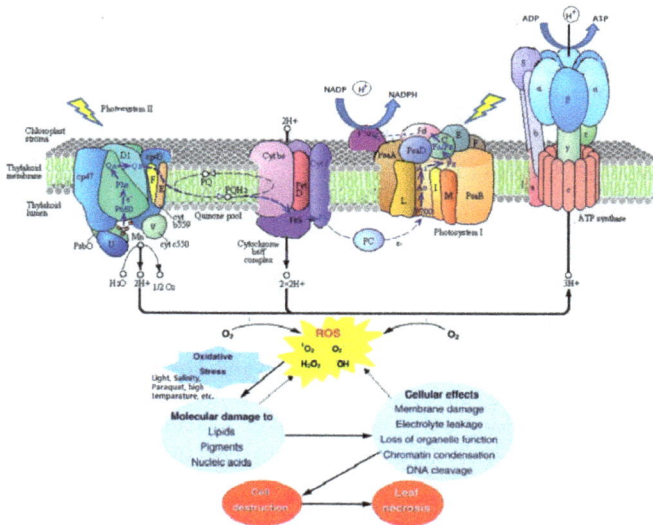

Figure 1. Schematic figure of oxidative stress effects on PSII and PSI

Photoinhibition of Photosystem I (PSI) was first reported by Jones and Kok [104], is the one who originally called them 'photoinhibition' [105]. The subsequent studies revealed that the activity of PSI could be photoinhibited in thylakoid membranes [106, 107] as well as in isolated PSI complexes [108, 109]. However, PSI-specific photoinhibition was never observed in intact leaves until 1994 [110]. The selective photoinhibition of PSI was first observed in cucumber leaves treated at chilling temperatures [111]. In contrast PSI is generally believed to be less sensitive to light stress and its photoprotection mechanisms were less investigated. Nevertheless, several recent evidences showed that PSI can also be targeted by photoinhibition, especially under chilling conditions and when the linear electron transport chain is unbalanced [111, 112]. PSI photoprotection has been suggested to be mainly mediated by oxygen scavenging enzymes (e.g., superoxide dismutase and ascorbate peroxidase) which efficiently detoxify reactive species produced at the reducing side of Photosystem I [113]. A decreased ability of these enzymes to scavenge ROS at low temperatures was proposed to be the reason of the major PSI photo-sensitivity in chilling conditions [111, 112, 114]. As a summary, processes during of PSI photoinhibition as follow;

1. Decreased rate of reducing power utilization by Calvin cycle enzymes (Rubisco) at low temperature;

2. Photoinduced electron transfer from PS2 and reduction of PS1 electron acceptors (FeS centres, Fd, NADP);

3. Cold-induced diminish of oxidative defense system (tAPX, sAPX etc.) capacity;

4. Recombination of separated charges in PS1 reaction centres between P700$^+$ and A$_0^-$ or A^{1-} and Chl triplet formation;

5. Energy migration from TChl to O^2 and production of singlet oxygen ^{1}O^2;

6. Superoxide anion radical and H2O2 production in Mehler reaction;

7. Fenton reaction (OH$^•$ formation as result of interaction of H_2O_2 with reduced FeS-clusters);

8. Destruction of FeS-clusters by OH;

9. Inactive FeS-clusters induce the conformational changes of PS1 core complex proteins facilitating its access for proteases;

10. Degradation of PsaB and PsaA gene products and release of 45 kDa and 51kDa proteins;

11. Processes (8) and (10) result PS1 photoinhibition

The eventual effect of the abiotic stress on plant growth and crop productivity is a result not only of the extent of the damage but also on the capacity for recovery after the damage has taken place. Although the recovery and repair of PSII after photoinhibition have been a subject of many studies [115, 116, 117], there is very little known about the recovery and repair of PSI Teicher et. al. [118] showed that PSI recovery is a very slow process, which may take several days even under optimal conditions in field-grown barley. A more recent study showed that PSI damage in cucumber is not even completely reversible [119]

The few previous studies have not shown whether PSI repair is similar to PSII repair where one particular subunit, D1, is specifically remade whereas the rest of the complex is reused [120]. Clearly, the PSI repair process must involve some protein turnover but it is not known whether the breakdown that is observed during photodamage is caused directly by the damage or is part of the recovery process.

Light are highly unpredictable resource for plants and the changes in growth irradiance induce several changes in biochemical and molecular composition of the plant cell. Murchie et al. [121] showed that there are 99-light responsive genes which were down regulated and 130 were up-regulated in rice during light treatment. Majority of these genes showed reduced levels of expression in response to high light, whereas stress related genes showed increased level of expression. In order to avoid over-excitation of chlorophyll protein complexes and photooxi-dation, a regulated degradation of LHC was observed in rice leaves along with a decline in CP-24, PSI genes and a 10 kD PSII gene was also noticed under high light [121].

PS I has long been reported to be less affected than PSII by high light [105]. PSI in isolated thylakoid membranes was inactivated by high light [122]. Since PSI is the terminal electron carrier in the chloroplast, it was identified as a major site producing ROS and shown to be closely associated with ROS-scavenging systems in the chloroplast [123]. The role of ROS inactivating PSI reaction center and degradation of psaA and psaB under high light conditions has been studied [124]. Very recently, Jiao et al [125] demonstrated that high light stress readily photoinhibited PSI, following the loss of psaC as well as degradation of PSI reaction center proteins (psaA and psaB). The findings suggest that PSI photoinhibition can be a limiting factor in crop productivity under high light.

Several studies demonstrated that thylakoid memebrane proteins were affected by salt stress. In *Synechococcus* cells, NaCl at 0.5 M concentration inactivated both PSII and PSI due to the changes in K/Na [102]. Lu and Voshak [126, 127] have also been demonstrated that salt stress itself has no direct effect on PSII activity in *Spirulina plantensis* preincubated in the dark, but the same salt stress in combination with Photosynthetically active radiation (PAR) led to block of electron transport between Q_A and Q_B (primer and secondary quinine electron acceptors of PSII), and the inhibition of PSII electron transport was proportional to the intensity of light. In addition, it has been revealed that cyclic electron flow around PSI was enhanced during salt stress in cyanobacteria [128,129]. Zhang et al. [103] observed that PSI activity in salt-stressed cells increased. Sudhir et al. [130] also reported that salt stress induced an increase in PSI activity. The increased PSI activity may be due to an increase in the content of P700 reaction centers[131,132]. An increase in PSI activity should increase cyclic electron transport. Several reports have shown that cyclic electron flow increases under salinity stress[130]. Hence, it seems that an increase in PSI activity in salt-adapted cells may protect PSII from excessive excitation energy under salt stress.

A few reports have shown the effects of metals on the activity of photosystem I (PSI) and some of them was controversial. Neelam and Rai [133] reported that cadmium treatment inhibits PSI activity in *Microcystis sp*. Zhou et al. [134] also suggested that the increase in PSI activity results from the increase of cyclic electron transport around PSI, which increases of ATP synthesis. At the same time, this cyclic electron transport around PSI is suggested to play an important role in the synthesis of more ATP, which would provide more energy to maintain proper defense system within the cell. In addition, Ivanov et al [135] showed that increased PSI mediated cyclic electron transport would be observed in plants which was grown under temperature / light stress conditions. It has been suggested that cyclic electron flow around PSI is required to supply sufficient proton motive force to initiate energy-dependent excitation quenching (q_E). Jin et al. [136] showed that cyclic electron flow around PSI plays an important role in the production of pH gradient across the thylakoid membrane (ΔpH) that leads to the effective dissipation of excess excitation energy under high temperature conditions. Thus, the increase in PSI activity resulting from the increase of cyclic electron transport around PSI could be one of the adaptive mechanisms to stress conditions [137, 138, 139].

Author details

Dilek Unal Ozakca

Address all correspondence to: dilek.unal@bilecik.edu.tr, dilek.unal@mail.ege.edu.tr

Bilecik Seyh Edebali University, Faculty of Science and Art, Department of Molecular Biology and Genetic, Bilecik, Turkey

References

[1] Chitnis P. R Photosystem I: Function and Physiology Annu Rev Plant Physiol Plant Mol Biol. 2001; 52:593-626.

[2] Scheller H.V, Jensen P. E, Haldrup. A, Lunde. C, Knoetzel J Role of subunits in eukaryotic Photosystem I. Biochim Biophys Acta 2001;1507: 41-60

[3] Manna P., Chitnis P.R Function and molecular genetics of Photosystem I, in: G.S. Singhal, G. Renger, S.K. Sapory (Eds.)i Concepts in Photobiology:Photosynthesis and Photomorphogenesis, 1998, pp.212-251

[4] Amunts A., Nelson N Fuctional organization of a plant Photosystem I:Evolution of a highly efficient photochemical machine. Plant Physiol. Biochem. 2008; 46:228-237

[5] Jensen P.E, Bassi R, Boekema E.J, Dekker J.P, Jansson S, Leister D, Robinson C, Scheller H.V Structure, function and regulation of plant photosystem I, Biochim. Biophys. Acta 2007;1767:335–352

[6] Ben-Shem A, Frolow F, Nelson N Crystal structure of plant photosystem I, Nature 2003;426:630-635

[7] Klukas O., Schubert WD., Jordan P., Krauss N., Framme P., Witt HT., Seanger W., 1999. Photosystem I, an improved model of the stromal subunits PsaC, PsaD, and PsaE. J. Biol. Chem. 274(11):7351-60.

[8] Ihnatowicz A., Pesaresi P., Varotto C., Richly E., Scheider A., Jahns P., Salamini F., Leister D Mutants of photosystem I subunit D of Arabidopsis thaliana : effects on photosynthesis, photosystem I stability and expression of nuclear genes for chloroplast functions. Plant J. 2004; 37, 839–852

[9] Jensen P. E., Haldrup A., Rosgaard L., Scheller, H. V Molecular dissection of photosystem I in higher plants: topology, structure and function. Physiol. Plant. 2003;119, 313–321

[10] Rochaix J. D, Fischer N, Hippler M Chloroplast site-directed mutagenesis of photosystem I in Chlamydomonas: Electron transfer reactions and light sensitivity Biochimie 82 2000; 635–645.

[11] Oh-oka H, Takahashi Y, Matsubara H Topological considerations of the 9-kDa polypeptide which contains centers A and B, associated with the 14- and 19-kDa polypeptides in the photosystem I complex of spinach. Plant Cell Physiol 1989; 30: 869-875

[12] Vallon O, Bogorad L Topological study of PSI-A and PSI-B, the large subunits of the photosystem-I reaction center. Eur J Biochem 1993; 214: 907-915

[13] Hoj P.B., Svendsen I., Scheller H.V., Moller B.L. Identification of a chloroplast-encoded 9-kDa polypeptide as a 2[4Fe- 4S] protein carrying centers A and B of photosystem I J. Biol. Chem. 1987; 262:12676-12684.

[14] Xia Z., Broadhurst R.W., Laue E.D., Bryant D.A., Golbeck J.H., Bendall D.S Structure and properties of PsaD in solution. Eur. J. Biochem. 1998; 255:309-316.

[15] Chitnis P.R., Xu Q., Chitnis V.P., Neshushtai R Function and organization of photosystem I polypeptides, Photosynth Res. 1995;44: 23-40

[16] Haldrup A., Simpson D.J, Scheller H.V Down-regulation of the PSI-F subunit of photosystem I in *Arabidopsis thaliana*. The PSI-F subunit is essential for photoautotrophic growth and antenna function, *J. Biol. Chem.* 2000;275:31211-31218.

[17] Varotto C., Pesaresi P., Jahns P., Lessnick A., Tizzano M., Schiavon F., Salamini F., Leister D Single and double knockouts of the genes for photosystem I subunits G, K, and H of Arabidopsis. Effects on photosystem I composition, photosynthetic electron flow, and state transitions, Plant Physiol. 2002;129:616–624.

[18] Allen JF. Redox control of gene expression and the function of chloroplast genomes —an hypothesis. Photosynth. Res. 1993; 36:95-102.

[19] Pfannschmidt T, Nilsson A, Allen JF Photosynthetic control of chloroplast gene expression, Nature 1999;397:625-628.

[20] Klukas O, Schubert W.D, Jordan P, Krauss N, Fromme P, Witt H.T, Saenger W., Photosystem I, an improved model of the stromal PsaC, psaD and psaE. J. Biol. Chem. 1999;274:7351-7360.

[21] Vassillev I.R, Jung Y.S, Yang F, Golbeck J.H PsaC subunit of photosystem I is oriented with iron-sulfur cluster F_B as the immediate electron donor to ferredoxin and flavodoxin. Biophys. J 1998; 74:2029-2035,

[22] Scheller H. V, Jensen P. E, Haldrup A, Lunde C, Knoetzel J. Role of subunits in eukaryotic photosystem I. Biochim. Biophys. Acta 2001;1507:41–60

[23] Jin P, Sun J, Chitnis P.RStructural features and assembly of the soluble overexpressed psaD subunit of Photosystem I. Biochm. Biophys. Acta 1999; 1410:7-18

[24] Krauß N, Schubert W-D, Klukas O, Fromme P, Witt HT, Saenger W Photosystem I at 4 Å resolution represents the first structural model of a joint photosynthetic reaction and core antenna system. *Nat Struct Biol* 1996; 3:965-973.

[25] Shubert WD, Klukas O, Krauß N, Saenger W, Framme P, Witt H. T Photosystem I of Synechoccus elongarus at 4 Å resolution:comprehensice structure analysis, J. Mol. Biol. 1997; 272:741-769

[26] Xu Q, Armbrust T.S, Guikema J.A, Chitnis P.R, Organization of Photosystem I polypeptides: A structural interaction between psaD and psaL subunits. Plant Physiol. 1994;106:1057-1063

[27] Yu J, Smart L.B, Yung Y.S, Golbeck J, McIntosh L, Absence of PsaC subunit allows assembly of photosystem I core but prevents the binding of PsaD and PsaE in Synechocyctis sp. PCC6803. Plant Mol. Biol. 1995; 29:331-342

[28] Chitnis P.R Photosytem I. Plant Physiol. 1996;111:661-669

[29] Lelong C, Setif P, Laqoutte B, Bottin H, Identification of the amino acids involved in the functional interaction between photosystem I and ferrodoxin from Synechocystis sp. PCC6803 by chemical cross-linking. J. Biol. Chem. 1994; 269:10034-10039

[30] Kowallik K.V, Stoebe B, Schaffran I, Kroth-Pancic P.G, Freier U The chloroplast genome of a chlorophyll a+c containing alga, Odontella sinensis. Plant Mol. Biol. Rep. 1995; 13:336-342

[31] Reith M, Munholland J Complete nucleotide sequence of the Porphyra purpurea chloroplast genome. Plant Mol. Biol. Rep. 1995; 13:333-335

[32] Falzone C.J, Kao Y-H, Zhao J, MacLaughlin K.L, Bryant D.A, Lecomte, J.T.J ^{1}H and ^{15}N NMR assignments of PsaE, a photosystem I subunit from the cyanobacterium *Synechococcus* sp. Strain PCC 7002. Biochemistry 1994;33: 6043–6051.

[33] Falzone C.J, Kao Y.H, Zhao J, Bryant D.A, Lecomte J.T Three-dimensional solution structure of PsaE from the Cyanobacterium Synechococcus sp. Strain PCC 7002, a photosystem I protein that shows structural homology with SH3 domains. Biochemistry 1994; 33(20):6052-62

[34] Jordan P, Fromme P, Witt H.T, Klukas O, Saenger W, Krauss N Three-dimensional structure of cyanobacterial photosystem I at 2.5 Å resolution. Nature 2001;411:909-917

[35] Andersen B., Scheller H.V, Møller B.L The PSI-E subunit of photosystem I binds ferredoxini NADP$^+$ oxidoreductase. FEBS Lett 1992; 311:169-173

[36] Schluchter WM., Bryant DA. Molecular characterization of ferrodoxin-NADP$^+$ oxidoreductase in cyanobacteria: cloning and sequence of the petH gene of *Synechococcus* sp PCC 7002 and studies on the gene product. Biochemistry 1992;31:3092-3102

[37] van Thor J.J, Geerlings T.H, Matthijs H.P, Hellingwerf K.J Kinetic evidence for the PsaE-dependent transient ternary complex photosystem I/Ferrodoxin/Ferrodoxin: NADP$^+$ reductase in a cayanobacterium. Biochemistry 1999; 38:12735-12746

[38] Xu Q, Odom W.R, Guikema J.A, Chitnis V.P, Chitnis P.R Targeted deletion of psaJ from the cyanobacterium Syneckocystis sp. PCC 6803 indicates structural interactions between the PsaJ and PsaF subunits of photosystem I. Plant Mol Biol 1994; 2624:291-30

[39] Jensen P.E, Gilpin M, Knoetzel J, Scheller H.V The PSI-K subunit of photosystem I is involved in the interaction between light-harvesting complex I and the photosystem I reaction center core, J. Biol. Chem. 2000;275:24701–24708.

[40] Schöttler M.A, Flügel C, Thiele W, Stegemann S, Bock R The plastome-encoded PsaJ subunit is required for efficient Photosystem I excitation, but not for plastocyanin oxidation in tobacco. Biochem. 2007;403:251-250

[41] Haldrup A, Naver H, Scheller H.V, The interaction between plastocyanin and photo-system I is inefficient in transgenic Arabidopsis plants lacking the PSI-N subunit of photosystem I. Plant J. 1999;17:689-98

[42] Haehnel W, Janse T, Gause K, Klosgen R.B, Stahl B, Michl D, Huvermann B, Karas M, Herrmann R.G Electron transfer from plastocyanin to photosystem I EMBO J. 1994;13:1028-1038.

[43] Farah J, Rappaport F, Choquet Y, Joliot P, Rochaix J.D Isolation of a psaF-deficient mutant of Chlamydomonas reinhardtii: efficient interaction of plastocyanin with the photosystem I reaction center is mediated by the PsaF subunit. EMBO J. 1995;14(20): 4976-84

[44] Hippler M., Drepper F., Farah J, Rochaix J.D Fast electron transfer from cytochrome c6 and plastocyanin to photosystem I of Chlamydomonas reinhardtii requires PsaF. Biochemistry 1997;36:6343-6349

[45] Filieger K, Tyagy A, Sopory S, Csepiö A, Hermann R.G, Oelmüller R A 42 bp frag-ment of the gene for subunit III of photosystem I (PsaF) is crucial for its activity, Plant J 1993;4:9-17

[46] Oelmüller R, Lübberstedt T, Bolle C, Sopory S, Tyagy A, Csepiö A, Filieger K, Her-mann R.G 1992. Promoter architecture of nuclear genes for thylakoid membrane pro-teins from spinach. In:Murata N (ed) Research in Photosynthesis Vol III. Kluwer Acad Publishers, Dordrecht pp 219-224.

[47] Mant A, Woolhead C.A, Moore M, Henry R, Robinson C Insertion of PsaK into the thylakoid membrane in a "horseshoe" conformation occurs in the absence of signal recognition particle, nucleoside triphosphates or functional Albino3. J. Biol. Chem. 2001;276:36200-36206

[48] Rosgaard L, Zygadlo A, Scheller H.V, Mant A, Jensen P.E Insertion of the plant pho-tosystem I subunit G into the thylakoid membrane in vitro and in vivo studies of wild-type and tagged versions of the protein. FEBS J. 2005; 272:4002-4010

[49] Ben-Shem A, Frolow F, Nelson N Crystal structure of plant photosystem I. Nature 2003; 426:630-635

[50] Jensen P.E, Bassi R, Boekema E.J, Dekker J.P, Jansson S, Leister D, Robinson C, Schel-ler H.V Structure, function and regulation of plant photosystem I. Biochim. Biophys. Acta 2007;176:335-352

[51] Jensen P.E, Glipin M, Knoetzel J, Scheller H.V The PSI-K subunit of photosystem I is involved in the interaction between light-harvesting complex I and the photosystem I reaction center core. J. Biol Chem 2000;275:24701-24708

[52] Jensen P.E, Rosgaard L, Knoetsel J, Scheller H.V Photosystem I activity is increased in the absence of the PSI-G subunit. J. Biol. Chem. 2002; 277:2798-2803

[53] Varotto C, Pesaresi P, Jahns P, Leβnick A, Tizzano M, Schiavon F, Salamini F, Leister D Single and double knockouts of the genes for photosystem I subunits G, K, and H of Arabidopsis, effects on photosystem I composition, photosynthetic electron flow, and state transitions. Plant Physiol. 2002; 129:616-624

[54] Scheller H.V, Jensen P.E, Haldrup A, Lunde C, Knoetzel J Role of subunits in eukaryotic Biochim Biophys Acta. 2001;1507(1-3):41-60.

[55] Andersen B, Koch B, Scheller H.VStructural and functional analysis of the reducing side of photosystem I. Physiol. Plant. 1992; 84:154-161

[56] Jansson S, Andersen B, Scheller H.V Neraest-neighbor analysis of higher-plant photosystem I holocomplex . Plant Physiol. 1996;112:409-4020

[57] Schubert W.D, Klukas O, Krauss N, Saenger W, Framme P, Witt H.T Photosystem I of Synechococcus elongates at 4 angstrom resolution. Comprehensive structure analysis. J. Mol. Biol. 1997;272:741-769

[58] Vonheijne G Membrane protein structure prediction, hydrophobicity analysis and the positive-inside rule. J. Mol. Biol. 1992; 225:487-494.

[59] Nielsen V.S, Mant A, Knoetzel J, Møller B.L, Robinson C Import of barley photosystem I subunit N into the thylakoid lumen is mediated by a bipartite presequence lacking an intermediate processing site. Role of the delta pH in translocation across the thylakoid membrane. J. Biol. Chem. 1994; 269(5):3762-3766

[60] Jensen P. E, Haldrup A, Zhang S, Scheller, H. V The PSI-O subunit of plant photosystem I is involved in balancing the excitation pressure between the two photosystems. J. Biol. Chem. 2004; 279, 24212–24217

[61] Hoshina S, Sue S, Kunishima N, Kamide K, Wada K, Itoh S Characterzation and N-terminal sequence of a 5 kDa polypeptide in the photosystem I core complex from spinach FEBS Lett 1989; 258:305-308

[62] Wyn R.N, Malkin R The photosystem I 5.5 kDa subunit (the psaK gene product) an intrinsic subunit of the PSI reaction center complex. FEBS Lett 1990; 262:45-48

[63] Ikeuchi M, Hirano A, Hiyama T, Inoue Y Polypeptide composition of higher plant photosystem I complex Identification of psaI, psaj and psaK gene products. FEBS 1990;263(2):274-278

[64] Kjaerulff S, Anderson B, Skovgaard N.V, Møller L.B, Okkels J.S The PSI-K subunit of the photosystem I from barley (Hordeum vulgare L.): Evidence for a gene duplication of an ancestral PSI-G/K gene. J. Biol Chem 1993; 268:18912-18916

[65] Nakamoto H, Hasegawa M Targeted inactivation of the gene psaK encoding a subunit of photosystem I from the cyanobacterium Synechocystis sp. PCC 6803. Plant Cell Physiol. 1999; 40:9-16.

[66] Morosinotto T, Breton J, Bassi R, Croce R The nature of chlorophyll ligand in Lhca proteins determines the far red fluorescence emission typical of photosystem I. J. Biol. Chem. 2003; 278(49):49223-49229

[67] Koziol A.G, Borza T, Ishida K, Keeling P, Lee L.W, Durnford D.G, Tracing the evolution of the light-harvesting antennae in chlorophyll a/b-containing organisms. Plant Physiol. 2007; 143:1802-1816.

[68] Zhang H, Goodman H.M, Jansson S, Antisense inhibition of the Photosystem I Antenna protein Lhca4 in *Arabidopsis thaliana*. Plant Physiol. 1997; 115(4):1525-31

[69] Haworth P, Watson J.L, Arntzen C.J, The detecting isolation and characterization of a light-harvesting complex which is specifically associated with Photosystem I. Biochim. Biophys. Acta 1983; 721:151-158

[70] Lam E, Ortiz W, Malkin R, Chlorophyll a/b proteins of photosystem I. FEBS Lett 1984; 168:10-14

[71] Gobets B, van Grondelle R, Energy transfer and trapping in Photosystem I. Biochim. Biophys. Acta 2001;1507:80-99

[72] Neilson J.A, Durnford D.G Structural and functional diversification of the light harvesting complex in photosynthetic eukaryotes. Photosyth. Res. 2010;106:54-71

[73] Ihalainen J. A, Jensen P. E, Haldrup A, van Stokkum I. H. M, van Grondelle R, Scheller H. V., Dekker J. P Pigment organization and energy transfer dynamics in isolated photosystem I (PSI) complexes from Arabidopsis thaliana depleted of the PSI-G, PSI-K, PSI-L, or PSI-N subunit. Biophys. J. 2002; 83, 2190–2201

[74] Mullet J.E, Burke J.J, Arntzen C.J, Chlorophyll proteins of Photosystem I. Plant Physiol. 1980; 65:814-822.

[75] Mozzo M, Morosinotto T, Bassi R, Croce R Probing the structure of Lhca3 by mutation analysis, Biochim. Biophys. Acta 2006;1757:1607-1613.

[76] Lelong C, Setif P, Lagoutte B, Bottin H Identification of the amino acids involved in the functional interaction between photosystem I and ferredoxin from Syneckocystis sp. PCC 6803 by chemical cross-linking. J Biol Chem 1994;269 10034-10039

[77] Xu Q, Jung Y.S, Chitnis V.P, Guikema J.A, Golbeck J.H, Chitnis P.R Mutational analysis of photosystem I polypeptides in Synechocystis sp. PCC 6803. Subunit requirements for reduction of NADP+ mediated by ferredoxin and flavodoxin. J Biol Chem. 1994; 269(34):21512-21518.

[78] Varotto C, Pesaresi P, Jahns P, Lessnick A, Tizzano M, Schiavon F, Salamini F, Leister D Single and double knock-outs of the genes for photosystem I subunits G, K and H of Arabidopsis. Effects on photosystem I composition, photosynthetic electron flow and state transtions. 2002; Plant Physiol. 129:616-624

[79] Lunde C, Jensen P.E, Haldrup A, Knoetzel J Scheller H.V The PS I-H subunit of Pho-
      tosystem I is essential for state transitions in plant photosynthesis. Nature 2000; 408:
      613–615

[80] Naver H., Haldrup A, Scheller H.V Cosuppression of photosystem I subunit PSI-H in
      Arabidopsis thaliana. Efficient electron transfer and stability of photosystem I is de-
      pendent upon the PSI-H subunit, J. Biol. Chem. 1999; 274:10784-10789

[81] Haldrup A, Simpson D.J., Scheller H.V Down-regulation of the PSI-F subunit of pho-
      tosystem I in *Arabidopsis thaliana*. The PSI-F subunit is essential for photoautotrophic
      growth and antenna function, J. Biol. Chem. 2000;275:31211-31218.

[82] Wöstemeyer A.; Oelmüller R The promoter of the spinach PSaF gene fort he subunit
      III of the photosystem I reaction center directs β-glucuronisade gene expression in
      transgenic tobacco roots, implication of the involment of phospholipases and protein
      kinase C in PsaF gene expression J.Plant Physiol. 2003;160: 503-508.

[83] Ihalainen J. A, Jensen P. E, Haldrup A, van Stokkum, I. H. M., van Grondelle R.,
      Scheller, H. V, Dekker, J. P Pigment organization and energy transfer dynamics in
      isolated photosystem I (PSI) complexes from Arabidopsis thaliana depleted of the
      PSI-G, PSI-K, PSI-L, or PSI-N subunit. Biophys. J. 2002; 83:2190–2201

[84] Oh-oka H, Takahaski Y, Kuriyama K, Saeki K, Matsubara H The protein responsible
      for center A/B in spinach photosystem I: isolation with iron-sulfur cluster(s) and
      complete sequence analysis. J. Biochem (Tokyo) 1988;103:962-68

[85] Okkels J.S, Scheller H.V, Jepsen L.B, Møller B.L A cDNA clone encoding the precur-
      sor for a 10.2 kDa photosystem I polypeptide of barley. FEBS Lett 1989; 250: 575–579

[86] Okkels J.S, Nielsen V.S, Scheller H.V, Møller B.L A cDNA clone from barley encod-
      ing the precursor from the photosystem I polypeptide PSI-G: sequence similarity to
      PSI-K. Plant Mol Biol 1992;18: 989–994

[87] Kusnetsov V, Bale C, Lübberstedt T, Hermann R.G, Oelmüller R Evidence that the
      plastid signal and light operate via the same cis-acting elements in the promoters of
      nuclear genes for plastid proteins. Mol. Gen. Genet. 1996; 252:631-639

[88] Ikeuchi M, Nyhus K.J, Inoue Y, Pakrasi H.B Identities of four low-molecular-mass
      subunits of the photosystem I complex from Anabaena variabilis ATCC 29413 evi-
      dence for the presence of the psaI gene product in cyanobacterial complex. FEBS
      1991; 287(1):5-9

[89] Chen S.C.G, Cheng M.C, Chen J, Hwang L.Y Organization of the rice chloroplast
      psaA-psaB-rps14 gene and the presence heterogeneity in this gene cluster. Plant Sci
      1990; 68:213-221

[90] Chen S.C.G, Cheng M.C, Chung K.R, Yu N.J, Chen M.C Expression of the rice chloro-
      plast of the rice chloroplast psaA-psaB-rps14 gene cluster. Plant Sci 1992;81:93-102

[91] Cheng M.C, Wu S.P, Chen L.F.O, Chen S.C.G Identification and prufication of a spinach chloroplast DNA-binding protein that interacts specifically with the plastid psaA-psaB-rps14 promoter region Planta 1997;203:373-380

[92] Kusnetsov V, Bale C, Lubberstedt T, Sapory S, Hermann R.G, Oelmüller R Evindence that the plastid signal and light operade via the same cis-acting elements in the promoters of nuclear genes for plastid proteins. Mol. Gen. Genetic 2006;252:631-639

[93] Kropat J, Oster U, Rudiger W, Beck C.I Chlorophyll precursors are signals of chloroplast origin involved in light induction of nuclear heat-schock genes Proc. Natl. Acad. Sci. USA 1997;94:14168-14172

[94] Mochizuki N, Brusslan J.A, Larkin R, Nagatani A, Chory J Arabidopsis genomes uncoupled 5 (GUN5) mutant reveals the involvement of Mg-chelatase H subunit in plastid-to-nucleus signal transduction. Proc. Natl. Acad. Sci. USA 2001; 98:2053-2058

[95] Kusnetsov V, Oelmüller R, Sarwat M.I, Porfirova S.A, Cherepneva G.M, Hormann R.G, Kulaeva O.N Cytokinins, abscisic acid and light affect on accumulation of chloroplast proteins in Lupinus luteus cothyledons without notable effect on steady state mRNA levels. Planta 1994;194:318-327

[96] Yamamoto Y.M, Tsuji H, Obokata J 5'-leader of a photosystem I gene in Nicotiana slyvestris, psaDb, contains a translational enhancer. J. Biol. Chem. 1995; 270(21): 12466-12470

[97] Kusnetsov V, Landsberger M, Neuer J, Oelmüller R, The assembly of the CAAT-box binding complex at the AtpC promoter is requlated by light, cytokinin and the stage of the plastids. J. Biol. Chem. 1999; 274:36009-36014

[98] Flieger K, Tyagi A, Sopary S, Cseplö A, Herrmann RG, Oelmüller R A 42 bp promoter fragment of the gene for subunit III of photosystem I (psaI) is crucial for its activity Plant J. 1993;4:9-17

[99] Oelmüller R, Lübberstedt T, Bolle C, Scpory S, Tyagy A, Cseplö A, Fliegr K, Herrmann R.G 1992. Promoter architecture of nuclear genes for thylakoid membrane proteins from spinach. In: Murata(ed) Research in Photosysthesis Vol III. Kluwer Acad. Publishers, Dordirecht pp. 219-224

[100] Gilmartin P.M, Memelink S, Hiratsuka K, Kay S.A, Chua N.HCharacterization of gene encoding a DNA binding protein with specificity for a light-responsive element. Plant Cell 1992;4:839-849

[101] Ihnatowicz A, Pesaresi P, Varotto C, Richly S, Scheider A, Jahns P, Salamini F, Leister D Mutants of photosystem I subunit D of Arabidopsis thaliana:effects on photosysthesis, photosystem I stability and expression of nuclear genes for chloroplast functions. Plant J. 2004;37:839-852.

[102] Allakhverdiev S. I, Sakamoto A, Nishiyama Y, Inaba M, Murata N Ionic and osmotic effects of NaCl-induced inactivation of photosystem I and II in *Synechococcus* sp. Plant Physiol 2000;123:1047-1056.

[103] Zhang T, Gong H, Wen X., Lu C Salt stress a decrase in excitation energy transfer from phycobilisomes to photosystem II but an increase to photosystem I in the cyanobacterium *Spirulina platensis*. J Plant Physiol. 2010;167:951-958.

[104] Jones L.W, Kok B Photoinhibition of chloroplast reaction I. Kinetics and action spectra. Plant Physiol 1966; 41:1037-1043

[105] Kok B On the inhibition of photosynthesis by intense light. Biochm. Biophys. Acta 1956; 21:234-244

[106] Sato K Mechanism of photoinactivation in photosynthetic systems. I. The dark reaction in photoinactivation. Plant Cell Physiol. 1970; 11:15-27

[107] Sato K Mechanism of photoinactivation in photosynthetic systems. II. The occurrence and properties of two different types of photoinactivation. Plant Cell Physiol. 1970; 11:29-38

[108] Inoue K, Sakurai H, Hiyama T Photoinactivation site of photosystem I in isolated chloroplasts. Plant Cell Physiol. 1986; 27:961-968

[109] Inoue K, Fujii T, Yakoyama E, Matsuura K, Hiyama T, Sakurai H The photoinhibition site of photosystem I in isolated chloroplasts under extremely reducing conditions. Plant Cell Physiol. 1989; 30:65-71

[110] Sonoike K Photoinhibition of photosystem I. Physiol. Plantarum 2011;142:56-64

[111] Teroshima I, Funayama S, Sonoike K The site of photoinhibition in leaves of Cucumis sativus L. at low temperature is photosystem I, not photosystem II. Planta 1994;193:300-306

[112] Sonoike K., Terashima Mechanism of the photosystem I photoinhibition in leaves of *Cucumis sativus* L. Planta 1994;194: 287-293

[113] Asada K. 1999 Radical production and scavenging in the chloroplasts. In: Baker NR(ed) photosynthesis and the environment Kluwer Academic Publishers, Dordrecht, pp 123-150

[114] Ivanov A.G, Morgan R.M, Gray G.R, Velitchkova M.Y, Huner N.P.A Temperature/light dependent development of selective resistance to photoinhibition of photosystem I. FEBS Lett. 1998;430:288-292

[115] Prasil O, Adir N, Ohad I 1992 Dynamics of photosystem I: Mechanism of photoinhibiton and recovery process. In Topics in Photosynthesis: the Photosystem Structure, Function and Molecular Biology, Edited by Barber J. pp. 295-348. Elsevier, Amsterdam

[116] Aro E.M, Virgin I, Anderson B, Photoinhibition of photosystem II. Inactivation, protein damage and turnover. Biochim. Biophys. Acta 1993;1143:113-134

[117] Andersson B, Barber J 1996. Mechanisms of photodamage and protein degradation during photoinhibition of photosystem II. In Photosynthesis and Enviroment. Edited by Baker NR, pp. 101-121. Kluwer Academic, Dordrecht

[118] Teicher H.B., Møller B.L, Scheller H.V photoinhibiton of photosystem I in field-grown barley (*Hordeum vulgare* L.): Induction, recovery and acclimation Photosynth. Res. 2000 64:53-61

[119] Kudah H, Sonoike K Irreversible damage to photosystem I by chilling in the light: cause of the degradation of chlorophyll after returaing to normal growth temperature. Planta 2002;215:541-548

[120] Zhang S, Scheller H.V Photoinhibition of photosystem I at chilling temperature and subsequent recovery in *Arabidopsis thalliana* Plant Cell. Physiol. 2004;45(11):1595-1602

[121] Murchie E. H, Hubbart S, Chen Y, Peng S, Horton P Acclimation of Rice Photosynthesis to Irradiance under Field Conditions. *Plant Physiol*. 2005;130, 1999-2010.

[122] Sonoike K Selective Photoinhibition of photosystem I In isolated thylakoid membrane from cucumber and spinach. *Plant Cell Physiol*. 1995; 36, 825-830.

[123] Ogawa K, Kanematsu S, Takabe K. Asada K. Attachment of Cu/Zn-superoxide dismutase to thylakoid membranes at the site of superoxide generation (PS1) in spinach chloroplasts: detection by immuno-gold labeling after rapid freezing and substitution method. *Plant cell Physiol*. 1995;36, 565-573.

[124] Tjus S. E, Møller B. L, Scheller H. V Photoinhibition of photosystem I damages both reaction center proteins PSI-A and PSI-B and acceptor-side located small photosystem I. *Photosynth. Res*. 1999; 60, 75-86

[125] Jiao D, Emmanuel H Guikema, J. A High light stress inducing photoinhibition and protein degradation of photosystem I in *Brassica rapa. Plant Sci*. 2004;167, 733-741.

[126] Lu C, Vonshak A, Characterization of PSII photochemistry in salt-adapted cells of cyanobacterium *Spirulina platensis*. New Phytol. 1999;141:231-239.

[127] Lu C, Vonshak A, Effects of salinity stres on photosystem II function in cyanobacterial *Spirulina platensis* cells. Physiol. Plant 2002;114:405-413.

[128] Gilmour D.J, Hipkins M.F, Boney A.D The effects of salt stres on the primary processes of photosynthesis in *Dunalielia tertioleca*. Plant Sci. Lett. 1982;26:325-330.

[129] Joset F, Jeanjean R, Hagemann M Dynamics of the response of cyanobacteria to salt stres:deciphering the molecular events. Physiol Plant. 1996; 738-744.

[130] Sudhir P. R, Pogoryelov D, Kovács L, garab G, Murthy S.D.S The effects of salt stres on photosynthetic electron transport and thylakoid memebrane proteins in the cyanobacterium *Spirulina platensis.* J Biochem Mol Biol. 2005;38:481-485.

[131] Hibino T, Lee B.H, Rai A.K, Ishikawa H, Kojima H, Tawada H, Shyimoyama H, Takebe T Salt enhances photosystem I content and cyclic electron flow via NAD(P)H dehydrogenase in the halotolerant cyanobacterium *Aphanothece halophytica.* Aust J Plant Physiol. 1996;23:321-330

[132] Jeanjean R, Matthijs H.C.P, Onana B, Havaux M, Joset F Exposure of the cyanobacterium Synechocystis PCC 6803 to salt stres induces concerted changes in respiration and photosynthesis. Plant Cell Physiol. 1993;34:1073-1079.

[133] Neelam A, Rai L. C Differantial responses of three cyanobacteria to UV-B and Cd, Microbiol. Biotechnol 2003;13:544-551.

[134] Zhou W, Juneau P, Qiu B Growth and photosynthetic responses of the bloom-forming cyanobacterium *Microcystis aeruginosa* to elevated lecels of cadmium, Chemosphere 2006;65:1738-1746.

[135] Ivanov A. G, Morgan R. M, Gray G.R, Velitchkova M. Y,. Huner N.P.A Temperature/ light dependent development of selective resistance to photoinhibition of photosystem I, FEBS Lett 1998;430:288–292.

[136] Jin S.H, Li X.Q, Hu J.Y, Wang J.G Cyclic electron flow around photosystem I is required for adaptation to high temperature in a subtropical forest tree, Ficus concinna. J Zhejiang Univ Sci B. 2009;10(10):784-790.

[137] Howitt C. A, Cooley J. W, Wiskich J. T, Vermaas W. F A strain of *Synechocystis* sp. PCC 6803 without photosynthetic oxygen evolution and respiratory oxygen consumption: implications for the study of cyclic photosynthetic electron transport, Planta 2001;214:46-56.

[138] Jeanjean R, Bedu S, Havaux M, Matthijs H. C. P, Joset F Salt-induced photosystem I cyclic electron transfer restores growth on low inorganic carbon in a type I NAD(P)H dehydrogenase deficient mutant of *Synechocystis* PCC6803, FEMS Microbiol. Lett 1998;167:131-137.

[139] Tanaka Y, Katada S, Ishikawa H, Ogawa T, Takabe T Electron flow front NAD(P)H dehydrogenase to photosystem I is required for adaptaion to salt shock in the Cyanobacterium *Synechocystis* sp. PCC 6803, Plant Cell Physiol 1997;38:1311-1318.

Hexavalent Chromium Induced Inhibition of Photosynthetic Electron Transport in Isolated Spinach Chloroplasts

Vivek Pandey, Vivek Dikshit and Radhey Shyam

Additional information is available at the end of the chapter

1. Introduction

Plants, being sessile, are often exposed to various kinds of harsh environmental conditions which adversely affect growth, metabolism and yield. Among various abiotic stresses, drought, salinity, gaseous pollutants and heavy metals are important environmental stressors which severely affect plant growth. Photosynthesis is essentially the only mechanism of energy input into living world. The decline in productivity in many plant species subjected to harsh environmental conditions is often associated with a reduction in photosynthetic capacity. Heavy metals have been increasing in the environment (air, water and soil) as a result of rapid industrialization, urbanization and agricultural runoff. Many of these metals have adverse effects on growth and metabolic processes in plants, including reduction in chlorophyll content, chloroplast degeneration, reduced photosynthesis and inhibition of enzyme activities.

Heavy metals influence photosynthesis by affecting pigments, electron transport activities and Calvin cycle enzymes. Of the various metals, Pb, Cu, Cd, Ni, Hg and Zn have been extensively studied in relation to their effects on plant photosynthesis [1-2]. Long term exposure of whole wheat plants to cadmium (Cd) affected chlorophyll and chloroplast development in young leaves [3]. However Pandey et al [4] reported significant increase in chlorophyll content in 20 μM Cr (VI) treated Indian mustard seedlings. Chloroplast membranes, particularly thylakoids, have been investigated as the sites of action of heavy metals. Photosynthetic electron transport within thylakoid membranes has been the primary target site of heavy metal action [5].

Light is shown to play a role in the binding of heavy metals to the chloroplast membranes and to inhibit photosynthetic electron transport. The accessibility of Cu (II) to site of inhibition in chloroplast membranes was much more rapid and to a greater extent in the light than in the dark [6]. A reduction in the inhibitory effects of Cu (II) upon leaving the membranes for short periods of time in the dark led them to presume that in the dark Cu (II) was being irreversibly bound to non-inhibitory site in the membranes, which prevented it from binding with inhibitory sites in the light [6]. The effect of Zn (II) on inhibition of oxygen evolution was somewhat identical in both light and dark [7]. However, Baker et al. [8] found a greater and tighter binding of zinc in the pea thylakoid membranes incubated in dark or exposed to low light as compared to those exposed to saturating light. Cu (II) inhibition increased in spinach chloroplasts with the time of light exposure [9]. Essentiality of light for the damage due to Cu (II) was also demonstrated in spinach PSII particles [10].

Chromium (Cr) is a toxic element that occurs in highly variable oxidation states. Chromium is found in all phases of the environment, including air, water and soil. Chromium content in naturally occurring soil ranges from 10 to 50 mg. kg^{-1} depending on the parental material. In ultramafic soils (serpentine), it can reach up to 125 g. kg^{-1} [11]. Cr and its compounds have multifarious industrial uses. They are extensively employed in leather processing and finishing [12], in the production of refractory steel, drilling muds, electroplating cleaning agents, catalytic manufacture and in the production of chromic acid and specialty chemicals. Hexavalent chromium compounds are used in industry for metal plating, cooling tower water treatment, hide tanning and, until recently, wood preservation. These anthropogenic activities have led to the widespread contamination of Cr in the environment. Very few studies have reported ameliorative measures for Cr toxicity in crop plants. *Pluchea indica* showed a good potential of phytoremediation, as it presented high levels of Cr accumulation and translocation to the leaves [13]. Mellem et al [14] found that *Amaranthus dubius* tolerate high Cr(VI) concentrations showing good potential for phytoremediation. Furthermore, Gardea-Torresdey et al [15] found that *Convolvulus arvensis* L. had capability to accumulate more than 3800 mg of Cr kg^{-1} dw tissue, showing that this specie can be used in phytoremediation of Cr(VI) contaminated soils. Khan (16) reported the potential of mycorrhizae in protecting tree species *Populus euroamericana, Acacia arabica and Dalbergia sisso* against the harmful effects of heavy metal and phytoremediation of Cr contamination in tannery effluent-polluted soils. The poor translocation of Cr from roots to shoots is a major hurdle in using plants and trees for phytoremediation.

Despite toxicity of chromium (Cr) to human and animals due to its instant exposure or via its incorporation in food chain, the mechanism of action of this element in the photosynthetic electron transport activity is not critically studied. Of the various species of Cr, trivalent Cr (III) and hexavalent Cr (VI) forms are of biological relevance due to their wide occurrence in the environment. While Cr (VI) is most biologically toxic oxidation state of chromium [17], Cr (III) is less toxic and is also an essential trace element in human nutrition [18]. The chemistry of Cr in respect to plant accumulation from the environment revolves around the reduction of Cr (VI) to Cr (III), the oxidation of Cr (III) to Cr (VI) and the relative stability of Cr (III) compound once formed [19].

Growth inhibition in Cr treated *Lemna gibba* was shown to be associated with an alteration of the PS II electron transport [20]. The net photosynthesis of primary and secondary leaves of *Phaseolus vulgaris* was decreased by Cr (VI) [21]. Decline in net CO_2 assimilation of leaves in Cr (VI) treated pea plants was found associated with depressed activities of both PS I and PS II. On the other hand, study on isolated pea chloroplasts showed that exogenously added Cr (VI) had markedly inhibited PS I activity while PS II activity was marginally affected [22]. This result was in contrast to the effects of most of the heavy metals (*Pb*, Cu, Cd, *Ni*, Hg and Zn) on photosynthetic electron transport where photosystem II was preferentially inhibited [23-25]. Cd treated pea and broad bean plants showed modified PS II activity due to damaged oxygen-evolving complex and caused disassembly of PS II itself [26]. Binding studies with radiolabel-led herbicide revealed that the Q_B pocket activity was also altered. Pätsikkä et al [27] while studying effect of Cu on photoinhibition of PS II in vitro in bean and pumpkin thylakoids concluded that the primary effect of excess Cu on the photoinhibition of PS II is caused by inhibition of electron donation to P680$^+$, which leads to donor-side photoinhibition. The *in vitro* study of Desmet et al [28] indicated reduction of Cr (VI) to a lower oxidation state of Cr by spinach chloroplasts in dark. The authors also showed that CrO_4^{2-} behave as an electron acceptor of photosynthesis in light.

The present study was undertaken to assess the impact of hexavalent Cr on isolated spinach chloroplasts, role of light in Cr binding and to resolve the site(s) of inhibition of Cr (VI) in photosynthetic electron transport.

2. Materials and methods

Chloroplasts were isolated from field grown spinach (*Beta vulgaris* L.) following the method of Navari-Izzo et al [29] with some modifications. Leaves were homogenized in a medium (1:3, w/v) containing 330 mM sucrose, 50 mM HEPES-KOH (pH 7.5), 5 mM $MgCl_2$, 10 mM NaCl. The homogenate was filtered through 4 layers of muslin cloth and centrifuged at 1100x *g* for 3 min. The pelleted chloroplasts were purified by layering on the top of a Percoll gradient (10 ml each of 40 and 80% Percoll prepared with isolation medium) and centrifuged for 30 min at 4000x *g*. After centrifugation the lower band containing intact chloroplasts was separated, washed gently and finally suspended in the isolation medium. The intactness of the chloro-plasts was estimated in the range of 90 to 95% by ferricyanide-dependent O_2 evolution [30]. Chloroplast fragments were prepared by giving osmotic shock to intact chloroplasts followed by their re-suspension in hypotonic medium [31].

PSII particles with high rate of oxygen evolution were prepared by the method of Berthold et al [32]. Chlorophyll was determined by method of Porra et al [33].

For inhibition study, thylakoid membranes or PSII particles (30 μg ml^{-1}) were incubated at 20° C with different concentration of $K_2Cr_2O_7$ in a reaction vessel for 5 min in the dark or in the presence of 10 μmol m^{-2}s^{-1} red light provided by LED, (Quantum Devices, Inc., USA) under continuous stirring. After treatment, the aliquots were centrifuged (at 5000x g for thylakoid and at 40000 g for PSII particles) and pellets after washing twice with assay buffer finally suspended in 1 ml assay medium to monitor electron transport activity. The conditions of assay for electron transport reactions using artificial electron donors and acceptors have been given under separate heads below. The oxygen evolution and consumption during different reactions were monitored polarographically in a total volume of 1 ml using a Clark type electrode (Hansatech, UK). A projector lamp (Kindermann-Germany) provided saturated light intensity under the condition of different electron transport assay.

2.1. Whole chain electron transport rate

The electron transport through the whole chain of photosynthesis, i.e. from H_2O to MV was measured as O_2 uptake. Assay medium, 1 ml, contained 30 mM HEPES, 10 mM NaCl, 5 mM NH_4Cl, 1 mM $MgCl_2$, 0.1 mM sodium azide, and 0.1 mM MV adjusted at pH 7.5. Thylakoid were added to the above reaction mixture to a final concentration of 30 µg.

2.2. PS I rate

PS I rate in lysed chloroplasts was monitored in a 1 ml reaction mix comprising of 50 mM HEPES (pH 7.5), 0.2 mM DCPIP, 2.0 mM ascorbate, 5 mM NH_4Cl, 2 µM DCMU, 0.1 mM MV, 0.1 mM Na azide and 30 µg thylakoid/ml. DCPIP reduction was measured spectrophotometerically at 600 nm in Shimadzu UV-VIS 1601 spectrophotometer.

2.3. PS II rate

PS II rate in lysed chloroplasts was monitored in 1 ml reaction mix consisting of 30 mM HEPES (pH 7.0), 10 mM NaCl, 1 mM $MgCl_2$, 0.5 mM DMBQ and 30 µg thylakoid/ml.

2.4. DQH_2 to MV and TMPD bypass

The photo reduction of MV with duroquinol as the donor was seen in reaction mix containing 30 mM HEPES (pH 7.5), 10 mM NaCl, 5 mM NH_4Cl, 1 mM $MgCl_2$, 0.1 mM sodium azide, 0.1 mM MV, 4 µM DCMU and 0.5 mM duroquinol (freshly prepared in ethanol) and 30 µg/ml thylakoids. For TMPD bypass, 30 µM TMPD and 1 µM DBMIB were added.

2.5. Reactions with PS II particles

The standard assay medium for PS II membranes consisted of 350 mM sucrose, 50 mM MES (pH 6.0), 2 mM $MgCl_2$, 15 mM NaCl, and 0.5 mM DMBQ or 0.2 mM $SiMO_4$ as electron acceptors. Ferricyanide was not used as it interfered with Cr. DMBQ was dissolved in ethanol and SiMO in 50% aqueous dimethyl sulfoxide.

2.6. Spectrophotometric measurements of PS II electron transfer activity in the presence and in the absence of Diphenylcarbazide (DPC)

In some experiments effect of Cr and light was seen on isolated spinach PS II particles by measuring the rate by which PS II particles reduce dichlorophenol-indophenol (DCPIP) in the presence and in the absence of the artificial electron donor diphenyl carbazide (DPC). DCPIP reduction was measured in the reaction water to DCPIP and DPC to DCPIP as a decrease in absorbance at 600 nm in a 2 ml reaction mixture containing 200 mM sucrose, 30 mM MES (pH 6.5), 1 mM $MgCl_2$, 10 mM NaCl, 60 µM DCPIP in the presence and absence of 0.5 mM DPC. Because Cr would react with DPC, PS II particles were washed twice with PS II buffer and pelleted. The light saturated rate of DCPIP reduction was measured by illuminating the PS II particles with the PPFD of 1000 µmol m^{-2} s^{-1} of red light in a spectrophotometer cuvette. The

absorbance reading was taken at 600 nm in Shimadzu UV-VIS 1601 spectrophotometer at 30-
s intervals during the 2-min assays.

2.7. Cr estimation in thylakoids

Chromium was determined in isolated chloroplasts. The chloroplast pellets were dried at 48
^{0}C for 48 h. Determination of Cr was made by atomic absorption spectromerty (AAnalyst 300,
Perkin Elmer, USA) on nitric acid: perchloric acid (3:1, v/v) digests of pelleted chloroplasts
using background correction. Stock chromium certified solutions (Sigma) and blanks were run
in parallel to validate the quality of metal analyses.

2.8. Hydroxyl radical estimation

Formation of hydroxyl radicals was measured under illumination in a stirred reaction vessel
with 30 μg chl ml^{-1} at 25° C in the presence of 0.7 M DMSO according to Babbs et al [34]. The
medium for illumination was 50 mM MES (pH 6.0), 2 mM $MgCl_2$ and 10 mM NaCl. After
centrifugation (3 min at 10,000 g), the pH of the supernatant was adjusted to 2.5 by adding
HCl. The colour reaction was started by adding 0.14 mM fast blue BB salt (prepared fresh) to
the chloroplast supernatant. The methane sulphinic acid (MSA) content was then calculated
from the absorbance at λ_{max} 425 nm using the extinction coefficient of 14.5 mM^{-1} cm^{-1}.

3. Results & discussion

Presence of chromium ions in the assay medium did not allow instant and accurate monitoring
of electron transport activities in the isolated spinach chloroplasts due to their reactivity with
the reagents (DMBQ, DCPIP, DPC and ascorbate) used in analyzing different electron
transport reactions (data not shown). The other uncertainty in monitoring of the electron
transport rates in the presence of chromium was that the chromium inhibition of electron
transport activity was more effective at acidic pH while optimal electron transport rates were
recorded at neutral to slightly basic pH of the assay medium. It was found therefore critical to
eliminate chromium from the assay medium. With a view to avoid interference of free Cr ions
during assay, we incubated spinach chloroplasts for different duration with varying concentrations of Cr and prior to monitoring electron transport activity, the treated thylakoids were
washed with the assay buffer to remove unbound Cr. The resulting Cr induced inhibition in
the electron transport activity was irreversible even after treatment with 5 mM EDTA (data
not shown).

Light is known to play a major role in heavy metal induced toxicity. Jagerschold et al [35] found
that strong illumination inhibited PS II electron transport and degraded D1 protein in chlorine
depleted medium. With a view to see the effects of light, we incubated spinach chloroplasts
under different per mutation combination of PAR and chromium concentrations (preliminary
data not shown). The light dependence of chromium toxicity in the chloroplasts incubated for
5 min with 10 μM Cr (VI) on whole chain electron transport activity is shown in Figure 1.

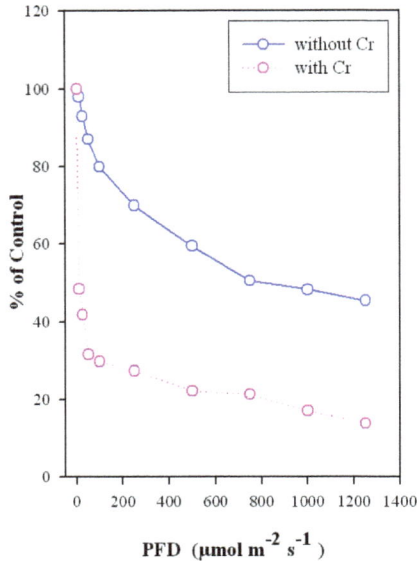

Figure 1. Whole chain electron transport rate (Water to MV) of spinach chloroplasts showing light dependence curve in absence and presence of Cr (10 μM)

Cr (VI) treatment of the chloroplasts in the light had a remarkable impact on photoreduction of methyl viologen. The inhibition of 8% whole chain electron transport activity observed in the chloroplasts incubated for 5 min with 10 μM Cr (VI) in dark, increased to ca 68% when chloroplasts were incubated for 5 min with 10 μM Cr (VI) at 50 μmol m^{-2} s^{-1} PPFD (Figure 2 A). Exposure of chloroplasts to and above 50 μmol m^{-2} s^{-1} in the presence of 10 μM Cr (VI) caused a gradual inhibition leading to 86% loss in whole chain electron transport activity at 1250 μmol m^{-2} s^{-1} (Figure 2 A). These results indicated that even low PFD during incubation of chloroplasts with chromium was substantially effective in inactivating electron transport reactions. In order to validate our finding we monitored rate of electron flow from water to MV, water to DMBQ (PSII) and ascorbate + DCPIP to MV (PSI) subsequent to incubation of chloroplasts at 10 μmol m^{-2} s^{-1} for 5 min with different concentrations of Cr (VI) (Figure 2 A-C).

The pretreatment of the chloroplasts with Cr (VI) at the PPFD as low as 10 μmol m^{-2} s^{-1} caused an abundant increase in the inhibition of whole chain as well as partial electron transport reactions. While the lowest Cr (VI) concentration (10 μM) in dark incubated chloroplasts decreased the whole chain, PSII and PSI electron transport activities by 8, 5 and 3%, the corresponding inhibition in the chloroplasts incubated at 10 μmol m^{-2} s^{-1} light amounted to 50, 32 and 30%, respectively (Figure 2 A-C). While the I_{50} value for inhibition in dark incubated chloroplasts was 1000 μM Cr (VI) for water to MV, it was only 10 μM Cr (VI) for corresponding reaction in the chloroplasts incubated at 10 μmol m^{-2} s^{-1}. The I_{50} Cr (VI) concentrations for PSII and PSI inhibition in the chloroplast incubated at 10 μmol m^{-2} s^{-1} was 100 and 1000 μM Cr (VI)

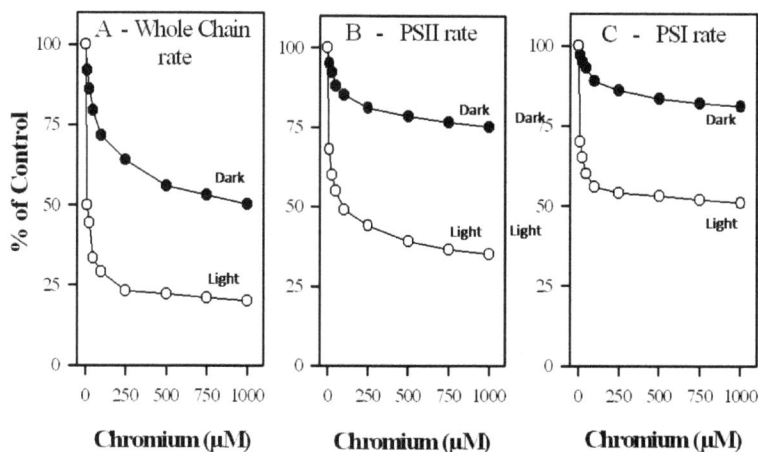

Figure 2. Effect of different concentrations of Cr on isolated spinach chloroplasts incubated either in dark or 10 μmol light. A: whole chain rate (Water to MV),B: PS II rate (Water to DMBQ) and C:. PS I rate (ASC/DCPIP to MV). Control rate of whole chain reaction: 200 μmol mg chl^{-1} h^{-1}; control rate of PS II reaction: 90 μmol mg chl^{-1} h^{-1}; control rate of PS I: 350 μmol mg chl^{-1} h^{-1}.

respectively which otherwise could not be achieved in chloroplasts incubated in dark even with 2000 μM Cr (VI).

In the past, most workers located the target of heavy metal inhibition of PSII to its oxidizing side [5]. In the present study, the assay of partial electron transport reaction was carried out with DCPIP as an electron acceptor. DCPIP accepts electrons from the plastoquinone pool of the electron transport chain and its photo-reduction is considered primarily a PS II reaction [7]. Cr (VI) treated PSII particles showed a gradual reduction in water to DCPIP rate. And DPC to DCPIP reaction of PS II did not show any significant recovery in rates from H$_2$O to DCPIP in the isolated PSII particles (Figure 3). These reactions ruled out the possibility of Cr effects on water oxidation complex.

As the DPC as electron donor could not reverse the PSII inhibition by Cr (VI), we may assume that the Cr (VI) if inactivate PSII donor side, that must be after DPC electron donating site i.e. the P680 and/or the acceptor side. It is possible that Cr impaired the PSII photochemistry by an interaction at or beyond the PSII primary electron carrier donor, TYR, (redox active Tyr of the D1 protein). It was, however, worth noting that in the present study Cr (VI) inhibited the DCMU-insensitive H$_2$O-SiMO$_4$ (silicomolybdate) activity. Although, silicomolybdate is an artificial electron acceptor with a very controversial binding site in PSII, this acceptor has been long considered the only specific one for testing the PSII donor side in a Hill reaction [36-38]. We found gradual reduction (5 to 32%) in water to silicomolybdate rate as the Cr concentration increased (Figure 4). Our results on above reaction carried out with Tris-washed PSII particles clearly excluded the Q$_B$ site and the water splitting system as the main Cr-inhibitory targets in PSII.

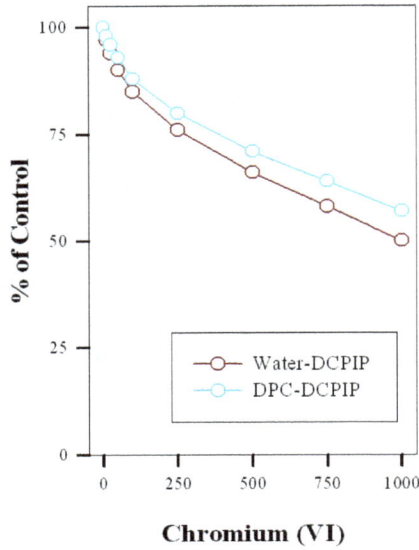

Figure 3. Effect of DPC on DCPIP reduction in Cr treated spinach PS II particles incubated in 10 μmol m^{-2} s^{-1} light for 5 min.

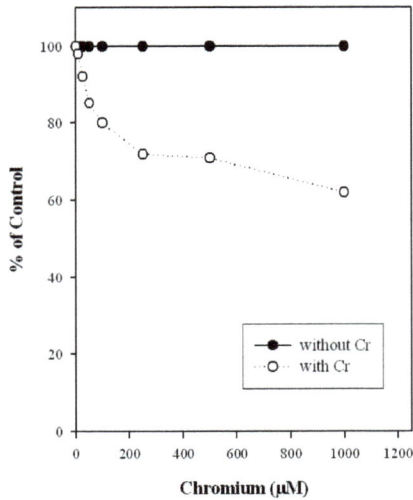

Figure 4. Effect of different concentrations of Cr on water to silicomolybdate rate of PS II particles isolated from spinach chloroplasts. 100% control activity was about150 μmol O$_2$ evolved mg chl^{-1} h^{-1}.

That the Cr (VI) inhibitory site is located before DCMU inhibitory site we studied the DCMU-inhibition pattern in the presence of different concentration of Cr by using Lineweaver-Burk plot. The data obtained were plotted 1/inhibition vs. 1/[DCMU] in the presence of different concentration of Cr (VI) and saturating amount of DMBQ. We obtained linear regression curves with the same intersect on the X axis, that implies that Cr (VI) is a non-competitive inhibitor with respect to DCMU (Figure 5). These results indicated that Cr (VI) binding site did not overlap with that of the DMBQ electron acceptor site and that of the DCMU binding site. These findings further suggested that the Cr (VI) binding site might be at the level of the Pheo-Q_A-Fe domain, separated from the Q_B niche (the DMBQ electron acceptor site). DCMU was shown to be competitive inhibitor with respect to Q_B is in agreement with the location of this herbicide-binding site at the level of Q_B niche.

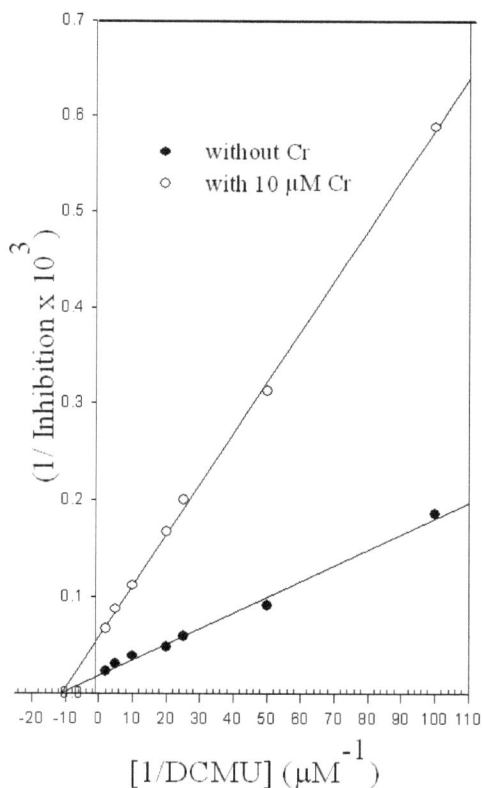

Figure 5. Reciprocal plot of the inhibitory effect of DCMU on the inhibition of oxygen evolution activity by Cr (VI). The DCMU-inhibition was measured in the absence of Cr (VI) and in the presence of 10 µM Cr (VI) and is given as µmol O_2 mg chl^{-1} h^{-1}. DMBQ was used as electron acceptor at a concentration of 0.5 mM.

In order to see if any component between the two photosystems was affected by Cr (VI), the electron transport from reduced plastoquinone to methyl viologen was monitored by using duroquinol as the donor for MV reduction At 10 – 1000 μM Cr (VI) a 30-70% inhibition was found (Figure 6). There was restoration of 17-24% of electron flow by addition of TMPD which has been shown to bypass the native plastohydroquinone site. Thus the results indicated that the components between the two photosystems were partly affected by Cr (VI).

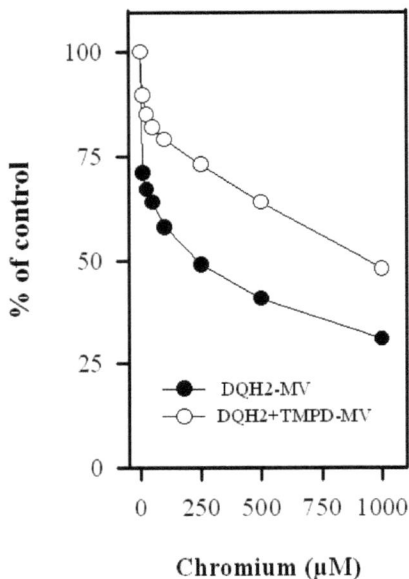

Figure 6. Effect of different concentrations of Cr on electron transport rate between the two photosystems in spinach chloroplasts

The possibility of Cr (VI) binding to the thylakoids during dark and light incubation might be a cause of different degree of inhibition, was examined by determining the amount of bound chromium. The membranes were washed thrice with incubation buffer containing $MgCl_2$, NaCl and MES at pH 6.5 to remove weakly associated Cr and then analyzed their Cr content. The amounts of Cr tightly bound to the thylakoid membranes treated with different concentrations of Cr (VI) in the dark and at low PPFD of 10 μmol $m^{-2}s^{-1}$ are presented in Figure 7. The results showed that thylakoid membranes treated in light had more bound Cr than those treated in dark. The differences in the binding of Cr to the thylakoid membranes were significantly higher in light than in the dark at all the Cr (VI) concentrations (Figure 7). Moreover, the binding of Cr to the membranes as a function of Cr (VI) concentrations both in dark and light showed a gradual increase from 3 and 12.3 μg Cr mg^{-1} Chl at 10 μM Cr (VI) to 9 and 21.4 μg Cr mg^{-1} Chl at 1000 μM Cr (VI) concentrations, respectively. It is possible that low PAR during Cr exposure to thylakoids may induce conformational and dynamic movements of the complete thylakoid membrane system that may facilitate transport of Cr into the granal region to bind with functional components of electron transport chain.

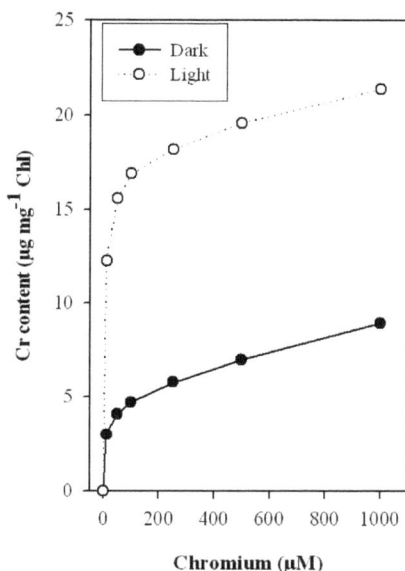

Figure 7. Binding of Cr to thylakoid membranes in light and dark

Chromium enhanced formation of lethal hydroxyl radical in light treated thylakoids (Figure 8). In order to investigate the mechanism of Cr (VI) mediated reactive oxygen species formation, we studied the effect of several oxygen radical scavengers on the oxygen evolution activity of Cr and light treated thylakoid membranes. With respect to treated thylakoids (10 μM Cr (VI) and 10 μmol m^{-2} s^{-1} light), where the activity was 53 % after 5 min of illumination, the activity was restored by ca. 10% by SOD and CAT (Table 1). We also used dimethyl sulfoxide (DMSO), a molecular probe for OH radical. It provided strong protection and activity was restored to 74% (Table 1). When DMSO, SOD and CAT were used together, they provided the best protection and activity was restored to 81% (Table 1). Ali et al [20] reported effect of Cr (VI) on *Lemna gibba* PS II photochemistry. The authors showed that Cr inhibitory site was located at oxygen evolving complex and Q_A reduction. The inhibition of PS II electron transport and formation of ROS by Cr were highly correlated with the decrease in the D1 protein and OEC proteins. There is also evidence that ROS may induce direct degradation of D1 peptide bonds [37]. Thus our results implicate superoxide in the inhibition mechanism and give evidence that hydroxyl radicals are formed via metal-catalyzed Haber-Weiss reaction and contribute to the inhibitory mechanism.

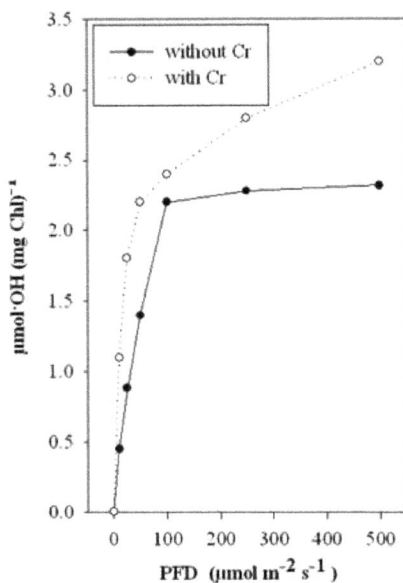

Figure 8. Effect of PFD on Cr (10 μM) induced generation of hydroxyl radical in isolated spinach thylakoids.

Protective system	Activity (% of control)
Without additions	58.81±1.8
Catalase (2000 U ml⁻¹)	64.26±1.6
SOD (300 U ml⁻¹)	62.84±1.1
7% DMSO	74.40±1.6
7% DMSO + Catalase (2000 U ml⁻¹) and SOD (300 U ml⁻¹)	81.52±2.4

Table 1. Effects of combined systems for protection against free radicals on the inhibition of electron transport activity (H_2O-MV) following Cr (10μM) +10μmol m^{-2} s^{-1} treatment of thylakoid membranes.

4. Conclusion

The present study showed that Cr (VI) inhibits both PS II and PS I electron transport of isolated spinach chloroplasts (Figure 9). Within the PS II, Pheo-Q_A region was more affected. Water oxidation complex was not affected by Cr (VI). Light as low as 10 μmol m^{-2} s^{-1} enhanced Cr

(VI) induced inhibition of electron transport. Our results also indicated that hydroxyl radical may be a major contributing factor towards decreased electron transport rate.

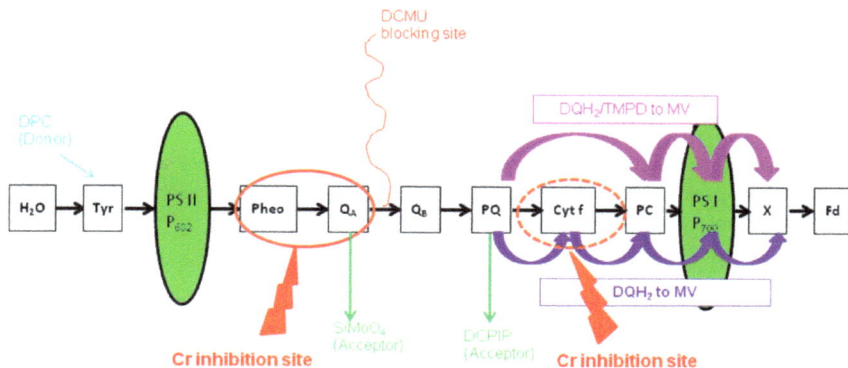

Figure 9. Schematic diagram of sites of Cr inhibition of photosynthetic electron transport in isolated spinach chloroplasts. The major sites of inhibition are at Pheo-Q_A domain and between two photosytems. Diagram also shows DQH_2 to MV and TMPD bypass rate. (for explanation please see the text). Electron transport chain components: Tyr tyrosine, P_{682} reaction centre chlorophyll for PS II, Pheo pheophytin, Q_A a plastoquinone tightly bound to PS II, Q_B another plastoquinone loosely bound to PS II, PQ plastoquinone, Cyt f cytochrome f, PC plastocyanin, P_{700} reaction centre chlorophyll for PS I, X primary electron acceptor of PS I, Fd ferredoxin

Author details

Vivek Pandey[1*], Vivek Dikshit[2] and Radhey Shyam[1,3]

*Address all correspondence to: v.pandey@nbri.res.in

1 Plant Ecology & Environmental Science, CSIR-National Botanical Research Institute, Lucknow, India

2 Biochemistry Division, Jain Biotech Lab, Jalgaon, Maharashtra, India

3 Vikas Khand, Gomtinagar, Lucknow, India

References

[1] Clijsters, H, & Van Assche, F. Inhibition of photosynthesis by heavy metals. Photosynthesis Research (1985). , 7, 31-40.

[2] Prasad MNV. Trace metals. In: Prasad MNV, editor. Plant Ecophysiology. New York: Wiley, (1997). , 207-249.

[3] Malik, D, Sheoran, I. S, & Singh, R. Carbon metabolism in leaves of cadmium treated wheat seedlings. Plant Physiology and Biochemistry (1992). , 30, 223-229.

[4] Pandey, V, Dixit, V, & Shyam, R. Antioxidative responses in relation to growth of mustard (*Brassica juncea* cv. Pusa Jaikisan) plants exposed to hexavalent chromium. Chemosphere (2005). , 61, 40-47.

[5] Krupa, Z, & Baszynski, T. Some aspects of heavy metals toxicity towards photosynthetic apparatus- direct and indirect effects on light and dark reactions. Acta Physiologia Plantarum (1995). , 17, 177-190.

[6] Cedeno-Maldonado, A, & Swader, J. A. The cupric ion as an inhibitor of photosynthetic electron transport in isolated chloroplasts. Plant Physiology (1972). , 50, 698-701.

[7] Tripathi, B. C, & Mohanty, P. Zinc inhibited electron transport of photosynthesis in isolated barley chloroplasts. Plant Physiology (1980). , 68, 1174-1178.

[8] Baker, N. R, Fernyhough, P, & Meek, I. T. Light dependent inhibition of photosynthetic electron transport by zinc. Physiologia Plantarum (1982). , 56, 217-222.

[9] Samuelsson, G, & Öquist, G. Effects of copper chloride on photosynthetic electron transport and chlorophyll-protein complexes of *Spinacia oleracea*. Plant and Cell Physiology (1980). , 21, 445-454.

[10] Arellano, J. B, Lazaro, J. J, Lopez-Gorge, J, & Baron, M. The donor side of photosystem II as copper-inhibitory binding site. Photosynthesis Research (1995). , 45, 127-134.

[11] Adriano, D. C. Trace Elements in the Terrestrial Environment. New York: Springer Verlag; (1986). , 105-123.

[12] Nriagu, J. O. Production and uses of chromium. Chromium in natural and human environment. New York, USA. John Wiley and Sons; (1988). , 81-105.

[13] Sampanpanish, P, Pongsapich, W, Khaodhiar, S, & Khan, E. Chromium removal from soil by phytoremediation with weed plant species in Thailand. Water, Air, and Soil Pollution, (2006). , 6, 191-206.

[14] Mellem, J. J, Baijnath, H, & Odhav, B. Bioaccumulation of Cr, Hg, As, Pb, Cu and Ni with the ability for hyperaccumulation by *Amaranthus dubius*. African Journal of Agricultural Research, (2012). , 7, 591-596.

[15] Gardea-Torresdey, J. L, Peralta-Videa, J. R, Montes, M, De La Rosa, G, & Corral-diaz, B. Bioaccumulation of cadmium, chromium and copper by *Convolvulus arvensis* L.: impact on plant growth and uptake of nutritional elements. Bioresource Technology, (2004). , 92, 229-235.

[16] Khan, A. G. Relationships between chromium bio magnification ratio, accumulation factor, and mycorrhizae in plants growing on tannery effluent-polluted soil. Environment International, (2001). , 26, 417-423.

[17] Von Burg, R, & Liu, D. Chromium and hexavalent chromium. Journal of Applied Toxicology (1993). , 13, 225-230.

[18] Mertz, W. Chromium occurrence and function in biological systems. Physiologial Review (1969). , 49, 163-239.

[19] Cary, E. E. Chromium in air, soil and natural waters. In: Langård S, editor. Biological and Environmental Aspects of Chromium. Elsevier Biomedical Press, (1982). , 50-63.

[20] Ali, N. A, Dewez, D, Didur, O, & Popovic, R. Inhibition of photosystem II photochemistry by Cr is caused by the alteration of both D1 protein and oxygen evolving complex. Photosynthesis Research (2006). , 89, 81-87.

[21] Austenfeld, F. A. The effect of Ni, Co and Cr on net photosynthesis of primary and secondary leaves of *Phaseolus vulgaris* L. Photosynthetica (1979). , 13, 434-438.

[22] Bishnoi, N. R, Chugh, L. K, & Sawhney, S. K. Effect of chromium on photosynthesis, respiration and nitrogen fixation in pea (*Pisum sativum* L.) seedlings. Journal of Plant Physiology (1993). , 42, 25-30.

[23] Hampp, R, Beulich, K, & Zeigler, H. Effects of zinc and cadmium on photosynthetic CO_2 fixation and Hill activity of isolated spinach chloroplasts. Zeitschrift für Pflanzenphysiologie (1976). , 77, 336-344.

[24] Bernier, M, Popovic, R, & Carpentier, R. Mercury inhibition at the donor side of photosystem II is reversed by chloride. FEBS Letters (1993). , 321, 19-23.

[25] Jegerschold, C, Arellano, J. B, Schroder, W. P, Kan, P. J, Maron, M, & Styring, S. Copper (II) inhibition of electron transfer through photosystem II studied by EPR spectroscopy. Biochemistry (1995). , 34, 12747-12754.

[26] Geiken, B, Masojidek, J, Rizzuto, M, Pompili, M. L, & Gjardi, M. T. Incorporation of [^{35}S] methionine in higher plants reveal that stimulation of D1 reaction centre II protein turnover accompanies tolerance to heavy metal stress. Plant Cell and Environment (1998). , 21, 1265-1273.

[27] Pätsikkä, E, Aro, E-M, & Tyystjärvi, E. Mechanism of copper-enhanced photoinhibition in thylakoid membranes. Physiologia Plantarum (2001). , 113, 142-150.

[28] Desmet, G, Ruyter, A. D, & Ringoet, R. Absorption and metabolism of $Cr_2O_4^-$ by isolated chloroplasts. Phytochemistry (1975). , 14, 2585-2588.

[29] Navari-Izzo, F, Quartacci, M. F, Pinzino, C, & Vecchia, F. D. Sgherri CLM. Thylakoid-bound and stromal antioxidative enzymes in wheat treated with excess copper. Physiologia Plantarum (1998). , 104, 630-638.

[30] Walker, D. A. Preparation of higher plant intact chloroplasts. Methods in Enzymology (1980). , 69, 94-105.

[31] Reeves, S. G, & Hall, D. O. Higher plant chloroplasts and grana: General preparative procedures (excluding high carbon dioxide fixation ability chloroplasts). Methods in Enzymology (1980). , 69, 85-94.

[32] Berthold, D. A, Babcock, G. T, & Yocum, C. A. A highly resolved, oxygen-evolving Photosystem II preparation from spinach thylakoid membranes. FEBS Letters (1981). , 134, 231-234.

[33] Porra, R. J, Thompson, W. A, & Kriedman, P. E. Determination of accurate extinction coefficients and simultaneous equations for assaying chlorophylls a and b extracted with four different solvents: Verification of the concentration of chlorophyll standards by atomic absorption spectroscopy. Biochemica et Biophysica Acta (1989). , 975, 384-394.

[34] Babbs, C. F, Pham, J. A, & Coolbaugh, R. C. Lethal hydroxyl radical production in paraquat treated plants. Plant Physiology (1989). , 90, 1267-1270.

[35] Jegerschold, C, Virgin, I, & Styring, S. Light-Dependent Degradation of the D1 Protein in Photosystem I1 Is Accelerated after Inhibition of the Water Splitting Reaction. Biochemistry (1990). , 29, 6179-6186.

[36] Barber, J, Chapman, D. J, & Telfer, A. Characterization of a PS II reaction centre isolated from the chloroplasts of Pisum sativum. FEBS Letters (1987). , 220, 67-73.

[37] Lubinkova, L, & Komenda, J. Oxidative modifications of the photosystem II D1 protein by reactive oxygen species: from isolated protein to cyanobacterial cells. Photochemistry and Photobiology (2004). , 79, 152-162.

[38] Takahashi, Y, Hansson, Ö, Mathis, P, & Satoh, K. Primary radical pair in the Photosystem II reaction centre. Biochemica et Biophysica Acta (1987). , 893, 49-59.

Methods

Chapter 10

Photoacoustics — A Novel Tool for the Study of Aquatic Photosynthesis

Yulia Pinchasov-Grinblat and Zvy Dubinsky

Additional information is available at the end of the chapter

1. Introduction

1.1. Photoacoustics

The photoacoustic method allows direct determination of the energy-storage efficiency of photosynthesis by relating the energy stored by it to the total light energy absorbed by the plant material (Canaani et al., 1988; Malkin & Cahen, 1979; Malkin et al., 1990). These authors applied the photoacoustic method to leaves in the gas phase, where brief pulses caused concomitant pulses of oxygen that caused a pressure transient detected by a microphone. This method is based on the conversion of absorbed light to heat. Depending on the efficiency of the photosynthetic system, a variable fraction of the absorbed light energy is stored, thereby affecting the heat evolved and the resulting photoacoustic signal. The higher the photosynthetic efficiency, the greater will be the difference between the stored energy with and without ongoing photosynthesis (Cha & Mauzerall, 1992). These authors collected microalgal cells onto a filter and studied them by an approach similar to that previously used with leaves. In both cases, the oxygen signal is combined with that of thermal expansion resulting from conversion of the fraction of the light energy in the pulse that is not stored by photochemistry.

In the case of liquid algal cultures, there is no signal due to photosynthetic oxygen evolution as gas; hence, the signal detectable by an immersed microphone is proportional to the heat generated by a laser pulse. The light absorbed by the photosynthetic pigments in the algal cells is, in part, stored by photochemistry as products of photosynthesis, while the remainder is converted to heat, causing an expansion of the culture medium. This expansion causes a pressure wave that propagates through the culture and is sensed by the hydrophone. By exposing the cells to continuous saturating background light, no storage of any of the pulse energy can take place, whereas in the absence of such light, a maximal fraction of the pulse energy is stored by photosynthesis. Thus, the maximal photosynthetic storage efficiency, PSmax, is determined from the difference between the signal obtained from a weak laser pulse under strong, continuous illumination (PAsat) and that obtained in the dark (PAdark). The above is then divided by PAsat.

$$PS\,max = (PAsat - PAdark)/PAsat \tag{1}$$

[For development of equations, see Cha and Mauzerall (1992)].

The experimental setup is shown schematically in Figure 1. The sample was placed in a sample cell. The beam of the brief laser pulse (5ns) is incident upon the suspension of algae, whose pigments absorb part of the laser light (Fig. 2). Depending on the experimental conditions, a variable fraction of the absorbed light pulse is stored in the products of photosynthesis. The remainder of the absorbed light is converted to heat, which causes a transient expansion of the surrounding water, producing an acoustic wave. This is intercepted by a submerged microphone containing a pressure-sensing ceramic disc.

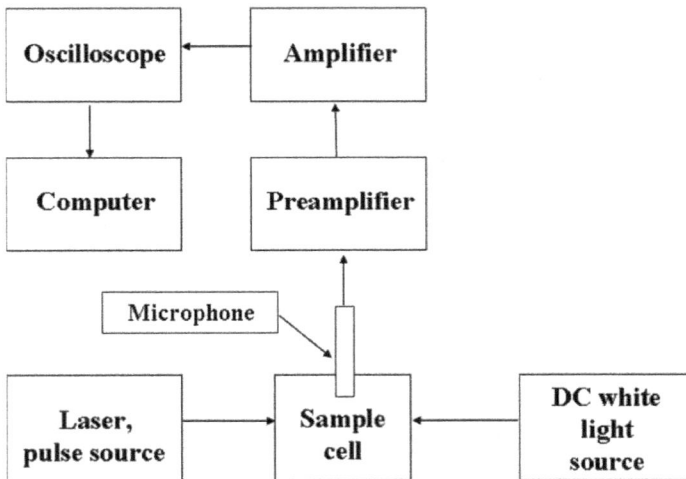

Figure 1. The photoacoustic setup.

A small portion of the laser pulse is used to trigger the Tektronix TDS 430A oscilloscope, where the amplified (Amptek A-250 Preamp and Stanford Research 560 Amp) photoacoustic signal is recorded (computer). An amount of 10 µs of the time scale was found by us as the optimal duration for quantification of the signal, beyond which the signal to noise ratio deteriorates. This time frame allowed us to fire the laser at 10 hz, thus averaging 128-256 pulses.

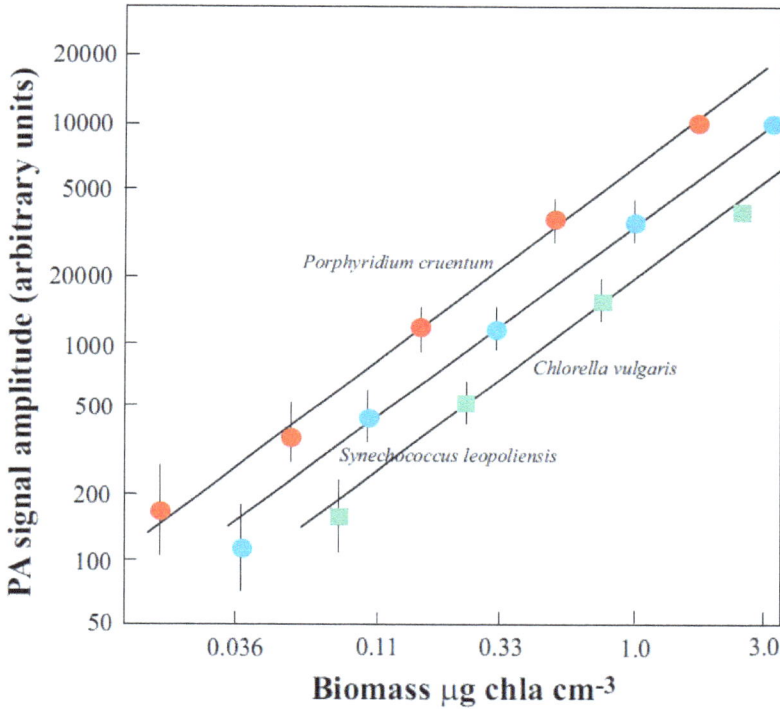

Figure 2. The photoacoustic signal vs Chl a content in 3-fold serially diluted laboratory cultures of three organisms ($\log_{10}$ scale: red circles – *Porphyridium cruentum*; blue circles – *Synechococcus leopoliensis*; green circles – *Chlorella vulgaris*. For biomass determination, the light pulse was 430 µJ at 532 nm [according to Mauzerall et al. (1998)]. The laser used was at 532 nm.

2. Quantification of biomass

Biomass detection by photoacoustics is based on the proportionality of the absorbed light to the amount of pigment (Dubinsky et al., 1998; Mauzerall et al., 1998). At high energies, the pulse saturates photosynthesis, and the photosynthetically stored energy becomes a negligible fraction of the absorbed energy. Under these conditions, the photoacoustic signal was proportional to the concentration of chlorophyll over the range of 14 mg to 8.5 µg chl a m^{-3} (Fig. 2) (Dubinsky et al., 1998; Mauzerall et al., 1998). Figure 3 shows the photoacoustic signal of a *Porphyridium cruentum* suspension and the same cells diluted 1:1 with medium (Mauzerall et al., 1998). The advantages of photoacoustics are the strict and exclusive proportionality to the light absorbed by the sample and the ease of obtaining photosynthetic efficiency.

3. Energy storage in photosynthesis

In a photosynthetic system at a given constant light intensity, a fraction of the reaction centers is closed at any time and only part of the light energy is stored (Dubinsky et al., 1998; Mauzerall et al., 1998). Figure 4 shows the photoacoustic signal in the dark (broken line) and saturating light conditions in a suspension of *Porphyridium cruentum*. With no background light ("in the dark"), all reactions are open and the very weak probe pulse causes no saturation, resulting in maximal photosynthesis and minimal heat release (Fig. 4) (Mauzerall et al., 1998).

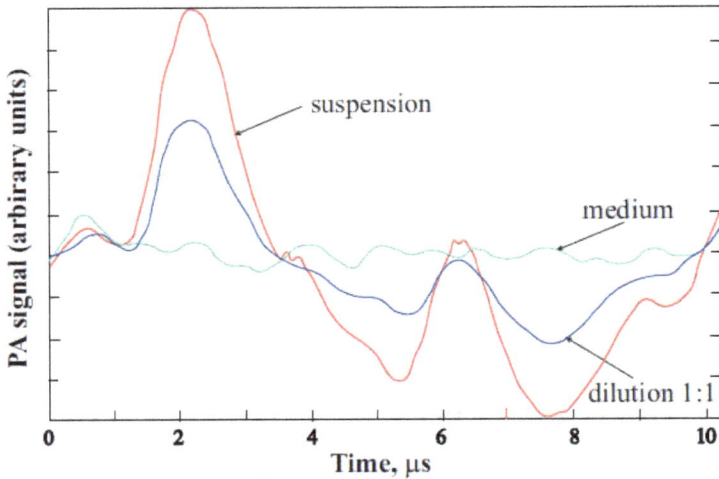

Figure 3. Photoacoustic-signal dependence on concentration of *Phorphyridium cruentum* in suspension. Dotted line is the signal from the (non-absorbing) medium. Red line is for ~0.5 µg/cm³ Chl a and purple is for 1:1 dilution of that solution [according to Mauzerall et al. (1998)].

By increasing the continuous background light intensity from zero to saturation of photosynthesis, an increasing fraction of the reaction centers is closed at any time, and a decreasing fraction of the probe laser pulse energy is stored. A corresponding increase in the fraction of the pulse energy in converted into heat and sensed by the photoacoustic detector. When all reaction centers are saturated, all the probe pulse energy is converted into heat (Fig. 4).

4. Demonstration of applications

We were able to follow the effects of the key environmental parameter, nutrient status, on the photosynthetic activity of phytoplankton and macroalgae. The nutrients examined were iron (Pinchasov et al., 2005), nitrogen, and phosphorus (Pinchasov-Grinblat et al., 2012).

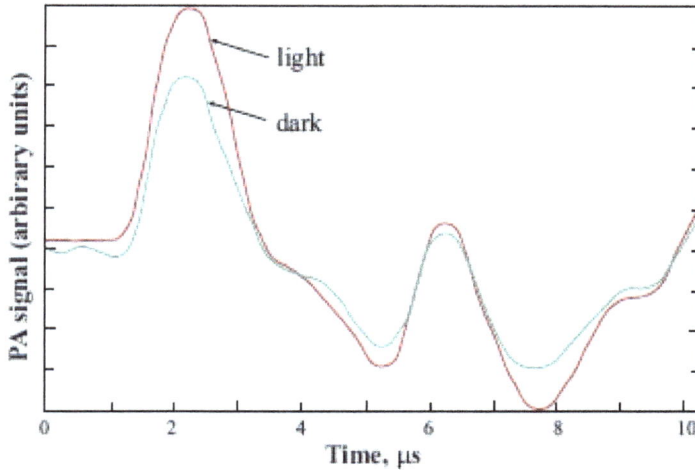

Figure 4. Photoacoustic signal from a *Phorphyridium cruentum* suspension with and without saturating background light (Mauzerall et al., 1998). The area between the curves is proportional to the fraction of the pulse energy stored by photosynthesis.

5. Iron limitation

Three algal species, the diatom *Phaeodactylum tricornutum*, a green alga *Nannochloropsis* sp., and the golden-brown flagellate *Isochrysis galbana*, were cultured in iron-replete media (artificial seawater medium Guillard's F/2) and grown at 24 °C under white fluorescent lights at ~220 μE m^{-2} s^{-1} PAR. These samples were subsequently transferred to the experimental media containing 0.00, 0.03, 0.09, 0.18, and 0.6 mg L^{-1} iron.

Each culture was diluted in the corresponding medium to chlorophyll *a* concentrations of 5.65 0.1 ± μg ml^{-1} in order to obtain similar absorptivity (Pinchasov et al., 2005).

The photoacoustic experiments were conducted after two weeks in these media. As the iron was progressively depleted, the ability of the three species to store energy decreased (Fig. 5). As seen in Figure 5, all three algal species showed a sharp decrease in efficiency.

6. Photoacclimation

Three species of marine phytoplankton, *Phaeodactylum tricornutum*, *Nannochloropsis* sp., and *Isochrysis galbana*, were studied. All cultures were grown in 250-mL Erlenmeyer flasks containing 200 mL enriched artificial seawater medium (Guillard's F/2) at 24 ± 0.5 °C, un-der white fluorescent lights at ~10 μmol q m^{-2} s^{-1} [low light (LL)] and ~500 μmol q m^{-2} s^{-1} [high light (HL)].

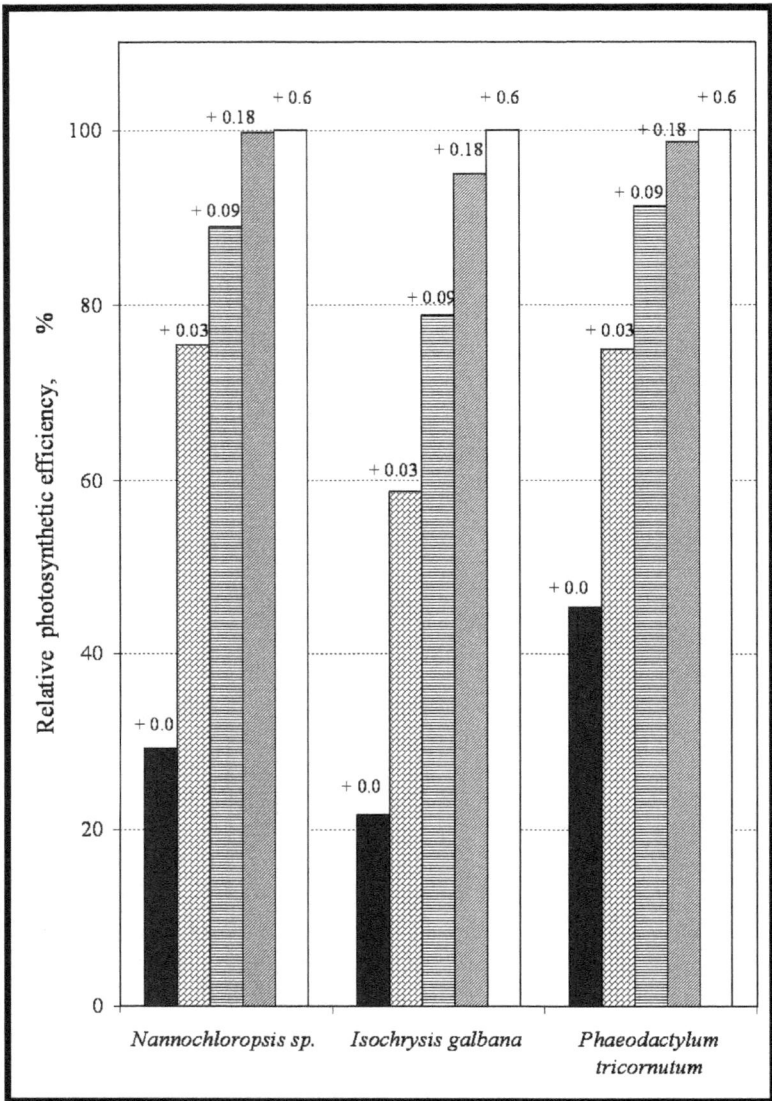

Figure 5. The effect of different iron concentrations on the relative photosynthetic efficiency in the three algae, *Nannochloropsis* sp., *Phaeodactylum tricornutum*, and *Isocrysis galbana*. For each species, the photosynthetic efficiency of the nutrient-replete control was taken as 100%. Controls (clear columns) were grown in iron-replete media contains 0.6 mgL^{-1}. The iron concentration in the iron-limited cultures (hatched columns) was 0 mg L^{-1}, 0.03 mg L^{-1}, 0.09 mg L^{-1}, and 0.18 mg L^{-1} (Pinchasov et al., 2005).

In these experiments, the photoacclimation of the three algal species, *Phaeodactylum tricornutum*, *Nannochloropsis* sp., and *Isochrysis galbana*, to low and high photon irradiances, was examined (Pinchasov-Grinblat et al., 2011). In general, photoacclimation to low light, results in increased cellular absorption due to a high concentration of light-harvesting pigments. In the numerous studies on the mechanism of photoacclimation in phytoplankton, a common trend of increase in chlorophyll and in thylakoid area as growth irradiance decreases (Dubinsky et al., 1986; Falkowski et al., 1986), was found. In addition to the changes in cellular chlorophyll, most other plant pigments also respond to ambient irradiance. All light-harvesting pigments increase under low light. These include the carotenoids fucoxantin and peridinin, in addition to all chlorophylls, phycoerythrin, and phycocyanin. The decrease of chlorophyll concentration under high-light growth conditions resulted in a parallel reduction in photosynthetic energy storage efficiency, as seen in Figure 6. All three species showed a decrease in efficiency for high-light acclimated algae compared to low-light grown conspecifics: by ~53% in *Isochrysis galbana*, and 33% and 31% in *Phaeodactylum tricornutum* and *Nannochloropsis* sp., respectively.

7. Lead exposure

In our experiments, the exposure of the cyanobacterium *S. leopoliensis* to different concentrations of lead resulted in major changes in photosynthesis (Pinchasov et al., 2006). Figure 7 shows the changes in photosynthetic efficiency following lead application. The reduction of photosynthesis reached ~50% and ~80% with 25 ppm and 200 ppm, respectively. Most of the decrease seen after the first 24 h already took place in the first 40 min.

With an increasing lead concentration and duration of exposure, the inhibition of photosynthesis increases. Since the photoacoustic method yields photosynthetic energy storage efficiency, the results are independent of chlorophyll concentration, which means that the observed decrease in efficiency is not due to the death of a fraction of the population, but rather due to the impairment of photosynthetic function in all cells, possibly due to the progressive inactivation of an increasing fraction of the photosynthetic units.

8. The effect of nutrient enrichment on seaweeds

Samples of the macroalga *Ulva rigida* were collected from the intertidal abrasion platforms in the Israeli Mediterranean during spring 2010. All samples were kept at 22 ± 0.1 °C in 100 mL Erlenmeyer flasks during 192 h under continuous irradiance at ~200 ± 5.0 μl m^{-2} s^{-1}.

The samples were exposed to 3 treatments: nitrogen (added as $NaNO_3$ at a concentration of 3.25 gL^{-1}), phosphorus (added as NaH_2PO_4 at a concentration of 0.025 gL^{-1}), and nitrogen and phosphorus together. Controls were kept in unenriched seawater.

Nutrient limitation, on the one hand, and anthropogenic eutrophication, on the other, are among the most important factors determining the overall ecological status of water bodies.

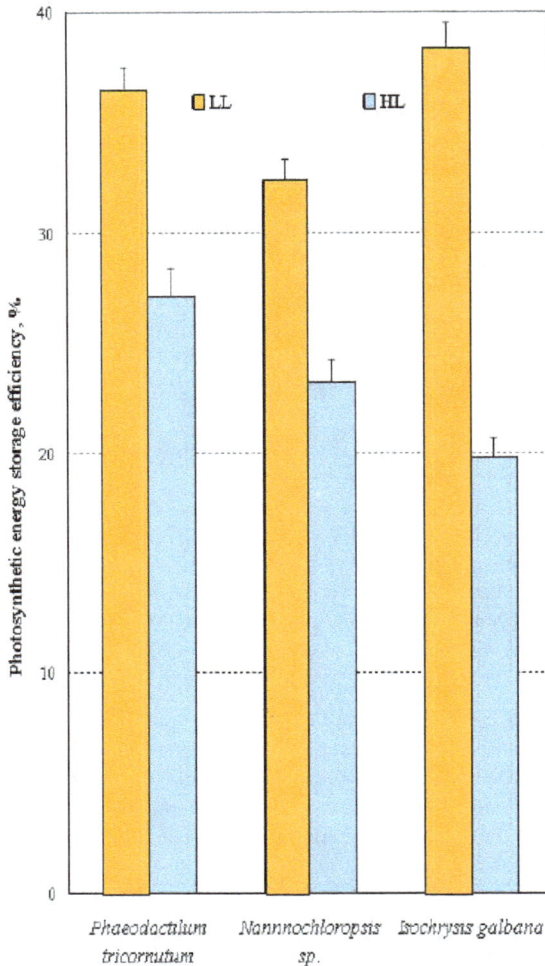

Figure 6. The effect of photoacclimation to high light (500 µmole qm^{-2} s^{-1}) and low light (10 µmole qm^{-2} s^{-1}) on photosynthetic energy storage efficiency for the three algae [according to Pinchasov et al. (2011)].

In general in all samples, photosynthetic efficiency and chlorophyll concentration (photoacoustic signal) decreased with time.

As is evident from Figure 8, macroalgae rapidly exhausted nutrients in the water, and within 190 h, the controls declined to approximately ~50% of the initial values in *U. rigida*. The addition of nutrients slowed down, but did not prevent, such decline (~20 % in *U. rigida*, Fig. 8).

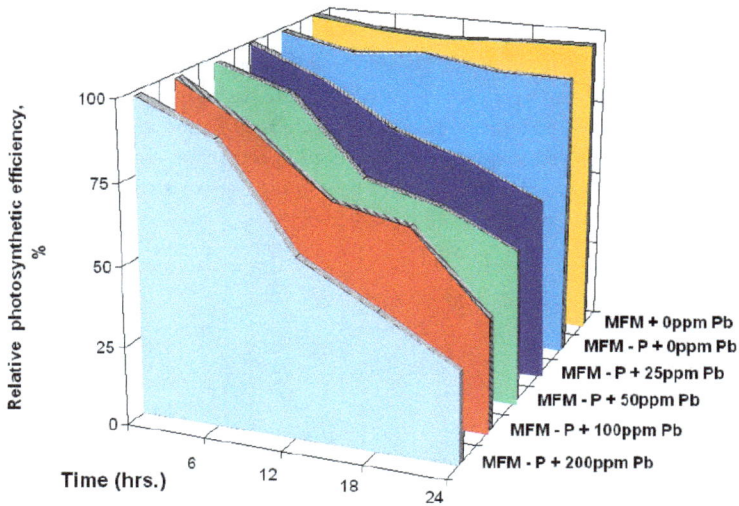

Figure 7. Relative photosynthetic efficiency following application of lead to *Synecococcus leopoliensis* [as per Pincha-sov et al. (2006)].

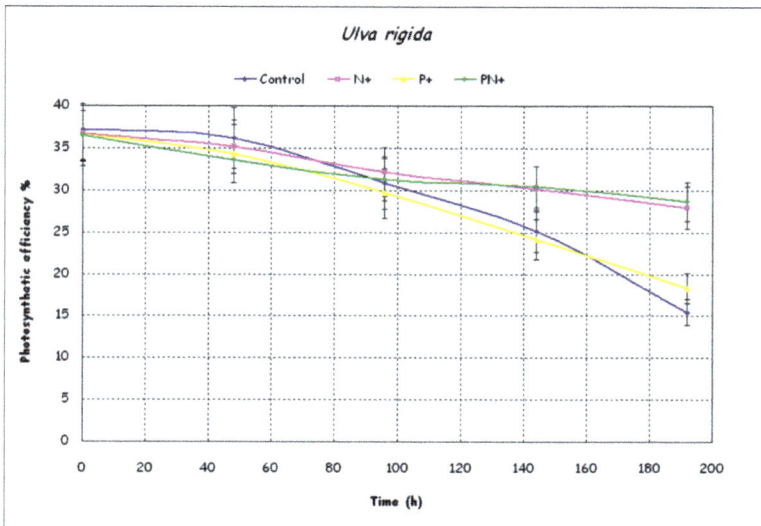

Figure 8. The effect of nutrient enrichment on photosynthetic efficiency of *Ulva rigida*, measured by photoacoustics [according to Pinchasov-Grinblat et al. (2012)].

Recently, Yan et al. (Yan et al., 2011), using a photoacoustic setup, measured thermal dissipation and energy storage in the intact cells of wild type *Chlamydomonas reinhardtii* and mutants lacking either PSI or PSII reaction centers. The photoacoustic signal from PSI-deficient mutants with open reaction centers had a positive phase at 25 °C but a negative phase at 4 °C. In contrast, PSII-deficient mutants showed large negative amplitude at 25 °C and an even larger effect at 4 °C. Kinetic analysis revealed that PSI and PSII reaction centers exhibit strikingly different photoacoustic signals, where PSI is characterized by a strong electrostriction signal and a weak thermal expansion component while PSII is characterized by a small electrostriction component and large thermal expansion (Yan et al., 2011).

9. Other applications
of photoacoustics

The thermal expansion of tissue, liquids, and gases due to light energy converted to heat, is termed the photothermal signal. This is always generated when photosynthetic tissue or cell is exposed to a light pulse, since plant tissue never absorbs all of the light stored as products of the process. The unused fraction of the absorbed light energy is converted to heat, resulting in measurable transient pressure (Cahen et al., 1980; Malkin, 1996). In addition to this thermal expansion signal, when a leaf is illuminated by a pulse of light, the resulting photosynthetic photolysis of water causes the evolution of a burst of gaseous oxygen. This process leads to an increase in pressure, a change which is readily detected by a microphone as the photobaric signal. For detailed definitions and description, see review by Malkin (1996).

The photoacoustic technique allows an investigation of energy conversion processes by photocalorimetry and direct measurement of the enthalpy change of photosynthetic reactions (Cahen & Garty, 1979; Malkin & Cahen, 1979). Oxygen evolution by leaf tissue can be measured photoacoustically with a time resolution that is difficult to achieve by other methods (Canaani et al., 1988; Cha & Mauzerall, 1992). A microphone can sense the photoacoustic waves via thermal expansion in the gas phase, thus allowing in-vivo measurements of the photosynthetic thermal efficiency and the optical cross section of the light harvesting systems. O_2 evolution in intact undetached leaves of dark adapted seedlings was measured during photosynthesis with the objective to detect genetic differences among cultivars (da Silva et al., 1995).

The rapid response of the phytoplankton populations to changes in environmental factors, such as temperature, light, nutrients, vertical mixing, and pollution, necessitates simple and frequent measurements. The photoacoustic method provides unique capabilities for ecological monitoring, photosynthesis research, and the optimization of algal mass cultures, such as those designed for the production of biofuel, aquaculture feed, and fine chemicals.

Acknowledgements

The authors wish to thank L.P.P. Ltd. Publishers for permission to use Figure 5 (The effect of different iron concentrations on the relative photosynthetic efficiency in the three algae *Nannochloropsis* sp., *Phaeodactylum tricornutum* and *Isocrysis galbana*).

This research was supported by European Research Council 2009 AdG – grant 249930 to Z.D. and by EU FP7 European Research Council 2012 – grant 309646 to Z.D.

Author details

Yulia Pinchasov-Grinblat and Zvy Dubinsky

The Mina and Everard Goodman Faculty of Life Sciences, Bar-Ilan University, Ramat-Gan, Israel

References

[1] Cahen, D, Bults, G, Garty, H, & Malkin, S. (1980). Photoacoustics in life sciences. Journal of Biochemical and Biophysical Methods, 0016-5022X, 3(5), 293-310.

[2] Cahen, D, & Garty, H. (1979). Sample cells for photoacoustic measurements. Analytical Chemistry, 0003-2700, 51(11), 1865-1867.

[3] Canaani, O, Malkin, S, & Mauzerall, D. (1988). Pulsed photoacoustic detection of flash-induced oxygen evolution from intact leaves and its oscillations. Proceedings of the National Academy of Sciences of the United States of America, , 85(13), 4725-4729.

[4] Cha, Y, & Mauzerall, D. C. (1992). Energy storage in linear and cyclic electron flows in photosynthesis. Plant Physiology, , 100(4), 1869-1877.

[5] da SilvaW.J., Prioli, L.M., Magalhaes, A.C.N., Pereira, A.C., Vargas, H., Mansanares, A.M., Cella, N., Miranda, L.C.M. & Alvaradogil, J. ((1995). Photosynthetic O_2 evolution in maize inbreds and their hybrids can be differentiated by open photoacoustic cell technique Plant Science, 0168-9452, 104(2), 177-181.

[6] Dubinsky, Z, Falkowski, P. G, & Wyman, K. (1986). Light harvesting and utilization by phytoplankton. Plant and Cell Physiology, 0032-0781, 27(7), 1335-1349.

[7] Dubinsky, Z, Feitelson, J, & Mauzerall, D. C. (1998). Listening to phytoplankton: Measuring biomass and photosynthesis by photoacoustics. Journal of Phycology, , 34(5), 888-892.

[8] Falkowski, P. G, Wyman, K, Ley, A, & Mauzerall, D. (1986). Relationship of steady state photosynthesis to fluorescence in eucaryotic algae. Biochimica et Biophysica Acta, , 849, 183-192.

[9] Malkin, S. (1996). The photoacoustic method in photosynthesis-monitoring and analysis of phenomena which lead to pressure changes following light excitation. In: Biophysical techniques in photosynthesis, J. Amesz and A. Hoff (Eds.), Kluwer Academic Publishers, Dordrecht, 191-206.

[10] Malkin, S, & Cahen, D. (1979). Photoacoustic spectroscopy and radiant energy conversion: theory of the effect with special emphasis on photosynthesis. Photochemistry and Photobiology, , 29, 803-813.

[11] Malkin, S, Herbert, S. K, & Fork, D. C. (1990). Light distribution, transfer and utilization in the marine red alga Porphyra perforata from photoacoustic energy-storage measurements. Biochimica et Biophysica Acta, , 1016(2), 177-189.

[12] Mauzerall, D. C, Feitelson, J, & Dubinsky, Z. (1998). Discriminating between phytoplankton taxa by photoacoustics. Israel Journal of Chemistry, , 38(3), 257-260.

[13] Pinchasov, Y, Kotliarevsky, D, Dubinsky, Z, Mauzerall, D. C, & Feitelson, J. (2005). Photoacoustics as a diagnostic tool for probing the physiological status of phytoplankton. Israel Journal of Plant Sciences, , 53(1), 1-10.

[14] Pinchasov, Y, Berner, T, & Dubinsky, Z. (2006). The effect of lead on photosynthesis, as determined by photoacoustics in Synechococcus leopoliensis (Cyanobacteria). Water Air and Soil Pollution, , 175, 117-125.

[15] Pinchasov-grinblat, Y, Hoffman, R, & Dubinsky, Z. (2011). The effect of photoacclimation on photosynthetic energy storage efficiency, determined by photoacoustics Open Journal of Marine Science, , 1(2), 43-49.

[16] Pinchasov-grinblat, Y, Hoffman, R, Goffredo, S, Falini, G, & Dubinsky, Z. (2012). The effect of nutrient enrichment on three species of macroalgae as determined by photoacoustics. Marine Science, , 2(6), 125-131.

[17] Yan, C. Y, Schofield, O, Dubinsky, Z, Mauzerall, D, Falkowski, P. G, & Gorbunov, M. Y. (2011). Photosynthetic energy storage efficiency in Chlamydomonas reinhardtii, based on microsecond photoacoustics. Photosynthesis Research, 0166-8595, 108(2-3), 215-224.

Chapter 11

EPR Spectroscopy — A Valuable Tool to Study Photosynthesizing Organisms Exposed to Abiotic Stresses

František Šeršeň and Katarína Kráľová

Additional information is available at the end of the chapter

1. Introduction

Abiotic environmental stresses, such as heat, cold, salt, drought, excess of photochemically active radiation (PAR) as well as UV-A and UV-B radiation, or presence of gaseous pollutants (e.g. ozone or SO_2), heavy metals or herbicides in the environment lead to inhibition of photosynthetic processes. Some of these abiotic stresses target specific cellular pathways, other ones have a broad cellular impact. They adversely affect photosynthetic apparatus of photosynthesizing organisms what ultimately results in negative effects on plant growth, productivity in agriculture, metabolic profile as well as plant nutritional potential. Therefore, plant abiotic stress has been a matter of concern for the maintenance of human life on earth and especially for the world economy [1].

The great power of EPR is its ability to identify the chemical nature of free radical species, and from the intensity of the signal to determine the number of radicals that have been formed in particular systems. Many components of the photosynthetic apparatus provide EPR signals at certain conditions (in detail see in subchapter 3). From the line widths and line shapes of the EPR spectra of radical species, frequently deliberately introduced to samples as spin-probes, various features of the local molecular environment may further be deduced [2].

Functional and undamaged thylakoid membrane is essential for successful process of photosynthesis. Based on EPR measurements using spin probes which are suitable to evaluate the effects of membrane-active compounds on the fluidity of PS 2 membranes or to study the fluidity of chloroplast thylakoid membranes of plants exposed to stressful conditions (e.g. herbicides, frost, etc.) changes in rotational mobility of lipids could be determined [3,4]. From the changes in EPR spectra relative membrane perturbation could be evaluated. On the other hand, rotational correlation time values (τ_c) can be used to monitor the effect of membrane active compounds on the rate of molecular reorientation of spin label in order to determine changes in the microviscosity of thylakoid membranes caused by these compounds [3].

Exposure of photosynthesizing organisms to stressful conditions such as drought, salinity, low temperature or heavy metals is connected with increased production of reactive oxygen species (ROS) that are generated due to the stepwise reduction of molecular oxygen by high-energy exposure or as a result of electron transfer chemical reactions. The enhanced production of ROS, including free radicals such as superoxide anion ($O_2^{-\bullet}$) and hydroxyl radical ($^\bullet OH$), as well as non-radical molecules H_2O_2, ozone and singlet oxygen ($^1O_2$), which occur during abiotic stresses, results in lipid peroxidation and oxidative damage of proteins, nucleic acids as well as in inhibition of enzyme activity [5]. EPR spectroscopy is a suitable method for qualitative and quantitative evaluation of ROS in photosynthesizing organisms. Experimental evidence of ROS generation in photosystem 2 particles can be directly obtained by EPR technique in combination with suitable spin trap after their irradiation by visible light [6]. This technique is also suitable to determine deleterious effects of UV-B and UV-A radiation on the photosynthetic apparatus [7-9].

2. Light reactions of photosynthesis

Photosynthesis is a process in which plants convert solar energy into chemical energy. In light reactions of photosynthesis the green algae and higher plants use two different reaction centres, called photosystem (PS) 1 and 2, while purple bacteria make do with a single reaction centre. The process of photosynthesis starts by the capture of photons in the antenna system of pigment-protein complexes and subsequent formation of excited forms of pigments. Excitation energy is then transported from the antenna to the cores of both photosystems where the primary charge separation occurs. Subsequent electron transfer steps prevent the primary charge separation from recombining by transferring the electron through the photosynthetic electron transport (PET) chain by a system of suitable electron acceptors and electron donors. The current concept of electron transport in the photosynthetic apparatus of green algae and higher plants is reflected in the following scheme:

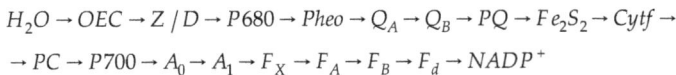

$$H_2O \rightarrow OEC \rightarrow Z \,/\, D \rightarrow P680 \rightarrow Pheo \rightarrow Q_A \rightarrow Q_B \rightarrow PQ \rightarrow Fe_2S_2 \rightarrow Cytf \rightarrow$$
$$\rightarrow PC \rightarrow P700 \rightarrow A_0 \rightarrow A_1 \rightarrow F_X \rightarrow F_A \rightarrow F_B \rightarrow F_d \rightarrow NADP^+$$

where OEC is oxygen evolving complex; Z/D are intermediates which participate at electron transport from OEC to P680, which is the core of PS 2 consisting from chlorophyll (Chl) a dimer; Pheo is pheophytin, the first electron acceptor in PS 2; Q_A and Q_B are the first and the second quinone acceptors of electron; PQ is plastoquinone pool consisting of a set of quinones; Fe_2S_2 is Rieske iron sulphur protein complex; cyt f is cytochrome f; PC is plastocyanin; P700 is the core of PS 1 consisting of Chl$a$ dimer; A_0, a Chl$a$ molecule, represents the primary electron acceptor in PS 1; A_1, a phylloquinone, is the secondary electron acceptor; F_X, F_A and F_B are the iron-sulphur centres; F_d is ferredoxin and NADP$^+$ is the final electron acceptor of PS 1 [10,11].

Photosystem 2 is the only known protein complex that can oxidize water, which results in the release of O_2 into the atmosphere. The catalytic cleaving of water occurs at a cluster consisting of four manganese atoms and one calcium atom which is situated at the luminal side of two key polypeptides of PS 2, D_1 and D_2. Water oxidation requires two molecules of water and involves four sequential turnovers of the reaction centre and manganese in the cluster undergoes light-induced oxidation. A kinetic model of oxygen evolution based on five S-states which was developed by Kok and co-workers postulated that each photochemical reaction creates an oxidant that removes one electron, driving the oxygen evolving complex to the next higher S-state [12]. The result is the creation of four oxidizing equivalents in the oxygen evolving complex. Electrons which were formed during photochemical cleavage of water are then transferred to P680 via Tyr_Z, a redox-active tyrosine residue on D_1 protein. From P680 the electrons are then transported to a mobile pool of plastoquinone molecules by subsequent redox reactions via Pheo, Q_A and Q_B. The electrons are further transmitted to the cytochrome complex and to the PS 1. The final electron acceptor of PS 1 is $NADP^+$ [10].

3. EPR signals of components of photosynthetic apparatus

The method of electron paramagnetic (spin) resonance (EPR or ESR) is suitable for detection of compounds which contain unpaired spins. During electron transport through the photosynthetic apparatus radicals are formed which can be recorded by EPR spectroscopy.

In the manganese cluster of OEC four manganese atoms, occurring in oxidation states II, III and IV, are bound to a 33 kDa protein. Due to spin-spin interactions the bound manganese atoms show no EPR signals at room temperature. After irradiance with light impulses four different states of OEC, in the literature known as S_0 to S_3, can be registered. In PS 2 particles prepared from spinach chloroplasts at cryogenic temperatures EPR spectra of some S states could be recorded.

The state S_0 consists of two signals. The first is similar to the multiline signal which is centred near g = 2.0 and spread over ~ 238 mT. It is constituted of 25 resolved lines spaced in the region of 6.5 – 9.5 mT. This signal was assigned to the antiferro-magnetically coupled S = 1/2 system in the ground state. The second signal is broad featureless signal at low field with g = 6 and g = 10 (S = 5/2) [13,14,15,16]. The integer spin of S_1 state (S = 1) exhibits two low field signals which were recorded by parallel-polarization mode EPR spectrometer. They are situated in low field region, the first at g = 12 with 18 lines (spacing 3.2 mT) and the second one at g = 4.8 (width 60 mT) [14,15,16]. S_1 state also exhibits a multiline EPR signal (S = 1/2) at g = 2.0 which is attributed to $S_1Q_A^-Fe^{2+}$ state [17] or to $S_1Y_Z^{\bullet}$ state [18]. Two EPR signals have been attributed to the manganese complex in the S_2 state. These signals are observable only at temperatures < 35 K. One of these signals is the multiline signal, ascribed to the ground state of S = 1/2, centred at g = 1.982 with a line width of 150-180 mT. The first signal consists of 18-20 partially resolved hyperfine lines (distance from each other 8-9 mT) which are often superimposed on broad Gaussian-shaped signal near g = 2. The other signal occurs at g = 4.1 (width ~ 36 mT, S = 5/2) [19,20]. The multiline EPR signal of $S_2Q_A^-Fe^{2+}$ state was also observed in PS 2 membranes [17].

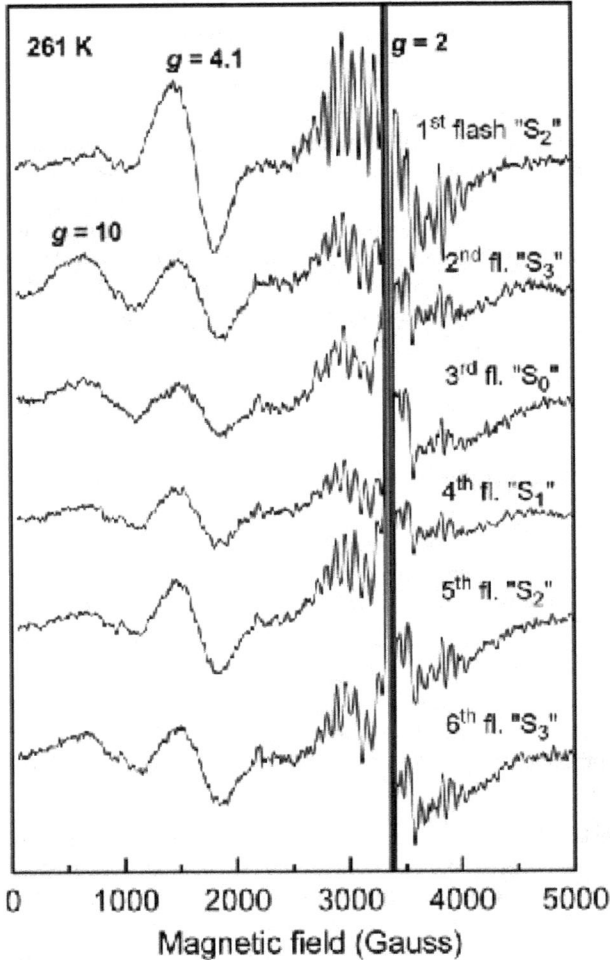

Figure 1. The EPR spectra of PS 2 particles during sequential flash illumination. The sample received each flash at 261 K, followed by rapid cooling to liquid helium temperatures to record the EPR spectrum. Source: [24].

Wider (~ 80 mT) EPR signal S_3 for $S = 1/2$ in Ca^{2+}-depleted PS 2 membranes was observed in the region of $g \sim 2.0$ [21]. The parallel polarization EPR method has been applied to investigate the manganese EPR signal of native S_3 state of OEC in PS 2 with $S = 1$, $g = 11$ and $g = 15$ with a width of 20 - 30 mT [15]. The EPR signal of S_3-state was also observed in the Boussac Works [20,22]. On the other hand, Hallahan et al. [23] assumed that the S_3 signal arises from the S_2-Y_Z^+ interaction. The typical EPR spectra of S states are presented in Fig. 1.

In EPR spectra of chloroplasts treated with some inhibitors of photosynthesis six line spectrum (at g ~ 2.0) originating from Mn^{2+} ions released from manganese cluster into interior of thylakoid membranes can be observed [25] (Fig. 2B).

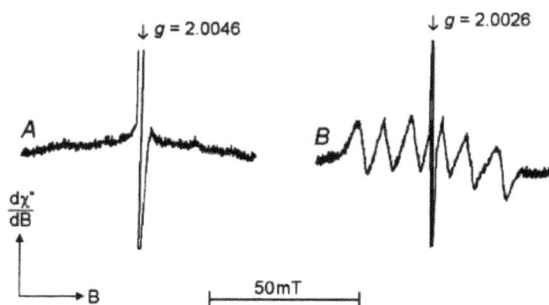

Figure 2. EPR spectra of spinach chloroplasts: A - control sample, B - chloroplasts treated with 0.05 mol dm^{-3} of $HgCl_2$. Source: [26].

EPR signals belonging to Z/D intermediates were observed for the first time in spinach chloroplasts already in 1956 [27]. They are long living radicals which could be recorded at room temperature by continual wave (cw) EPR spectrometer. These signals are situated in the region of free radicals and they are referred in the literature as signal II_{slow} (Fig. 3A, full line) with g = 2.0046 and ΔB_{PP} = 1.9 mT [28] and signal $II_{very\ fast}$ (Fig. 3A, dotted line) with g = 2.0046 and ΔB_{PP} = 1.9 mT [29]. These signals were associated with some components of PS 2 [30]. Signal II_{slow} can be registered even in darkness, while signal $II_{very\ fast}$ is observable only in the light. At the end of 80 years of the last century it was found that signal II_{slow} corresponds to the oxidized intermediate D•, i.e. to the tyrosine radical occurring in the 161st position of D_2 protein on the donor side of PS 2 [31]. The EPR signal $II_{very\ fast}$ is observable as an increase of signal II in the light and it belongs to the intermediate Z•, i.e. to the tyrosine radical occurring in the 161st position of D_1 protein [32].

EPR signals originating from P680 can be observed in two states: as oxidized P680+ with g = 2.0030 and line width of 0.85 mT [17,33] or as a triplet state of P680 (^{3}P680) [34].

Pheophytin is the first acceptor of electrons on the acceptor side of PS 2. After its reduction a radical with short life time (Pheo-) is formed and EPR spectrum of this radical (g = 2.0032, ΔB_{PP} = 1.2 mT) can be recorded at cryogenic temperature by the methods of electron spin polarization [35] or by time resolved EPR spectroscopy [36].

Due to the interaction with paramagnetic iron atom, the EPR spectra of Q_A^- and Q_B^- are registered as very broad signals at g < 2.0. Q_A in triplet state was also recorded by optical-ly detected magnetic resonance spectroscopy (ODMR) [37]. Q-band spectrometer was used for better resolution of anisotropic g tensor for Q_A^- •and the estimated values of g-tensor were g_{xx} = 2.0073, g_{yy} = 2.0054 and g_{zz} = 2.0023 [38]. Moreover, the EPR signal from the radical Q_A^--Fe^{2+} with g = 1.67 and 1.82 was registered by Jegerschöld and Styring [39]. On the other hand, van Mieghem et al. [34] registered this signal with g = 1.9.

The EPR spectrum of oxidized core of PS 1 ($P700^+$) belongs to the first observed EPR signals. Commoner and Heise [27] called it as signal I with g = 2.0026 and line width 0.8 mT. This signal is well visible in Fig. 3C (both lines). Later, Warden and Bolton [40] found that this signal belongs to chlorophyll a dimer in the oxidized core of PS 1.

The EPR signal of the $P700^+$-A_1^- radical pair was observed by time resolved EPR technique [41,42]. Snyder et al. [43] were the first who attributed A_1^- to vitamin K_1 on the basis of EPR measurements. Evans et al. [44] found the presence of a bound electron transport component in spinach chloroplasts showing EPR spectrum characteristic for ferredoxins in PS 1 (g = 1.95, g = 1.93 and g = 1.87 at 77 K). Later, Sonoike et al. [45] identified EPR signals from ferredoxins in PS 1: g = 1.94 for F_A, g = 1.92 for F_B, g = 1.89 for F_A /F_B mixture and g = 1.78 for F_X. These EPR spectra were recorded at 20 and 8 K. The EPR signal from F_X with g = 1.77 was observed in PS 1 particles at 10 K [46].

Moreover, EPR signals from other components of photosynthetic apparatus, namely the large signal in the g = 2.00 region from the chlorophyll free radical [35,44,47], EPR signal of oxidized carotene (Car^+) with g = 2.0033 and line width 1.1 mT and EPR signal of protochlorophyllide [48] were registered as well.

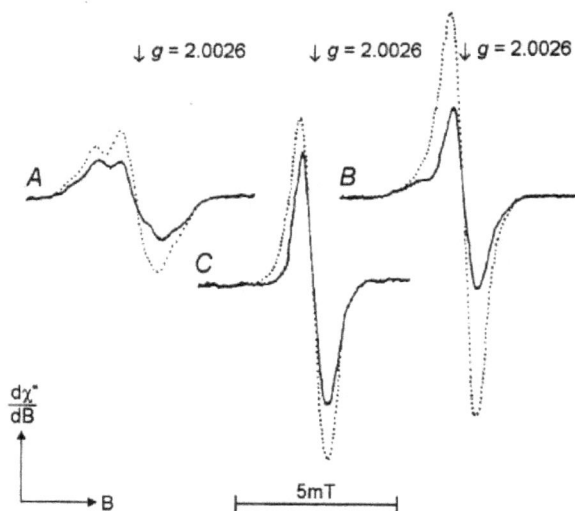

Figure 3. EPR spectra of spinach chloroplasts: control sample (A) and chloroplasts treated with 8 mmol dm^{-3} (B) or 40 mmol dm^{-3} (C) of $HgCl_2$. The full lines were recorded in the dark and the dotted ones in the light. Source: [26].

4. Investigation of photosynthesizing organisms exposed to toxic metal stress

Several transient metals belong to very effective inhibitors of photosynthesis due to their ability to interact with amino acids occurring in proteins of photosynthetic apparatus or to release manganese ions from the water splitting complex. These processes can also be examined by EPR spectroscopy.

4.1. Copper

Copper is essential bioelement which occurs directly in photosynthetic electron transport chain, namely in the plastocyanin on the donor side of PS 1 [49] and in the light harvesting complex of PS 2 [50]. However, higher Cu concentrations result in the inhibition of photosynthesis due to interaction of Cu^{2+} with several parts of photosynthetic apparatus [51]. These interactions can be observable by EPR spectroscopy. It was found that Cu^{2+} ions at concentrations ~ 10 mmol dm^{-3} decrease the EPR signal intensity of $Tyr_Z{}^{\bullet}$ in spinach chloroplasts treated with Cu^{2+} ions [52-54]. The disappearance of both signals belonging to $Tyr_Z{}^{\bullet}$ and $Tyr_D{}^{\bullet}$ (shown in Fig. 3B and 3C for $HgCl_2$-treated chloroplasts [26]) was observed at higher Cu^{2+} concentrations (~ 50 mmol dm^{-3}) as well [54-57]. Moreover, incubation of chloroplasts with Cu^{2+} resulted in loss of the normal EPR signal from $Q_A{}^-$ which is coupled to the non-heme Fe^{2+} on the acceptor side of PS 2 (the $Q_A{}^-$-Fe^{2+} EPR signal). In the presence of excess Cu reduction of Q_A results in the formation of a free radical spectrum which is 0.95 mT wide and centred at g = 2.0044. This signal is attributed to $Q_A{}^{-\bullet}$ which is magnetically decoupled from the non-heme iron. This suggests that Cu^{2+} displaces the Fe^{2+} or severely alters its binding properties [53,58]. Moreover, application of higher Cu^{2+} concentrations resulted in displacement of Mn^{2+} ions from the manganese cluster and their release into interior of thylakoid membranes. This was documented by the appearing of six lines of hyperfine structure in EPR spectra of Cu^{2+}-treated spinach chloroplasts [54-57]. Similar effect for $HgCl_2$-treated chloroplasts is shown in Fig. 2B.

The Cu^{2+} ions appear to be predominantly associated with PS 2 proteins. In Cu-treated chloroplasts the formation of Cu(II)-protein complexes was confirmed by changes in EPR spectra of the applied Cu(II) compounds [55,57,59]. Using EPR spectroscopy it was found that Cu^{2+} is bound on two different sites of PS 2: one of them is situated near the Zn site that modulates electron transport between the quinones Q_A and Q_B and the second one occurs at the Fe site [60].

Interaction of copper with plastocyanin was presented by Bohner et al. [61] who found that in copper-treated *Scenedesmus* the content of this electron carrier dramatically varied with increasing external copper concentration.

4.2. Mercury

Mercury is a potential environmental contaminant which strongly inhibits photosynthetic processes in algae and higher plants [62]. Several sites of mercury action in both photosynthetic centres were determined. It was found that Hg^{2+} ions inhibit PET through PS 1 by interactions with: i/ plastocyanin on the donor side of PS 1 [63-65]; ii/ ferredoxin [66,67]; iii/ F_B iron-sulphur cluster [68]. Hg^{2+} ions also damage PET through PS 2 [63,65,66] by interactions: i/ with OEC on the oxidizing side of PS 2 [26, 67, 69-72]; ii/ with the core of PS 2 (P680) [73]; iii/ with both quinone acceptors (Q_A and Q_B) on the reducing side of PS 2 [74,75]. Due to strong affinity of Hg^{2+} to CO, CN, CS and CSH groups the formation of organo-mercury compounds with amino acid residues in photosynthetic proteins was proposed as possible mechanism of the Hg^{2+} action [72,76,77].

Despite known Hg^{2+} action sites in the photosynthetic electron transport chain, EPR studies of mercury effect on photosynthetic apparatus were reported only by Sakurai et al. [78], Jung et al. [68], Šeršeň et al. [26] and Šeršeň and Kráľová [77]. Šeršeň`s group found that Hg^{2+} ions interact with the intermediates $Z^•/D^•$ what was reflected in reduction of both components of the EPR signal II (Fig. 3B and 3C). Moreover, EPR spectra of Mn^{2+} ions (Fig. 2B) which were released from manganese cluster into interior of thylakoid membranes confirmed interaction of Hg^{2+} with OEC [26]. A damage of the F_B iron-sulphur cluster in PS 1 after $HgCl_2$ treatment was demonstrated by EPR spectroscopy, while the EPR spectra of F_A and F_X remain unchanged [68,78]. The decay of EPR signal I after switching off the light in Hg-treated chloroplasts indicated the damage of direct cyclic and non-cyclic electron flow through PS 1 [77].

4.3. Cadmium

Cadmium belongs to the major heavy metal pollutants which have toxic effects on living organisms [79]. Cadmium exhibits several toxic effects on higher plants, which are caused by direct and indirect mechanisms of its action on plant photosynthetic apparatus [80]. The site of Cd action in photosynthetic apparatus was found to be situated on several sites of photosynthetic electron transfer chain within PS 2: i/ particularly in the OEC or in its vicinity on the donor side of PS 2 [67,77,81-84]; ii/ in the site of Q_A or Q_B on the acceptor side of PS 2 [69,85,86].

Some of the above mentioned sites of Cd^{2+} action were supported by EPR spectroscopy. It was found that Cd^{2+} decreased signal intensity of intermediates $Z^•/D^•$ and released Mn^{2+} ions from OEC into interior of thylakoid membranes [77]. These interactions of Cd^{2+} with OEC and $Z^•/D^•$ intermediates resulted in great increase of the signal I intensity [77]. Similarly to Hg-treated chloroplasts, application of Cd resulted in the damage of direct cyclic and non-cyclic electron flow through PS 1 what was demonstrated by kinetic behaviour of EPR signal I after switching off the light (Fig. 4) [77]. Addition of Cd^{2+} to Ca^{2+}-extracted PS 2 particles, which exhibited neither the multiline EPR signal nor g = 4.1 signal, did not restore these signals unlike Ca^{2+} addition when only the EPR g = 4.1 signal remained lost [85,87]. Ono and Inoue [88] found that Cd^{2+} substitution in Ca^{2+}-extracted PS 2 particles restored neither the multiline EPR nor g = 4.1 signals of S_2 state. A decrease in unstable free radical level in the leaves of wheat seedlings (*Triticum aestivum* L.) treated with low Cd concentrations (less than 3.3 mg kg^{-1} soil), followed by their significant enhancement with increasing Cd concentrations, were determined by EPR spectroscopy using spin trap [89].

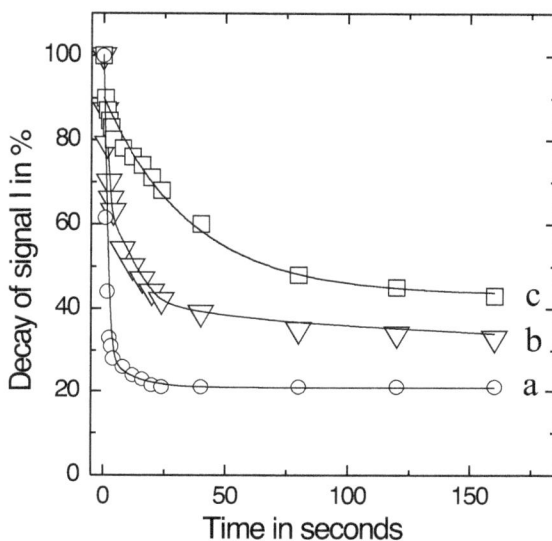

Figure 4. The kinetics of signal I of spinach chloroplasts after switch off the light in control chloroplasts treated with: 5 mmol dm^{-3} of DCMU (a), 0.05 mol dm^{-3} of CdCl$_2$ (b) and 0.05 mol dm^{-3} of HgCl$_2$ (c). Source: [77].

4.4. Zinc, chromium, selenium, nickel and iron

Karavaev et al. [90] investigated light-induced changes in the EPR signal I from oxidized reaction centres P700$^+$ of the photosynthetic apparatus of broad beans grown in aqueous solutions of zinc chloride. High concentrations of ZnCl$_2$ in the hydroponic medium slowed down the plant development and inhibited the light-induced production of oxygen and kinetics of redox transients of P700 induced by ZnCl$_2$ correlated with the changes in photosynthetic activity.

Frontasyeva et al. [91] studied the interaction of various chromium forms (Cr(III) and Cr(VI)) with *Spirulina platensis* biomass and found that from a nutrient medium the cells of this cyanobacterium mainly accumulated vitally essential form Cr(III) rather than toxic Cr(VI). Using EPR spectroscopy they demonstrated that the *Spirulina platensis* biomass enriched with Cr(III) was free from other toxic chromium forms. The toxicity of hexavalent chromium to photosynthesizing organisms is closely connected with generation of reactive oxygen species [92,93]. In the root tissues of some plants that were exposed to high concentrations of Cr(VI), the presence of intermediate Cr species, i.e., Cr(IV) or Cr(V) was confirmed by EPR spectroscopy studies by Micera and Dessi [94] as well as by low-frequency EPR experiments [95].

Labanowska et al. [96] used EPR spectroscopy to examine the alteration of radicals in wheat seedlings exposed for 2 days to selenium stress in two genotypes of Polish and one of Finnish wheat, differing in their tolerance to long-term stress treatment. The action of reactive oxygen species in short-term action of Se stress was confirmed by the reduction of PS 2 and PS 1 system activities. EPR studies showed changes in redox status (especially connected with Mn(II)/ Mn(III), and semiquinone/quinone ratios) in wheat cell after Se treatment. Finnish wheat was recognized as the genotype more sensitive to short-term Se stress than the Polish varieties.

Using EPR spectroscopy it was found that Ni(II) complexes with N-donor ligands of the type NiX_2L_y (where X = Cl, Br, I, $ClCH_2COO$ or Cl_2CHCOO, L = nicotinamide or ronicol and y = 2 or 4) interact with Z^+/D^+ intermediates, i.e. with tyrosine cation-radicals $Tyr_Z^{+\bullet}$, and $Tyr_D^{+\bullet}$ situated in D_1 and D_2 proteins on the donor side of photosystem 2 and with the manganese cluster in the oxygen evolving complex as well [97]. Similarly, the interaction with tyrosine radicals $Tyr_Z^{+\bullet}$ and $Tyr_D^{+\bullet}$ was also confirmed for Fe(III) complexes $[Fe(nia)_3Cl_3]$ and $[Fe(nia)_3(H_2O)_2](ClO_4)$, however release of Mn(II) from the oxygen evolving complex was observed only after treatment with $Fe(nia)_3(H_2O)_2](ClO_4)$ [98].

5. Investigation of photosynthesizing organisms exposed to herbicides

Herbicides are compounds used to kill weeds and unwanted plants. They also intervene in photosynthetic machinery and restrict photosynthetic electron transport due to their interaction with several components of photosynthetic apparatus. Therefore, EPR is very useful method for determination of the site of herbicide action.

Electron transfer from Q_A to Q_B is inhibited by a wide variety of plastoquinone (PQ) analogues that compete with PQ at the Q_B site. The most widely studied classes of inhibitors are the urea and triazine herbicides, such as 3-(3,4-dichlorophenyl)-1,1-dimethylurea (DCMU) and atrazine. Although their binding domains are likely to overlap with the urea/triazine herbicides on the Q_B site, "phenolic" inhibitors, which include bromoxynil, ioxynil, dinoseb and 2,4,6-trinitrophenol, appear to inhibit in a more complex fashion [99]. As mentioned above, different classes of herbicides apparently have different binding sites. DCMU-type herbicides bind to a 32 kDa polypeptide thought to be a regulatory protein associated with Q_A to Q_B electron transfer while phenolic herbicides bind to a 42 kDa polypeptide, which is probably a reaction centre protein. Herbicide-induced perturbations of these polypeptides might be expected to modify the EPR signals arising from reaction centre components. In particular, changes in the signal arising from Q_AFe might be expected since it arises from an interaction between two components, the semiquinone and the iron, which are located close to the herbicide binding sites [100].

Incubation of PS 2 membranes with herbicides results in changes in EPR signals arising from reaction centre components. Dinoseb, a phenolic herbicide which binds to the reaction centre polypeptide, changes the width and form of the EPR signal arising from photoreduced Q_A^-Fe at $g = 1.82$. Orthophenanthroline slightly broadens the Q_A^-Fe signal. These effects are attributed to changes in the interaction between the semiquinone and the iron. Herbicide effects can also be seen when Q_AFe is chemically reduced what is reflected by changes in splitting and amplitude of the split Pheo⁻ signal. Dinoseb application also results in the loss of signal II$_{slow}$, in the conversion of reduced high-potential cytochrome b_{559} to its oxidized low-potential form and in the presence of transiently photooxidized carotenoid after a flash at 25 °C. These effects indicate that dinoseb may also deactivate OEC by accelerating the deactivation reactions of the water splitting enzyme [100].

Some commercial herbicides are known to bind to the site of the exchangeable quinone Q_B in the PS 2 reaction centre, thus blocking the electron transfer to Q_B what results in PET interruption. This effect can be demonstrated in EPR spectra of chloroplasts treated with DCMU recorded by continuous wave (cw) EPR apparatus at room temperature (Fig. 5). In Fig. 5 an increase of signal I intensity due to PET interruption from PS 2 is presented.

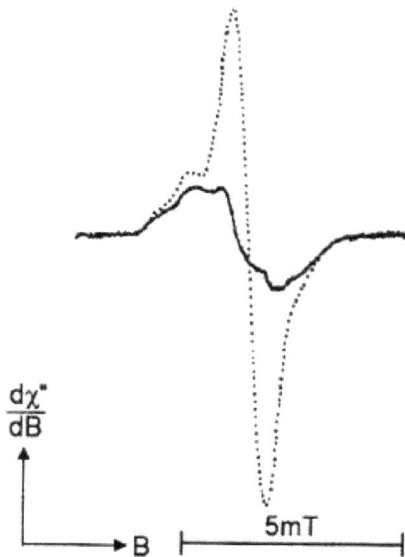

Figure 5. EPR spectrum of spinach chloroplasts treated with 100 μmol dm⁻³ of DCMU recorded in the dark (full line) and in the light (dotted line). Source: [24].

Some authors suggested that the herbicide-induced toxicity requires light and it is connected with chlorophyll-mediated $^1O_2$ generation (photooxidative stress). It is generally believed that as the energy is not being used for photosynthetic electron transfer, the chances of forming ^{3}Chl are increased, leading to $^1O_2$ formation and protein damage, similarly to the photodamage mechanism. This mechanism is also likely to apply to other herbicides. When electron transfer is blocked by herbicide binding, the level of the $S_2Q_A^{\bullet-}$ charge pair decays by a charge recombination pathway that involves formation of a chlorophyll triplet in the heart of the reaction centre. This triplet is thus able to react with $^3O_2$ to give $^1O_2$ [101]. All above mentioned intermediates ($S_2Q_A{}^{\bullet-}$, ^3Chl and $^1O_2$) can be observed by EPR spectroscopy.

Tenuazonic acid is natural PS 2 inhibitor with several action sites [102,103]. Treatment of PS 2 particles with this bioherbicide resulted in generation of ROS such as $^1O_2$, $O_2{}^{-\bullet}$ and $^\bullet OH$ which can be detected by EPR spectroscopy. Singlet oxygen was recorded as 2,2,6,6-tetramethylpiperidinoxyl radical (TEMPO) by EPR and production of $O_2{}^{-\bullet}$ and $^\bullet OH$ was estimated by spin trap 5-diethoxyphosphoryl-5-methyl-1-pyrroline N-oxide (DEPMPO) [104].

Photodamage of PS 2 in strong illumination of thylakoid membranes was documented by changes in the extent of the $Q_A{}^-Fe^{2+}$ and chlorophyll triplet EPR signals. In the presence of DCMU, a decrease of the $Q_A{}^-Fe^{2+}$ EPR signal (corresponding to the inhibition of oxygen evolution) and an increase of the chlorophyll triplet EPR signal indicated a possible overreduction of Q_A [105].

Multiphasic kinetic curve of light-induced EPR signal I from $P700^+$ was observed in the cyanobacterium *Synechocystis* sp PCC 6803 whereas in those treated with DCMU the PS 1 kinetics was monophasic [106].

Illumination of native NH_3-treated PS 2 membranes results in the appearance of an EPR signal at g = 2. It was suggested that this signal arises from perturbated S_3 state of manganese cluster [107]. González-Pérez et al. [108] observed the EPR signal with a g value of approximately 2.0043 and a line width of 0.1 mT which was induced under continuous illumination in the presence of peroxynitrite. In the absence of magnetic interaction with the non-heme Fe^{2+} this new EPR signal corresponds with the semireduced plastoquinone Q_A and it can be concluded that peroxynitrite impairs PS 2 electron transport in the Q_AFe^{2+} niche.

EPR spectroscopy was also used to determine the site of inhibitory action in the photosynthetic electron transport chain of many aromatic, heteroaromatic and amphiphilic compounds (in detail see in [109,110]).

6. Monitoring of the effect of gaseous environmental pollutants on photosynthetic apparatus

6.1. Ozone

Ozone is now regarded as the most important phytotoxic air pollutant, with long-term background concentrations increasing progressively. It is well known to have a negative impact on the photosynthetic apparatus of plant leaves and is thought to act via the formation of other ROS forms. Mehlhorn et al. [111] directly detected elevated levels of free radicals in plants exposed to ozone by EPR spectroscopy. It is generally thought that free radical generation in ozone stressed plants is enhanced by the direct reaction of ozone with biomolecules. This mechanism was supported by EPR measurements with freeze-dried samples of leaves from ozone-exposed wheat plants, showing a positive relationship between the free radical signal and the stress history of the plant [112].

In a C_3 grass, *Poa pratensis* L. and a C_4 grass, *Setaria viridis* Beauv., during exposure to O_3 both signal I (from P700+ in PS 1) and signal II (from Tyr_D in the D_2 protein of PS 2) were stimulated by O_3. However, the fact that signal I observed in white light rose to the level of signal I in far-red light indicated reduced electron flow through PS 1 [113]. Reichena-uer and Goodmann [114] found that after exposure to ozone in EPR spectra of freeze-dried samples from the same batches of plants an unidentified stable free radical appeared. The intensity of this radical signal increased with the duration of ozone exposure in leaves that received an additional ozone treatment.

Direct observations of radical signals obtained by EPR spectroscopy of intact, attached leaves of bluegrass *(Poa pratensis* L.) and ryegrass *(Lolium perenne* L.) and leaf pieces of radish *(Raphanus sativus* L.) during exposure to O_3 in air flowing through the spectrometer cavity have revealed the appearance of a signal with the characteristics of $O_2^{-\bullet}$. In each species, the signal only appeared after about 1 h of exposure to O_3, and then increased steadily over the next 4 h [115]. After prolonged exposure to ozone a free-radical signal with parameters similar to the superoxide anion free-radical signal was also formed in plant leaves of Kentucky bluegrass [116].

6.2. SO$_2$

The oxygen evolving enzyme of chloroplasts responsible for reduction of the PET carriers has been shown to be sensitive to SO_2 absorption by plants [117]. In SO_2-fumigated spinach leaves the time course of EPR signal I indicated that reduction of P700 by white light illumination was inhibited but dark reduction of P700 was not significantly affected. Photosynthetic O_2 evolution was also inhibited by SO_2 fumigation but all of these effects were reversible after removal of SO_2 [118].

Radish, Kentucky bluegrass and perennial ryegrass leaves subjected to high levels of sulphur dioxide (10–500 ppm) revealed the formation of signal I upon irradiation with broad-band white or 650 nm light, thereby indicating an interruption of normal electron flow from PS 2 to PS 1. Damage to the oxygen evolving complex and reaction centre of PS 2 was also detected through changes in signal II and Mn^{2+} signal [116]. These changes in the normal EPR signals were dose-dependent. Leaves subjected to low levels of sulphur dioxide (600–2000 ppb) revealed the disappearance of signal I after 3 hours of fumigation and the formation of a new free-radical signal with parameters similar to the sulphur trioxide free-radical signal. After prolonged exposure to sulphur dioxide a free-radical signal with parameters similar to the superoxide anion free-radical signal was formed in plant leaves [116].

7. Study of the thylakoid membrane arrangement under abiotic stress

The EPR spin label technique provides useful information about polarity of the local micro-environment near the spin label and the dynamic properties of the labelled site, which reflect its conformational state. As spin probes are usually used the stable nitroxide radicals which are built into thylakoid membranes. From the line widths and line shapes of the EPR spectra of such built-in spin probe, various features of the local molecular environment and phase-transitions may further be deduced.

Using spin-labelled phosphatidylglycerol incorporated into thylakoid suspension it was found that Cu, Pb and Zn increased the surfaces available for lipid-protein interaction by dissociating membrane protein complexes [119]. EPR spectra of the lipid vesicles spin probed with n-doxylstearic acid (n-DSA) were used to explore the lipid rotational free-dom at different depth of the bilayer. The EPR measurements indicated that the copper stress resulted in more tightly packed bilayers of the photosynthetic membranes with reduced acyl chain motion. Moreover, investigations of nitroxide radicals probing the bilayer at different depth suggested that changes in mobility associated with stress involve different parts of the bilayer in a similar way [120].

Line shape analysis of EPR spectra recorded as a function of temperature on concentrat-ed suspensions of thylakoids with labelled 5-doxylstearic acid (5-DSA) allowed getting information about the fluidity of differently treated membranes. The immobilization of the spin probes in the hydrophobic part of the membranes was supported by experiments of Calucci et al. [121]. Another example of the use of spin labels 16-DSA (16-doxylstearic acid) or CAT-16 (N-hexadecyl-N-tempoyl-N,N-dimethyl-ammonium bromide) was carried out by Šeršeň et al. [3] who determined the inhibitory mechanism of N-alkyl-N,N-dimethyl amine oxides (ADAO) action upon PET. Above mentioned authors found that interactions of ADAO with thylakoid lipids of chloroplasts resulted in the decrease of the ordering and the microviscosity of lipid phase. The motion of a spin label after its incorporation into membranes will be limited and consequently changes in its EPR spectra occur. From these changes in EPR spectrum order parameter S characterizing the arrangement of the membrane can be calculated (in detail see in [3]). Application of membrane-active compounds results in perturbation in membrane structure. Consequently, the value of S parameter will be affected depending on the extent of membrane damage. From the dependences of the order parameter of thylakoid membranes S (determined from EPR spectra of CAT 16) on the concentration of ADAO for hexadecyl-, dodecyl- and hexyl-derivatives (Fig. 6A) and on the alkyl chain length of ADAO at the constant ADAO concentration (Fig. 6B) it is evident that the membrane perturbation increases with increasing concentration of membrane active compounds and that the most effective compound was dodecyl derivative.

Addition of membrane-active compound to chloroplasts containing spin labels results in an increase in the rate of molecular reorientation of spin label. Therefore, changes in the rotational correlation time τ_c (which is linearly proportional to the microviscosity of the environment in which the spin label is located) due to addition of membrane-active compound can also be used to characterize alterations in membrane arrangement (Fig. 7A, 7B).

It was shown that ADAO decreased the microviscosity of thylakoid membranes and the course of τ_c of 16 DSA spin label located in the thylakoid membrane on the alkyl chain length of ADAO at constant compound concentration 50 μmol dm^{-3} (Fig. 7B) was similar to that obtained for order parameter S (Fig. 6B), i.e. the lowest τ_c exhibited dodecyl derivate.

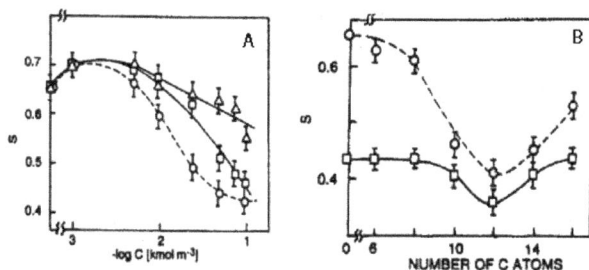

Figure 6. Dependences of the order parameter of thylakoid membranes S (determined from EPR spectra of CAT 16) on the concentration of ADAO for hexadecyl- (□), dodecyl- (o) and hexyl- (Δ) derivatives (A) and on the alkyl chain length of ADAO at the constant concentration of ADAO 50 μmol dm^{-3} (B); S was evaluated from EPR spectra of CAT 16 (o) and 16 DSA (□) spin labeles. Source: [3].

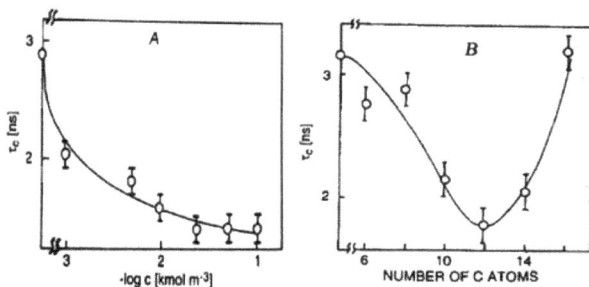

Figure 7. Dependences of rotational time τ_c of 16 DSA spin label located in the thylakoid membranes on the concentration of N-dodecyl-N,N-dimethylamine oxide (A) and on the alkyl chain length of ADAO at the constant concentration of ADAO 50 μmol dm^{-3} (B). Source: [3].

Alteration of membrane structure can also be caused by ionic amphiphilic compounds, including alkyl substituted quaternary ammonium salts [122-124]. Using EPR spectroscopy and spin label CAT 16 an increase of order parameter S was observed after application of low concentrations of 1-dodecyl-1-ethylpiperidinium bromide (DEPBr), followed with its decrease at higher DEPBr concentrations (Fig. 8A) [125]. These findings were in agreement with the results of oxygen evolution rate (OER) measurements which confirmed that OER was stimulated by low DEPBr concentrations (Fig, 8A). Because in the OER and in EPR experiments different chlorophyll content in chloroplast suspensions was applied, DEPBr concentration within chloroplast organelles was calculated using DEPBr partition coefficient between chloroplast organelles and aqueous environment. From Fig. 8B in which the dependences of OER and order parameter S on the DEPBr concentration in chloroplast organelles are presented it is evident that OER stimulation and increase of order parameter S occurred in the same concentration range. Consequently, it could be assumed that OER stimulation is caused by changes in the arrangement of thylakoid membranes.

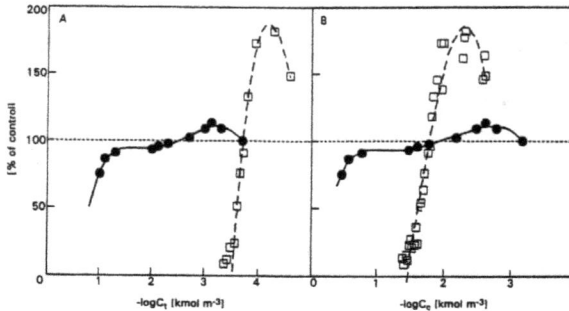

Figure 8. The dependence of the 2,6-dichlorophenolindophenol reduction (□) and the order parameter S (•) expressed as % of control sample upon concentration of 1-dodecyl-1-ethylpiperidinium bromide (DEPBr) in chloroplast suspension (A) and in chloroplast organelles (B). Source: [121].

Quartacci et al. [126] studied the effect of copper on the fluidity of PS 2 membranes by EPR measurements, using spin-probed fatty acids as probes. They found that due to treatment of PS 2 membranes (spin probed by means of 5- and 16-doxylstearic acids) with 50 µmol dm^{-3} Cu only the fluidity of the surface region of the bilayer close to the polar head group was reduced, while the fluidity of the inner membrane region of the bilayer did not show any change.

8. Study of photoinhibition in photosynthesizing organisms

When organisms that perform oxygenic photosynthesis are exposed to strong visible or UV light, inactivation of photosynthetic apparatus occurs. Under high light intensities the components of PET chain get damaged in a process called photoinhibition. It has been shown that PS 2 is the most susceptible pigment protein complex to photoinhibition. However, such organisms are able rapidly to repair the photoinactivated PS 2. Photoinhibition can be invoked by impairment of PS 2 acceptor side electron transport, and by the damage of PS 2 donor side [127]. Acceptor side induced photoinhibition takes place when reoxidation of quinones at acceptor side of PS 2 reaction centres is limited, e.g. when the plastoquinone pool is highly reduced. Under these conditions a charge recombination reaction between P680$^+$ and Q_A^- can take place, leading to a re-population of the primary charge pair P680$^+$Pheo$^-$. Recombination of this charge pair leads to the formation of the excited state of P680, both in its singlet and triplet state. ^{3}P680 reacts with O_2 leading to the formation of the highly oxidizing species $^1O_2$ [128]. In isolated thylakoids methylviologen (paraquat) was used to $O_2^{-\bullet}$ generation in the light in PS 1. $O_2^{-\bullet}$ was trapped to the spin trap by 5-diethoxyphosphoryl-5-methyl-1-pyrroline N-oxide (DEPMPO) and this spin adduct was recorded by EPR spectrometer [128].

High-intensity illumination of thylakoids results in the impairment of PS 2 electron transport, followed by the degradation of the D_1 reaction centre protein. This impairment is caused by reactive oxygen species (ROS), which are formed in photosynthetic apparatus by high-intensity illumination. They are mainly superoxide anion radical ($O_2^{-\bullet}$) and singlet oxygen ($^1O_2$). The formation of both above-mentioned ROS was confirmed by EPR spectroscopy. $^1O_2$ is generated by interaction of molecular oxygen with excited triplet of chlorophyll formed via charge recombination of radical pair ^{3}[P680$^+$ Pheo$^-$] [129]. Singlet oxygen was detected by following the formation of 2,2,6,6-tetramethylpiperidine-1-oxyl, a stable nitroxide radical yielded in the reaction of singlet oxygen with the sterically hindered amine 2,2,6,6-tetramethylpiperidine. Singlet oxygen, a non-radical form of active oxygen, was detectable only in samples undergoing acceptor-side-induced photodamage [130]. During both types of photo-inhibition also other free radicals were detected as spin adducts of the spin trap 5,5-dimethyl-1-pyrroline N-oxide, and identified on the basis of hyperfine splitting constants of the EPR spectra.: i/ The acceptor-side induced process was accompanied by the oxygen dependent production of carbon centred (alkyl or hydroxyalkyl) radicals, probably from the reaction of singlet oxygen with histidine residues. (ii) Donor-side induced photoinhibition was dominated by hydroxyl radicals, which were produced in anaerobic samples, too. The production rate of these radicals, as well as D_1 protein degradation, was dependent on the possibility of electron donation from manganese ions to PS 2. The marked distinction between the active oxygen forms produced in acceptor- and donor-side induced photoinhibition are in agreement with earlier reports on the different mechanism of these processes [130]. The two types of trapped ROS radicals are presented in Fig. 9.

Ogami et al. [131] observed that after 4 hours cultivation of cyanobacterium *Thermosynecho-coccus elongatus* cells under high light conditions, the S_2 multiline signal was undetectable. Ivanov et al. [132] confirmed the impairment of PS 1 by measuring of EPR signal intensity of oxidized P700 under high light stress.

The UV-A (320-400 nm) component of sunlight is a significant damaging factor of plant photosynthesis, which targets the PS 2 complex. UV-A irradiation results in the rapid inhibition of oxygen evolution accompanied by the loss of the multiline EPR signal from the S_2 state of the water-oxidizing complex. Gradual decrease of EPR signals arising from the $Q_A^-Fe^{2+}$ acceptor complex, Tyr$_D$, and the ferricyanide-induced oxidation of the non-heme Fe^{2+} to Fe^{3+} was also observed, but at a significantly slower rate than the inhibition of oxygen evolution and the reduction of the multiline signal. The amplitude of signal II$_{fast}$, arising from Tyr$_Z$ in the absence of fast electron donation from the Mn cluster, was gradually increased during UV-A treatment. However, the amount of functional Tyr$_Z$ decreased to a similar extent as Tyr$_D$ as shown by the loss of amplitude of signal II$_{fast}$ that could be measured in the UV-A-treated particles after Tris washing. It was concluded that the primary damage site of UV-A irradiation is the catalytic manganese cluster of the water-oxidizing complex, where electron transfer to

Tyr_Z and P680$^+$ becomes inhibited. This damaging mechanism is very similar to that induced by the shorter wavelength UV-B (280-320 nm) radiation, but different from that induced by the longer wavelength photosynthetically active light (400-700 nm) [7].

Strong UV-A light from a laser inactivated the oxygen-evolving machinery and the photochemical reaction centre of PS 2. The release of Mn^{2+} ions from PS 2 during incubation of thylakoid membranes in very strong UV-A light was documented by recording of EPR spectra of Mn^{2+} ions in thylakoid membranes from *Synechocystis* cells [9].

Pospisil et al. [134] indicated that $^{\bullet}$OH is produced on the electron acceptor side of PS 2 by two different routes: i) $O_2^{-\bullet}$, which is generated by oxygen reduction on the acceptor side of PS 2, interacts with a PS 2 metal centre, probably the non-heme iron, to form an iron-peroxide species that is further reduced to $^{\bullet}$OH by an electron from PS 2, presumably via $Q_A^{-\bullet}$; ii) $O_2^{-\bullet}$ dismutates to form free H_2O_2 that is then reduced to $^{\bullet}$OH via the Fenton reaction in the presence of metal ions, the most likely being Mn^{2+} and Fe^{2+} released from photodamaged PS 2. H_2O_2 causes extraction of manganese from OEC, inhibits its activity and photosynthetic electron transfer, and leads to the destruction of the photosynthetic apparatus. EPR spectroscopy documented an increase in the level of P700 photooxidation, a decrease of the rate of its subsequent reduction in the dark and an increase of free Mn^{2+} ions after addition of H_2O_2 [135].

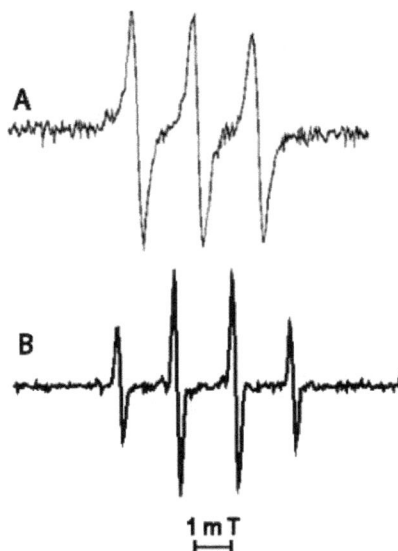

Figure 9. EPR detection of singlet oxygen trapped by TEMP (2,2,6,6-tetramethylpiperidine) (A), and hydroxyl radicals trapped by DMPO (5,5-dimethyl-pyrroline N-oxide) (B) in thylakoid membranes exposed to the light. Source: [133].

The other concept of the monitoring of detrimental action of strong light or other unfavourable living conditions is inspection of the typical hyperfine structure of monodehydroascorbate (MDA). Oxygen is a natural electron acceptor in the PET chain, during which the superoxide anion radical ($O_2^{-\bullet}$) is formed in the thylakoids. $O_2^{-\bullet}$ is reduced to H_2O_2 by CuZn superoxid dismutase (SOD). During the reduction of H_2O_2 by ascorbate peroxidase, MDA is formed, which is detectable by EPR in photoactive conditions. Plants, which are exposed to an excess of radiation, cannot completely utilize it in photosynthesis. This excess radiation can exhibit a damaging effect on photosynthetic apparatus. Leaves are equipped with several protective mechanisms involved in preventing oxidative and photoinhibitory damage. The resulting negative effects on plants depend on the capacity of cellular systems to scavenge ROS and to prevent or to repair harmful effects of light on the PET components. MDA is a long lasting anion radical, it can be detected by EPR spectroscopy at room temperature and serves as an endogenous probe for oxidative stress. Under optimal conditions the concentration of MDA in the leaves is too low to be detected in the light or darkness. Under environmental stress when the rate of oxygen activated species surpasses the MDA reducing capacity, MDA can be detected by EPR, as was the case in leaves treated with paraquat or aminotriazole (Fig. 10). In illuminated leaves paraquat is photoreduced to the paraquat radical that rapidly reacts with O_2 to $O_2^{-\bullet}$, $O_2^{-\bullet}$ is disprotonated to H_2O_2 by SOD at the site of its production, leading to an increase of H_2O_2 level in the chloroplasts. Scavenging of H_2O_2 gives rise to MDA signal that is light dependent [136]. Impact of air pollutants, chemicals, herbicides, photooxidants and unfavourable environmental conditions like drought, high temperatures and even mechanically induced injuries lead to increase of the MDA concentration which can be monitored by EPR spectroscopy. Thus, by simple measurements a proof of oxidative stress can easily be performed. Therefore, the MDA EPR signals can be used as a general marker of stress situations [137]. Light-induced MDA radical production was not detectable by EPR spectroscopy in untreated broad bean leaves, but it was observed after exposing the leaves to UV-B irradiation. After this pretreatment, a low level of MDA radicals was also detectable without illumination [138].

However, we would like to note that beside of investigations focused on the effects of strong visible light and UV irradiation on photosynthesizing organisms, effects of low light on plants were investigated as well. In these experiments mainly chlorophyll fluorescence characteristics were estimated [139,140], however there are some papers in which the use of EPR technique is reported. Paddock et al. [141] investigated reaction centres from *Rhodobacter sphaeroides* by EPR at high and low light intensities. They found that decay kinetics of EPR signal after switch off light exhibited two phases. The fast decay with a time constant $\tau = 30$ ms belongs to the decay of $D^{+\bullet}Q_A^{-\bullet} \rightarrow DQ_A$, where D is the reaction centre in *Rhodobacter sphaeroides*. Slow phase had a longer time constant $\tau = 6$ s. However, when the sample was illuminated at lower light intensity, the relative amplitude of the slow phase was larger indicating that the slow descending component is connected with the decay of the $D^{+\bullet}Q_B^{-\bullet}$ state. EPR technique was also used to study light-induced alteration of low-temperature interprotein electron transfer between PS 1 and flavodoxin [142] utilizing the fact that deuteration of flavodoxin enables the signals of the reduced flavin acceptor and oxidized primary donor, P_{700}^+, to be well-resolved at X- and D-band EPR. While in dark-adapted samples photoinitiated interprotein electron transfer does not occur at 5 K, for samples prepared in dim light significant interprotein

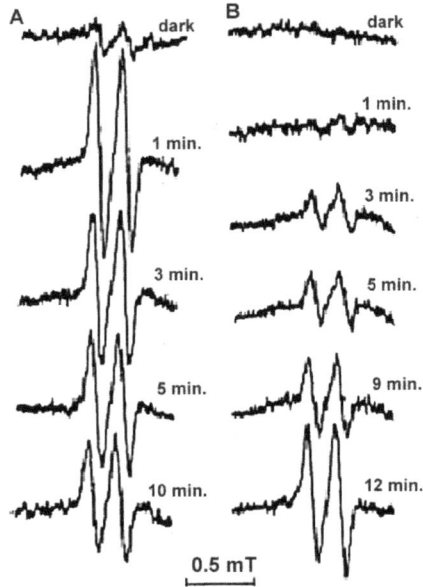

Figure 10. EPR spectra of MDA in *Vicia faba* leaves treated with paraquat (A) or aminotriazole (B). The irradiance was 1000 W m⁻². Numbers in A and B mark the time in light when EPR spectra were recorded. Source: [136].

electron transfer occurred at this temperature and a concomitant loss of the spin-correlated radical pair $P_{700}^+A_{1A}^-$ signal was observed. This indicated a light-induced reorientation of flavodoxin in the PS 1 docking site that allows high quantum yield efficiency for the interprotein electron transfer reaction.

9. Study of photosynthesizing organisms exposed to drought and chilling stress

9.1. Drought

Water deficits cause a reduction in the rate of photosynthesis. Limitation of carbon dioxide fixation results in exposure of chloroplasts to excess excitation energy. When carbon dioxide fixation is limited by water deficit, the rate of active oxygen formation increases in chloroplasts as excess excitation energy, not dissipated by the photoprotective mechanisms, is used to form superoxide and singlet oxygen which can be detected by EPR spectroscopy. Superoxide formation leads to changes suggestive of oxidative damage including lipid peroxidation and a decrease in ascorbate level [143].

Elevated levels of free radicals were detected in leaves of drought-stressed barley plants during and after release of drought stress compared with those seen in the controls. However, they returned rapidly to the control levels after release of the stress. On the other hand, a sizeable increase in the level of a mononuclear Fe(III) complex was seen in the droughted samples (compared with the levels in the watered controls), and these elevated levels remained after release of the stress [144].

EPR quantification of superoxide radicals revealed that drought acclimation treatment led to 2-fold increase in superoxide radical accumulation in leaf and roots of wheat (*Triticum aestivum*) cv. C306 with no apparent membrane damage. However, under subsequent severe water stress condition, the leaf and roots of non-acclimated plants accumulated significantly higher amount of superoxide radicals and showed higher membrane damage than that of acclimated plants indicating that acclimation-induced restriction of superoxide radical accumulation is one of the cellular processes that confers enhanced water stress tolerance to the acclimated wheat seedlings [145].

When germinating *Zea mays* L. seeds were rapidly desiccated, free radical-mediated lipid peroxidation and phospholipid deesterification was accompanied by a desiccation-induced generation of a stable free radical associated with rapid loss of desiccation tolerance, which was detected by EPR spectroscopy. At the subcellular level, the radical was associated with the hydrophilic fraction resulting from lipid extraction. Modulation of respiration using a range of inhibitors resulted in broadly similar modulation of the build-up of the stable free radical [146].

EPR measurements showed that also microsomes isolated from wheat leaves exposed to drought, and from leaves exposed to drought followed by watering, generated significantly higher amount of hydroxyl radical as compared to microsomes isolated from control leaves, suggesting higher production of $^{\bullet}OH$ in the cellular water-soluble phase after drought and watering, as compared to control values. Lipid radicals combined with the spin trap α-(4-pyridyl-1-oxide)-*N-tert*-butylnitrone (4-POBN) resulted in adducts that gave a characteristic EPR spectrum with hyperfine coupling constants of $a^N = 1.58$ mT and $a^H = 0.26$ mT but no significant effect on lipid radical content was measured after drought and drought followed by watering, as compared to controls [147].

The values of signal II for drought-stressed *Sorghum bicolor* and *Pennisetum glaucum* plants were found to be lower by 3 to 9%, similar to non-drought-stressed plants after light stress. However, after a combination of light and drought stress, signal II was decreased by 11 to 32%, indicating that the donor side of PS 2 is also affected by drought stress and high irradiance [148].

9.2. Chilling

At chilling temperature and low light intensity PS 1 is selectively inhibited while PS 2 remains practically unchanged [149]. The activity of PS 1 in cucumber leaves was selectively inhibited by weak illumination at chilling temperatures with almost no loss of P700 content and PS 2 activity. The sites of inactivation in the reducing side of PS 1 were determined by EPR and flash photolysis. EPR measurements showed the destruction of iron-sulphur centres, F_x, F_A

and F_B, in parallel with the loss of quantum yield of electron transfer from diaminodurene to NADP+ [45].

EPR spectra of stearic acid spin labels incorporated into spinach thylakoids can be used to monitor membrane changes during freezing and changes in the EPR parameters can be directly correlated to the extent of functional freeze damage. Jensen et al. [4] used stearic acid spin labels to study the effect of freeze damage to thylakoid membranes microviscosity. They determined changes in EPR parameters either as a function of temperature or during freezing at −15 °C as a function of time and found that an empirical parameter h_+/h_0 (ratio of height of a low field line component h_+ over height of the central line h_0) proved to be very sensitive to minute changes in membrane structure. The observed changes in line shapes were interpreted as an increase in mobility and/or orientation of the lipids following the swelling of thylakoids, however, they did not indicate a disorganization of the lipid phase. Freeze-induced changes in the EPR parameters were found to be strongly dependent on the osmotic conditions of the incubation medium and they were similar to changes observed by transferring thylakoids from an isotonic to a hypotonic medium, i.e., by swelling osmotically flattened thylakoids [150].

Broadening of the EPR signals of 16-doxyl stearic acid in chloroplast membranes of frost-sensitive needles of *Pinus sylvestris* L. and changes in the amplitudes of the peaks were observed upon a decrease in temperature from +30 °C to −10 °C, indicating a drastic loss in rotational mobility. The EPR spectrum of thylakoids from frost-tolerant needles at −10 °C was typical of a spin label in highly fluid surroundings. However, an additional peak in the low-field range appeared in the subzero temperature range for the chloroplast membranes of frost-sensitive needles, which represents spin-label molecules in a motionally restricted surrounding. The domains with restricted mobility could be attributed to protein-lipid interactions in the membranes [151].

EPR study of phospholipid multibilayers, obtained from two cultivars of thermally acclimated wheat of different frost revealed two breaks in the motion of the spin-labelled fatty acid 2-(14-carboxyte-tradecyl)-2-ethyl-4,4-dimethyl-3-oxazolidinyloxyl, for both cultivars (+3 °C, −17 °C and +5 °C, −18 °C, respectively) when grown at 22 °C. Exposure of the resistant cultivar to cold (+2 °C) resulted in the shift of the onset of the apparent phase-separation temperature from +3 °C to −16 °C. However, the sensitive cultivar was unable to do so [152].

Intact tissues and microsome preparations from root tips of coffee seedlings subjected for 6 days to temperatures of 10, 15, 20 and 25 °C in darkness were investigated by EPR using fatty acid spin probes 5-, 12- and 16-doxylstearic acid. It was found that at the depth of the 5th and 16th carbon atom of the alkyl chains the nitroxide radical detected more rigid membranes in seedlings subjected to 10 °C compared with 15 and 25 °C. Membrane rigidity induced by chilling was interpreted as due to lipid peroxidation that could have been facilitated by higher density of peroxidizable chains below the membrane phase transition. At the C-12 position of the chains the probe showed very restricted motion and was insensitive to chilling induced membrane alterations [153].

Incubation at 5 °C and a moderate photon flux density (PFD) decreased the rate of $O_2^{-\bullet}$ production by 40% and 15% in thylakoids from *Spinacia oleracea* L. and 20 °C grown *Nerium oleander* L. (chilling-insensitive plants), but increased the rate by 56% and 5% in thylakoids from *Cucumis sativus* L. and 45 °C grown *N. oleander* (chilling-sensitive plants). The rate of $O_2^{-\bullet}$ production increased in thylakoids when the rate of electron transfer to NADP was reduced what could explain differences in the susceptibility of thylakoids from chilling-sensitive and chilling-insensitive plants to chilling at a moderate PFD, and is consistent with the proposal that $O_2^{-\bullet}$ production is involved in the injury leading to the inhibition of photosynthesis induced under these conditions [154].

10. Other potential uses of EPR in photosynthesis

10.1. Oxymetry

Oxygen concentration in thylakoids can be monitored by EPR using some spin probes (nitroxide probes, phtalocyanine, etc.). Determination of oxygen concentration is based on physical phenomenon of an oxygen-induced line broadening in the EPR spectrum of selected stable free radicals due to their collisions with molecular oxygen. EPR oxymetry offers high sensitivity (typically 10^{-12} mol dm^{-3} at 100 μmol dm^{-3} nitroxide during 1 s), small sample volume and permits monitoring of time-resolved changes in oxygen concentration on millisecond time-scale [155-157].

10.2. Measurements of pH in thylakoids

There are two most frequently used EPR methods for ΔpH measurements in chloroplasts. They are based on pH-indicating spin probes. The first method consists in the calculation of Δ pH from the partitioning of permeable amine spin probes (usually TEMPAMINE; 2,2,6,6-tetra-methylpiperidine-*N*-oxyl-4-amine) between the thylakoid lumen and the suspending medium. By addition of a membrane-impermeable paramagnetic compound, chromium oxalate (which broadens the EPR signal from TEMPAMINE in the external medium), into chloroplast suspension, the probe molecules occurring outside and inside of the thylakoid can be visualized. The second method is based on the measurement of EPR spectra of pH-sensitive spin probes (imidazoline nitroxide radicals) loaded into the vesicles [158].

11. Conclusion

Electron paramagnetic resonance (EPR) spectroscopy is a useful method to study photosynthetic processes because it can monitor the presence of compounds with unpaired spins in the photosynthetic apparatus. Intermediate compounds with unpaired spins are formed directly in photosynthetic apparatus during photosynthetic electron transport in light reactions of photosynthesis and more than 20 different EPR signals covering all electron transfer components can be observed. Moreover, the formation of reactive oxygen species (ROS) during photosynthetic processes in plant chloroplasts can be recorded by EPR spectroscopy as well. This method enables to study the effects of various abiotic stresses (e.g. gaseous pollutants such as ozone and SO$_2$, heavy metals, herbicides, heat, cold, salt, drought, excess of the light as well as UV-A and UV-B radiation and dim light) on photosynthetic apparatus. The formation of complexes between toxic metal ions and photosynthetic proteins as well as excessive formation of ROS caused by various abiotic stress factors can also be monitored by EPR spectroscopy. Moreover, EPR spin probe technique is suitable to study changes in the arrangement and viscosity of photosynthetic membranes in the presence of an abiotic stressor. EPR spectroscopy is a valuable tool to study photodynamic processes (short living radicals or radical pairs, as well as kinetics of their creation or decay) in photosynthesis. Beside this, EPR spin probe method can be used to determine the concentration of photosynthetically released oxygen (EPR oxymetry) or to measure ΔpH in chloroplasts.

Acknowledgements. This chapter is dedicated to the memory of Prof. Ľudovít Krasnec (1913–1990), former leader of our research team. This paper was financially supported by the grant VEGA of the Ministry of Education of the Slovak Republic and the Slovak Academy of Sciences, No. 1/0612/11, and Sanofi Aventis Pharma Slovakia. The authors wish to thank the journals *Biochimica et Biophysica Acta, Jugoslav Physiological and Pharmacological Acta* and *Photosynthetica* for providing us with the original data and figures published in these journals.

Author details

František Šeršeň and Katarína Kráľová

*Address all correspondence to: sersen@fns.uniba.sk

*Address all correspondence to: kralova@fns.uniba.sk

Institute of Chemistry, Faculty of Natural Sciences, Comenius University, Bratislava, Slovakia

References

[1] Macedo AF. Abiotic Stress Responses in Plants: Metabolism to Productivity. In: Ahmad P., Prasad MNV. (eds.) Abiotic Stress Responses in Plants: Metabolism, Productivity and Sustainability. New York: Springer; 2012. p41-61.

[2] Rhodes CJ. Electron Spin Resonance. Part Two: a Diagnostic Method in the Environmental Sciences. Science Progress 2011;94(Pt 4) 339-413.

[3] Šeršeň F., Gabunia G., Krejčíová E., Kráľová K. The Relationship between Lipophilicity of N-alkyl-N,N-dimethylamine Oxides and Their Effects on the Thylakoid Membranes of Chloroplasts. Photosynthetica 1992;26(2) 205-212.

[4] Jensen M., Heber U., Oettmeier W. Chloroplast Membrane Damage During Freezing: The Lipid Phase. Cryobiology 1981;18(3) 322-335.

[5] Bhattacharjee S. Reactive Oxygen Species and Oxidative Burst: Roles in Stress, Senescence and Signal Transduction in Plants. Current Science 2005;89(7) 1113-1121.

[6] Hideg E., Kálai T., Hideg K. Direct Detection of Free Radicals and Reactive Oxygen Species in Thylakoids. In: Carpentier R. (ed.) Photosynthesis Research Protocols, 2nd edition. Book Series: Methods in Molecular Biology, Vol. 684. Totowa, USA: Humana Press Inc; 2011. p.187-200.

[7] Vass I., Turcsanyi E., Touloupakis E., Ghanotakis D., Petroluleas V. The Mechanism of UV-A Radiation-Induced Inhibition of Photosystem II Electron Transport Studied by EPR and Chlorophyll Fluorescence. Biochemistry 2002;41(32) 10200-10208.

[8] Vass I., Szilárd A., Sicora, C. Adverse Effects of UV-B Light on the Structure and Function of the Photosynthetic Apparatus.. In: Pessrakli M. (ed.) Handbook of Photosynthesis, 2nd edition. New York: Marcel Dekker INC; 2005. p827-845.

[9] Zsiros O., Allakhverdiev SI., Higashi S., Watanabe M., Nishiyama Y., Murata N. Very Strong UV-A Light Temporally Separates the Photoinhibition of Photosystem II into Light-Induced Inactivation and Repair. Biochimica et Biophysica Acta-Bioenergetics 2006;1757(2) 123-129.

[10] Whitmarsh J. Electron Transport and Energy Transduction. In: Raghavendra AS. (ed.) Photosynthesis: A Comprehensive Treatise. Cambridge: University Press; 1998. p87-110.

[11] Govindjee. Milestones in Photosynthesis Research. In: Yunus M., Pathre U., Mohanty P. (eds.) Probing Photosynthesis: Mechanisms, Regulation and Adaptation. London: Taylor & Francis; 2000. p9-39.

[12] Kok B., Forbush B., McGloin M. Cooperation of Charges in Photosynthetic O_2 Evolution. I. A Linear Four-Step Mechanism. Photochemistry and Photobiology 1970;11 467-475.

[13] Boussac A., Kuhl H., Ghibaudi E., Rögner M., Rutheford AW. Detection of an Electron Paramagnetic Resonance Signal in the S_0 State of the Manganese Complex of Photosystem II from *Synechococcus elongatus*. Biochemistry 1999;38(37) 11942-11948.

[14] Britt RD., Peloquin JM. Pulsed and Parallel-Polarization EPR Characterization of the Photosystem II Oxygen-Evolving Complex. Annual Review of Biophysics and Biomolecular Structure 2000;29 463-495.

[15] Mino H., Kawamori A. EPR Studies of Water Oxidizing Complex in S_1 and the Higher S States: the Manganese Cluster and Y_Z Radical. Biochimica et Biophysica Acta-Bioenergetics 2001;1503(1-2) 112-122.

[16] Haddy A. EPR Spectroscopy of the Manganese Cluster of Photosystem II. Photosynthesis Research 2007;92(3) 357-368.

[17] Kodera Z., Takura K., Kawamori A. Distance of P680 from the Manganese Complex in Photosystem II Studied by Time Resolved EPR. Biochimica et Biophysica Acta-Bioenergetics 1992;1101(1) 23-32.

[18] Peloquin JM., Britt RD. EPR/ENDOR Characterization of the Physical and Electronic Structure of the OEC Mn Cluster. Biochimica et Biophysica Acta-Bioenergetics 2001;1503(1-2) 96-111.

[19] Debus RJ. The Manganese and Calcium Ions of Photosynthetic Oxygen Evolution. Biochimica et Biophysica Acta-Bioenergetics 1992;1102(3) 269-352.

[20] Boussac A. Quantification of the Number of Spins in the S_2- and S_3-States of Ca^{2+} Depleted Photosystem II by Pulsed-EPR Spectroscopy. Biochimica et Biophysica Acta-Bioenergetics 1996;1277(3) 253-265.

[21] Booth PJ., Rutherford AW., Boussac A. Location of the Calcium Binding Site in Photosystem II: A Mn^{2+} Substitution Study. Biochimica and Biophysica Acta-Bioenergetics 1996;1277(1-2) 127-134.

[22] Boussac A., Rutheford AW. Does the Formation of the S_3-State in Ca^{2+}-Depleted Photosystem II Correspond to an Oxidation of Tyrosine Z Detectable by cw–EPR at Room Temeperature? Biochimica et Biophysica Acta-Bioenergetics 1995;1230(3) 195-201.

[23] Hallahan BJ., Nugent JHA., Warden JT., Evans MCW. Investigation of the "S_3" EPR Signal from the Oxygen-Evolving Complex of Photosystem 2: The Role of Tyrosine Z. Biochemistry 1992;31(19) 4562-4573.

[24] Chrysina M., Zahariou G., Ioannidis N., Petruleas V. Conversion of the g = 4.1 EPR Signal to the Multiline Conformation During the S-2 to S-3 Transition of the Oxygen Evolving Complex of Photosystem II. Biochimica et Biophysica Acta-Bioenergetics 2010;1797(4) 487-493.

[25] Blankenship RE., Sauer K. Manganese in Photosynthetic Oxygen Evolution. Electron Paramagnetic Resonance Study of the Environment of Mn in Tris-Washed Chloroplasts. Biochimica et Biophysica Acta-Bioenergetics 1974;357(2) 252-266.

[26] Šeršeň F., Kráľová K., Bumbálová A. Action of Mercury on the Photosynthetic Apparatus of Spinach Chloroplasts. Photosynthetica 1998;35(4) 551-559.

[27] Commoner BJ., Heise JJ., Townsend J. Light-Induced Paramagnetism in Chloroplasts. Proceedings of the National Academy of Sciences of the United States of America 1956;42(10) 710-718.

[28] Babcock GT., Sauer K. Electron Paramagnetic Resonance Signal II in Spinach Chloroplasts. Biochimica et Biophysica Acta-Bioenergetics 1973;325(3) 483-503.

[29] Blankenship RE., Babcock GT., Warden JT., Sauer K. Observation of a New EPR Transient in Chloroplasts that May Reflect Electron-Donor to Photosystem 2 at Room-Temperature. FEBS Letters 1975;51(1) 287-293.

[30] Hoff, AJ. Application of ESR in Photosynthesis. Physics Reports 1979;54(2) 75-200.

[31] Debus RJ., Barry DA., Babcock GT., McIntosh L. Site-Directed Mutagenesis Identifies a Tyrosine Radical Involved in the Photosynthetic Oxygen-Evolving System. Proceedings of the National Academy of Sciences of the United States of America 1988;85(2) 427-430.

[32] Debus RJ., Barry DA., Sithole I., Babcock GT., McIntosh L. Direct Mutagenesis Indicates that Donor to P680$^+$ in Photosystem II is Tyrosine-161 of the D1 Polypeptide. Biochemistry 1988;27(26) 9071-9074.

[33] Hoganson CW., Babcock GT. Redox Cofactor Interaction in Photosystem II. Electron Spin Resonance Spectrum P680$^+$ is Broadened in Presence of Y_Z^+. Biochemistry 1989;28(4) 1448-1454.

[34] van Mieghem F., Brettel K., Hillmann B., Kamlowski A., Rutherford AW., Schloder E. Charge Recombination Reactions in Photosystem II. 1. Yields, Recombination Pathways, and Kinetics of the Primary Pair. Biochemistry 1995;34(14) 4798-4813.

[35] Hoff AJ., Proskuryakov II. Electron Spin Polarization (CIDEP) of a Primary Acceptor in Photosystem II. Biochimica et Biophysica Acta-Bioenergetics 1985;808(2) 343-347.

[36] Bock CH., Gerken S., Stehlik D., Witt HT. Time Resolved EPR on Photosystem II Particles after Irreversible and Reversible Inhibition of Water Cleavage with High Concentrations of Acetate. FEBS Letters 1988;227(2) 141-146.

[37] Angerhofer A., Bittl R. Radicals and Radical Pairs in Photosynthesis. Photochemistry and Photobiology 1996;63(1) 11-38.

[38] MacMillan F., Lendzian F., Renger G., Lubitz W. EPR and ENDOR Investigation of the Primary Electron Acceptor Radical Anion $Q_A^{\bullet-}$ in Iron-depleted Photosystem II Membrane Fragments. Biochemistry 1995;34(25) 8144-8156.

[39] Jegerschöld C., Styring S. Spectroscopic Characterization of Intermediate Steps involved in Donor-Side Induced Photoinhibition of Photosystem II. Biochemistry 1996;35(24) 7794-7801.

[40] Warden JT., Bolton JR. Falsh-Photolysis ESR Studies of the Dynamics of PS1 in Green Plant Photosynthesis. Photochemistry and Photobiology 1974;20(2) 263-269.

[41] Sieckman I., van der Est A., Botin H., Sétif P., Stehlik D. Nanosecond Electron Transfer Kinetics in Photosystem I Following Substitution of Quinones for K_1 Vitamin as Studied by Time Resolved EPR. FEBS Letters 1991;284(1) 98-102.

[42] Rustandi RR., Snyder SW., Biggins J., Norris JR., Thurnauer MC. Reconstitution and Exchange of Quinones in A1 Site of Photosystem I. An Electron Spin Polarization

Electron Paramagnetic Resonance Study. Biochimica et Biophysica Acta-Bioenergetics 1992;1101(3) 311-320.

[43] Snyder SW., Rustandi RR., Biggins J., Norris JR., Thurnauer MC. Direct Assignment of Vitamin K_1 as the Secondary Acceptor A_1 in Photosystem I. Proceedings of the National Academy of Sciences of the United States of America 1991;88(21) 9895-9896.

[44] Evans MCW., Telfer A., Lord AV. Evidence for Role of a Bound Ferredoxin as the Primary Electron Acceptor of Photosystem I in Spinach Chloroplasts. Biochimica et Biophysica Acta-Bioenergetics 1972;267(3) 530-537.

[45] Sonoike K., Terashima I., Iwaki M., Itoh S. Destruction of Photosystem I Iron-Sulphur Centers in Leaves of Cucumis sativus L. by Weak Illumination at Chilling Temperatures. FEBS Letters 1995;362(2) 235-238.

[46] Setif P., Brettel K. Photosystem I Photochemistry under Highly Reducing Condition: Study of the P700 Triplet State Formation from the Secondary Radical Pair (P700$^+$-$A_1$$^-$). Biochimica et Biophysica Acta-Bioenergetics 1990;1020(3) 232-238.

[47] Nixon PJ., Diner BA. Asparate 170 of the Photosystem II Reaction Center Polypeptide D1 is Involved in Assembly of the Oxygen-Evolving Manganese-Cluster. Biochemistry 1992;31(3) 942-948.

[48] Bellyaeva OB., Timofeev KN., Litvin FF. The Primary Reaction in Protochlorophyll(ide) Photoreduction as Investigated by Optical and ESR Spectroscopy. Photosynthesis Research 1988;15(3) 247-256.

[49] Kaim W., Schwederski B., editors. Bioinorganic Chemistry: Inorganic Elements in the Chemistry of Life. New York: Wiley; 1994.

[50] Sibbald PR., Green BR. Copper in Photosystem II – Association with LHC II. Photosynthesis Research 1987;14(3) 201-209.

[51] Yurela I. Copper in Plants. Brazilian Journal of Plant Physiology 2005;17(2) 145-156.

[52] Schröder WP., Arellano JB., Bittner T., Barón M., Eckert HJ., Renger G. Flash Induced Absorption Spectroscopy Studies of Copper Interaction with Photosystem II in Higher Plants. Journal of Biological Chemistry 1994;269(52) 32865-32870.

[53] Jegerschöld C., Arellano JB., Schröder WP., van Kan PJM., Barón M., Styring S. Copper (II) Inhibition of Electron Transfer through Photosystem II Studied by EPR Spectroscopy. Biochemistry (USA) 1995;34(39) 12747-12754.

[54] Šeršeň F., Kráľová K., Bumbálová A., Švajlenová O. The Effect of Cu(II) Ions Bound with Tridentate Schiff Base Ligands upon Photosynthetic Apparatus. Journal of Plant Physiology 1997;151(3) 299-305.

[55] Kráľová K., Šeršeň F., Blahová M. Effects of Cu(II) Complexes on Photosynthesis in Spinach Chloroplasts. Aqua(aryloxyacetato)copper(II) Complexes. General Physiology and Biophysics 1994;3(6) 483-491.

[56] Kráľová K., Šeršeň F., Melník M. Inhibition of Photosynthesis in *Chlorella vulgaris* by Cu(II) Complexes with Biologically Active Ligands. Journal of Trace and Microprobe Techniques 1998;16(4) 491-500.

[57] Šeršeň F., Kráľová K., Blahová M. Photosynthesis of *Chlorella vulgaris* as Affected by 4-Chloro-2- Methylphenoxyacetato)Copper(II) Complex. Biologia Plantarum 1996;38(1) 71-75.

[58] Jegerschöld C., MacMillan F., Lubitz W., Rutherford AW. Effects of Copper and Zinc Ions on Photosystem II Studied by EPR Spectroscopy. Biochemistry 1999;38(38) 12439-12445.

[59] Vierke G., Struckmeier PZ. Binding of Copper(II) to Proteins of the Photosynthetic Membrane and its Correlation with Inhibition of Electron Transport in Class II Chloroplasts of Spinach. Zeitschrift für Naturforschung-Biosciences 1977;32c(7/8) 605-610.

[60] Utschig LM., Poluektov O., Tiede DM., Thurnaer MC. EPR Evidence of Cu^{2+} Substituted Photosynthetic Bacterial Reaction Centers: Evidence for Histidine Ligation at the Surface Metal Site. Biochemistry 2000;39(11) 2961-2969.

[61] Bohner H., Böhme H., Böger P. Reciprocal Formation of Plastocyanin and Cytochrome c − 553 and the Influence of Cupric Ions on Photosynthetic Electron Transport. Biochimica et Biophysica Acta-Bioenergetics 1980;592(1) 103-112.

[62] Patra M., Sharma A. Mercury Toxicity in Plants. Botanical Review 2000;66(3) 379-422.

[63] Kimura M., Katoh S. Studies on Electron Transport Associated with Photosystem I. Functional Site of Plastocyanin: Inhibitory Effects of $HgCl_2$ on Electron Transport and Plastocyanin in Chloroplasts. Biochimica et Biophysica Acta-Bioenergetics 1972;283(2) 279-292.

[64] Radmer R., Kok B. Kinetic Observation of the System II Electron Acceptor Pool Isolated by Mercuric Ions. Biochimica et Biophysica Acta-Bioenergetics 1974;357(2) 177-180.

[65] Rai LC., Singh AK., Mallick N. Studies on Photosynthesis, the Associated Electron Transport System and Some Physiological Variables of *Chlorella vulgaris* under Heavy Metals Stress. Journal of Plant Physiology 1991;137(4) 419-424.

[66] Honeycutt RC., Krogmann DW. Inhibition of Chloroplast Reactions with Phenylmercury Acetate. Plant Physiology 1972;49(3) 376-380.

[67] De Filippis LF., Hampp R., Ziegler H. The Effect of Sublethal Concentration of Zinc, Cadmium and Mercury on *Euglena*. Adenylates and Energy Charge. Zeitschrift für Pflanzenphysiologie 1981;103(1) 1-7.

[68] Jung YS., Yu L., Golbeck JH. Reconstitution of Iron-Sulfur Center F_B Results in Complete Restoration of $NADP^+$ Photoreduction in Hg-treated Photosystem I Complexes from *Synechococus* sp. PCC631. Photosynthesis Research 1995;46(1-2) 249-255.

[69] Singh DP., Singh SP. Action of Heavy Metals on Hill Activity and O_2 Evolution in *Anacystis nidulans*. Plant Physiology 1987;83(1) 12-14.

[70] Samson G., Morissette JC., Popovic F. Determination of Four Apparent Mercury Interaction Sites in Photosystem II by Using a New Modification of the Stern-Volmer Analysis. Biochemical and Biophysical Research Communications 1990;166(2) 873-878.

[71] Bernier M., Popovic R., Carpentier R. Mercury Inhibition at the Donor Side of Photosystem II is Reversed by Chloride. FEBS Letters 1993;321(1) 19-23.

[72] Bernier M., Carpentier R. The Action of Mercury on Binding of the Extrinsic Poypeptides Associated with the Water Oxidizing Complex of Photosystem II. FEBS Letters 1995; 360(3) 251-254.

[73] Murthy SDS., Mohanty N., Mohanty P. Prolonged Incubation with Low Concentrations of Mercury Alters Energy Transfer and Chlorophyll (Chl) *a* Protein Complexes in *Synechococccus* 6301: Changes in Chl *a* Absorption and Emission Characteristics Loss of the F695 Emission Band. BioMetals 1995;8(3) 237-242.

[74] Miles D., Bolen P., Farag S., Goodin R., Lutz J., Moustafa A., Rodriguez B., Weil C. Hg++ - A DCMU Independent Electron Acceptor of Photosystem II. Biochemical and Biophysical Research Communications 1973;50(4) 1113-1119.

[75] Prokowski Z. Effects of $HgCl_2$ Long-lived Delayed Luminescence in *Scenedesmus quadricauda*. Photosynthetica 1993;28(4) 563-566.

[76] Nahar S., Tajmir-Riahi HA. A Comparative Study of Fe(II) and Fe(III) Ions Complexation with Proteins of the Light-Harvesting Complex of Chloroplast Thylakoid Membranes. Journal of Inorganic Biochemistry 1994;54(2) 79-90.

[77] Šeršeň F., Kráľová K. New Facts about $CdCl_2$ Action on the Photosynthetic Apparatus of Spinach Chloroplasts and its Comparison with $HgCl_2$ Action. Photosynthetica 2001;39(4) 575-580.

[78] Sakurai H., Inoue K., Fujii T., Mathis P. Effect of Selective Destruction of Iron-Sulphur Center B on Electron Transfer and Charge Recombination in Photosystem I. Photosynthesis Research 1991;27(1) 65-71.

[79] Siedlecka A., Krupa Z. Interaction Between Cadmium and Iron and its Effect on Photosynthetic Capacity of Primary Leaves of *Phaseolus vulgaris*. Plant Physiology and Biochemistry 1996;34(6) 833-841.

[80] Krupa Z. Cadmium Against Higher Plant Photosynthesis – a Variety of Effects and Where Do They Possibly Come From? Zeitschrift für Naturforschung-A Journal of Biosciences 1999;54c(9-10) 723-729.

[81] Bazzaz NB., Govindjee. Effects of Cadmium Nitrate on Spectral Characteristics and Light Reaction of Chloroplasts. Environmental Letters 1974;6(1) 1-12.

[82] Van Duijvendijk-Mateoli MA., Desmet GM. On the Inhibitory Action of Cadmium on the Donor Side of Photosystem II in Isolated Chloroplasts. Biochimica et Biophysica Acta-Bioenergetics 1975;408(2) 164-169.

[83] Baszynski T., Wajda L., Król M., Wolińska D., Krupa Z., Tukendorf A. Photosynthetic Activities of Cadmium-Treated Tomato Plants. Physiologia Plantarum 1980;48(3) 365-370.

[84] Atal N., Saradhi PP., Mohanty P. Inhibition of Chloroplast Photochemical Reactions by Treatment of Wheat Seedlings with Low Concentrations of Cadmium: Analysis of Electron Transport Activities and Changes in Fluorescence Yield. Plant and Cell Physiology 1991;32(7) 943-951.

[85] Sigfridsson KGV., Bernát G., Mamedov F., Styring S. Molecular Interference of Cd^{2+} with Photosystem II. Biochimica et Biophysica Acta-Bioenergetics 2004;1659(1) 19-31.

[86] Fodor F., Sárvári E., Láng F., Szigeti Z., Cseh E. Effects of Pb and Cd on Cucumber Depending on the Fe-Complex in the Culture Solution. Journal of Plant Physiology 1996;148(3-4) 434-439.

[87] Pagliano C., Raviolo M., Dalla Vecchia F., Gabbrielli R., Gonnelli C., Rascio N., Barbato R., La Rocca N. Evidence for PSII Donor-Side Damage and Photoinhibition induced by Cadmium Treatment on Rice (Oryza sativa L.). Journal of Photochemistry and Photobiology B: Biology 2006;84(1) 70-78.

[88] Ono TA., Inoue Y. Roles of Ca^{2+} in O_2 Evolution in Higher Plant Photosystem II: Effects of Replacement of Ca^{2+} Site by other Cations. Archives of Biochemistry and Biophysics 1989;275(2) 440-448.

[89] Lin R., Wang X., Luo Y., Du W., Guo H., Yin D. Effects of Soil Cadmium on Growth, Oxidative Stress and Antioxidant System in Wheat Seedlings (Triticum aestivum L.). Chemosphere 2007;69(1) 89-98.

[90] Karavaev VA., Baulin AM., Gordienko TV., Dovydkov SA., Tikhonov AN. Changes in the Photosynthetic Apparatus of Broad Bean Leaves as Dependent on the Content of Heavy Metals in the Growth Medium. Russian Journal of Plant Physiology 2001;48(1) 38-44.

[91] Frontasyeva MV., Pavlov SS., Mosulishvili L., Kirkesali E., Ginturi E., Kuchava N. Accumulation of Trace Elements by Biological Matrice of Spirulina platensis. Ecological Chemistry and Engineering S-Chemia i inzynieria ekologiczna S 2009;16(3) 277-285.

[92] Pandey V., Dixit V., Shyam R. Chromium Effect on ROS Generation and Detoxification in Pea (Pisum sativum) Leaf Chloroplasts. Protoplasma 2009;236(1-4) 85-95.

[93] Pandey V., Dikshit V., Shyam R. Hexavalent Chromium Induced Inhibition of Photosynthetic Electron Transport in Isolated Spinach Chloroplasts. In: Dubinsky Z. (ed.) Photosynthesis. Rjeka: Intech; 2013; in this book.

[94] Micera G., Dessi A. Chromium Adsorption by Plant Roots and Formation of Long-Lived Cr(V) Species: An Ecological Hazard? Journal of Inorganic Biochemistry 1988;34 157-166.

[95] Liu KJ., Jiang J., Shi X., Gabrys H., Walczak T., Swartz HM. Low-Frequency EPR Study of Cr(V) Formation from Cr(VI) in Living Plants. Biochemical and Biophysical Research Communications 1995;206(3) 829-834.

[96] Labanowska M., Filek M., Košcielniak J., Kurdziel M., Kuliš E., Hartikainen H. The Effects of Short-Term Selenium Stress on Polish and Finnish Wheat Seedlings—EPR, Enzymatic and Fluorescence Studies. Journal of Plant Physiology 2012;169(3) 275-284.

[97] Šeršeň F., Kráľová K., Jóna E., Sirota A. Effects of Ni(II) Complexes with N-donor Ligands on Photosynthetic Electron Transport in Spinach Chloroplasts. Chemické Listy 1997;91(9) 685-686.

[98] Kráľová K., Masarovičová E., Šeršeň F., Ondrejkovičová I. Effect of Different Fe(III) Compounds on Photosynthetic Electron Transport in Spinach Chloroplasts and on Iron Accumulation in Maize Plants. Chemical Papers 2008;62(4) 358-363.

[99] Roberts AG., Gregor W., Britt RD., Kramer DM. Acceptor and Donor-Side Interactions of Phenolic Inhibitors in Photosystem II. Biochimica et Biophysica Acta-Bioenergetics 2003;1604(1) 23-32.

[100] Rutherford AW., Zimmermann JL., Mathis P. The Effect of Herbicides on Components of the PS II Reaction Centre Measured by EPR. FEBS Letters 1984;1077 156-162.

[101] Rutherford AW., Krieger-Liszkay A. Herbicide-Induced Oxidative Stress in Photosystem II. Trends in Biochemical Sciences 2001;26(11) 648-653.

[102] Chen SG., Xu XM., Dai XB., Yang CL., Qiang S. Identification of Tenuazonic Acid as a Novel Type of Natural Photosystem II Inhibitor Binding in Q_B-Site of Chlamydomonas reinhardtii. Biochimica et Biophysica Acta-Bioenergetics 2007;1767(4) 306-318.

[103] Chen SG., Yin CY., Dai XB., Qiang S., Xu XM. Action of Tenuazonic Acid, a Natural Phytotoxin, on Photosystem II of Spinach. Environmental and Experimental Botany 2008;62(3) 279-289.

[104] Chen SG., Yin CY., Qiang S., Zhou FY., Dai XB. Chloroplastic Oxidative Burst Induced by Tenuazonic Acid, a Natural Photosynthesis Inhibitor, Triggers Cell Necrosis in Eupatorium adenophorum Spreng. Biochimica et Biophysica Acta-Bioenergetics 2010;1797(3) 391-405.

[105] Kirilovsky D., Rutherford AW., Etienne AL. Influence of DCMU and Ferricyanide on Photodamage in Photosystem II. Biochemistry 1994;33(10) 3087-3095.

[106] Timofeev KN., Kuznetsova GV., Elanskaya IV. Effects of Dark Adaptation on Light-Induced Electron Transport through Photosystem I in the Cyanobacterium *Synechocystis* sp. PCC 6803. Biochemistry-Moscow 2005;70(12) 1390-1395.

[107] Andreasson LE., Lindberg K. The Inhibition of Photosynthetic Oxygen Evolution by Ammonia Probed by EPR. Biochimica et Biophysica Acta-Bioenergetics 1992;1100(2) 177-183.

[108] González-Pérez S., Quijano C., Romero N., Melo TB., Radi R., Arellano JB. Peroxonitrite Inhibits Electron Transport on the Acceptor Side of Higher Plant Photosystem II. Archives of Biochemistry and Biophysics 2008;473(1) 25-33.

[109] Kráľová K., Šeršeň F. Effects of Bioactive Natural and Synthetic Compounds with Different Alkyl Chain Length on Photosynthetic Apparatus. In: Najafpour MM. (ed.) Applied Photosynthesis. Rjeka: Intech; 2012. p165-190.

[110] Doležal M., Kráľová K. Synthesis and Evaluation of Pyrazine Derivatives with Herbicidal Activity. In: Soloneski S., Larrmendy M. (eds.) Herbicides, Theory and Application. Rjeka: Intech; 2011. p581-610.

[111] Mehlhorn H., Tabner B., Wellburn A. Electron Spin Resonance Evidence for the Formation of Free Radicals in Plants Exposed to Ozone. Physiologia Plantarum 1990;79(2) 377-383.

[112] Pirker KF., Reichenauer TG., Pascual EC., Kiefer S., Soja G., Goodman BA. Steady State Levels of Free Radicals in Tomato Fruit Exposed to Drought and Ozone Stress in a Field Experiment. Plant Physiology and Biochemistry 2003;41(10) 921-927.

[113] Mazarura U., Runeckles VC. EPR Studies on the Effect of Sequences of Fluctuating Levels of Ozone on *Poa pratensis* L., and *Setaria viridis* Beauv. 2001. http://abstracts.aspb.org/pb2001/public/P37/0065.htmlPoster: oxidation (accessed 12 July 2012).

[114] Reichenauer TG., Goodman BA. Stable Free Radicals in Ozone-Damaged Wheat Leaves. Free Radical Research 2001;35(2) 93-101.

[115] Runeckles VC., Vaartnou M. EPR Evidence for Superoxide Anion Formation in Leaves During Exposure to Low Levels of Ozone. Plant Cell and Environment 1997;20(3) 306-314.

[116] Vaartnou M. EPR investigation of free radicals in excised and attached leaves subjected to ozone and sulphur dioxide air pollution. PhD thesis. The University of British Columbia Vancouver; 1988.

[117] Beauregard M. Involvement of Sulfite and Sulfate Anions in the SO_2-Induced Inhibition of the Oxygen Evolving Enzyme Photosystem II in Chloroplasts: A Review. Environmental and Experimental Botany 1991;31(1) 11-21.

[118] Shimazaki K., Ito K., Kondo N., Sugahara K. Reversible Inhibition of the Photosyn-
       thetic Water-Splitting Enzyme System by SO_2-Fumigation Assayed by Chlorophyll
       Fluorescence and EPR Signal *in vivo*. Plant and Cell Physiology 1984; 25(5) 795-803.

[119] Szalontai B., Horváth LI., Debreczeny M., Droppa M., Horváth G. Molecular Rear-
       rangements of Thylakoids after Heavy Metal Poisoning, as Seen by Fourier Trans-
       form Infrared (FTIR) and Electron Spin Resonance (ESR) Spectroscopy.
       Photosynthesis Research 1999;61(3) 241-252.

[120] Berglund AH., Quartacci MF., Calucci L., Navari-Izo F., Pinzino C., Liljenberg CB.
       Alterations of Wheat Root Plasma Membrane Lipid Composition Induced by Copper
       Stress Result in Changed Physicochemical Properties of Plasma Membrane Lipid
       Vesicles. Biochimica et Biophysica Acta-Biomembranes 2002;1564(2) 466-472.

[121] Calucci L., Navari-Izzo F., Pinzino C., Sgherri CLM. Fluidity Changes in Thylakoid
       Membranes of Durum Wheat Induced by Oxidative Stress: A Spin Probe EPR Study.
       Journal of Physical Chemistry B 2001;105 (15) 3127-3134.

[122] Apostolova EL. Effect of Detergent Treatment and Glutaraldehyde Modification on
       the Photochemical Activity of Chloroplasts. Comptes Rendus de l'Académie Bulgaire
       des Sciences 1988;41(7) 117-120.

[123] Devínsky F., Kopecká-Leitmanová A., Šeršeň F., Balgavý P. Interaction of Surfactants
       with Model and Biological Membranes. 8. Amine Oxides. 24. Cut off Effect in Anti-
       microbial Activity and in Membrane Perturbation Efficiency of the Homologous Ser-
       ies of N,N-Dimethylalkylamine Oxides. Journal of Pharmacy and Pharmacology
       1990;42(11) 790-794.

[124] Balgavý P., Devínsky F. Cut-off Effects in Biological Activities of Surfactants. Advan-
       ces in Colloid and Interface Science 1996;66 23-63.

[125] Šeršeň F., Lacko I. Stimulation of Photosynthetic Electron Transport by Quaternary
       Ammonium Salts. Photosynthetica 1995;31(1) 153-156.

[126] Quartacci MF., Pinzino C., Sgherri CLM., Dalla Vecchia F., Navari-Izzo F. Growth in
       Excess Copper Induces Changes in the Lipid Composition and Fluidity of PSII-En-
       riched Membranes in Wheat. Physiologia Plantarum 2000;108(1) 87-93.

[127] Aro E.-M, Virgin I., Andersson B. Photoinhibition of Photosystem II. Inactivation,
       Protein Damage and Turnover. Biochimica et Biophysica Acta-Bioenergetics
       1993;1143(2) 113-124.

[128] Krieger-Liszkay A., Kós PB., Hideg E. Superoxide Anion Radicals Generated by
       Methylviologen in Photosystem I Damage Photosystem II. Physiologia Plantarum
       2011;142(1) 17-25.

[129] Krieger-Liszkay A., Fufezan C., Trebst A. Singlet Oxygen Production in Photosystem
       II and Related Protection Mechanism. Photosynthesis Research 2008;98(1-3) 551-564.

[130] Hideg E., Spetea C., Vass I. Singlet Oxygen and Free Radical Production During Acceptor-Induced and Donor-Induced Photoinhibition – Studies with Spin-Trapping EPR Spectroscopy. Biochimica et Biophysica Acta-Bioenergetics 1994;1186(3) 143-152.

[131] Ogami S., Boussac A., Sugiura M. Deactivation Processes in PsbA1-Photosystem II and PsbA3-Photosystem II under Photoinhibitory Conditions in the Cyanobacterium *Thermosynechococcus elongatus*. Biochimica et Biophysica Acta-Bioenergetics 2012;1817(8):1322-1330.

[132] Ivanov AG., Morgan RM., Gray GR., Velitchkova MY., Huner NPA. Temperature/ Light Dependent Development of Selective Resistance to Photoinhibition of Photosystem I. FEBS Letters 1998;439(3) 288-292.

[133] Hideg E., Vass I. EPR Detection of Reactive Oxygen in the Photosynthetic Apparatus of higher Plants Under Light Stress. 1998. http://www.photobiology.com/v1/hideg/ (accessed 14 August 2012).

[134] Pospisil P., Arato A., Krieger-Liszkay A., Rutherford AW. Hydroxyl Radical Generation by Photosystem II. Biochemistry 2004;43(21) 6783-6792.

[135] Samuilov VD., Timofeev KN., Sinitsyn SV., Bezryadnov DV. H_2O_2-Induced Inhibition of Photosynthetic O_2 Evolution by *Anabaena variabilis* Cells. Biochemistry-Moscow 2004;69(8) 926-933.

[136] Veljović-Jovanović S. Active Oxygen Species and Photosynthesis: Mehler and Ascorbate Peroxidase Reactions. Jugoslav Physiological and Pharmacological Acta 1998;34(2) 503-522.

[137] Stegmann HB., Schuler P. Environmental Stress of Higher Plants Monitored by the Formation of Monodehydroascorbate Radicals with Continuous Wave EPR. Applied Magnetic Resonance 2000;19(1) 69-76.

[138] Hideg E., Mano J., Ohno C., Asada K. Increased Levels of Monodehydroascorbate Radical in UV-B-Irradiated Broad Bean Leaves. Plant and Cell Physiology 1997;38(6) 684-690.

[139] Sui XL., Mao SL., Wang LH., Zhang BX., Zhang ZX. Effect of Low Light on the Characteristics of Photosynthesis and Chlorophyll *a* Fluorescence during Leaf Development of Sweet Pepper. Journal of Integrative Agriculture 2012;11(10) 1633-1643.

[140] Zhang YJ., Xie ZK., Wang YJ., Su PX., An LP., Gao H. Light Intensity Affects Dry Matter, Photosynthesis and Chlorophyll Fluorescence of Oriental Lily. Philippine Agricultural Scientist 2011;94(3) 232-238.

[141] Paddock ML., Isaacson RA., Abresch EC., Okamura MY. Light Induced EPR Spectra of Reaction Centers from *Rhodobacter sphaeroides* at 80 K: Evidence for Reduction of Q_B by B-Branch Electron Transfer in Native Reaction Centers. Applied Magnetic Resonance 2007; 31(1-2) 29-43.

[142] Utschig LM., Tiede DM., Poluektov OG. Light-Induced Alteration of Low-Temperature Interprotein Electron Transfer between Photosystem I and Flavodoxin. Biochemistry 2010;49(45) 9682-9684.

[143] Smirnoff N. Tansley Review. 52. The Role of Active Oxygen in the Response of Plants to Water Deficit and Desiccation. New Phytologist 1993;125(1) 27-58.

[144] Goodman BA., Newton AC. Effects of Drought Stress and Its Sudden Relief on Free Radical Processes in Barley. Journal of the Science of Food and Agriculture 2005;85(1) 47-53.

[145] Selote DS., Bharti S., Khanna-Chopra R. Drought Acclimation Reduces $O_2{}^{\bullet-}$ Accumulation and Lipid Peroxidation in Wheat Seedlings. Biochemical and Biophysical Research Communications 2004;314(3) 724-729.

[146] Leprince O., Atherton NM., Deltour R., Hendry CAF. The Involvement of Respiration in Free Radical Processes During Loss of Desiccation Tolerance in Germinating *Zea mays* L. Plant Physiology 1994;104(4) 1333-1339.

[147] Bartoli CG., Simontacchi M., Tambussi E., Beltrano J., Montaldi E., Puntarulo S. Drought and Watering-Dependent Oxidative Stress: Effect on Antioxidant Content in *Triticum aestivum* L. Leaves. Journal of Experimental Botany 1999;50(332) 375-383.

[148] Masojldek J., Trivedi S., Halshaw L., Alexiou A., Hall DO. The Synergistic Effect of Drought and Light Stresses in Sorghum and Pearl Millet. Plant Physiology 1991;96(1) 198-207.

[149] Yordanov I., Velikova V. Photoinhibition of Photosystem 1. Bulgarian Journal of Plant Physiology 2000;26(1-2) 70-92.

[150] Jensen M., Oettmeier W. Effects of Freezing on the Structure of Chloroplast Membranes. Cryobiology 1984;21(4) 465-473.

[151] Vogg G., Heim R., Gotschy B., Beck E., Hansen J. Frost Hardening and Photosynthetic Performance of Scots Pine (*Pinus sylvestris* L.). II. Seasonal Changes in the Fluidity of Thylakoid Membranes. Planta 1998;204(2) 201-206.

[152] Vígh L., Horváth I., Woltjes J., Farkas T., Hassett P., Kuiper PJC. Combined Electron Spin Resonance, X-Ray Diffraction Studies on Phospholipid Vesicles Obtained from Cold-Hardened Wheats. I. An Attempt to Correlate Electron-Spin Resonance Spectral Characteristics with Frost Resistance. Planta 1987;170(1) 14-19.

[153] Alonso A., Queiroz CS., Magalhães AC. Chilling Stress Leads to Increased Cell Membrane Rigidity in Roots of Coffee (*Coffea arabica* L.) Seedlings. Biochimica et Biophysica Acta-Biomembranes 1997;1323(1) 75-84.

[154] Hodgson RAJ., Raison JK. Superoxide Production by Thylakoids During Chilling and Its Implication in the Susceptibility of Plants to Chilling-Induced Photoinhibition. Planta 1991;183(2) 222-228.

[155] Strzalka K., Sarna T., Hyde JS. Electron Spin Resonance Oxymetry – Measurement of Photosynthetic Oxygen Evolution by Spin Probe Technique. Photochemistry and Photobiophysics 1986;12(1-2) 67-71.

[156] Tang XS., Moussavi M., Dismukes GC. Monitoring Oxygen Concentration in Solution by ESR Oximetry Using Lithium Phtalocyanine – Application to Photosynthesis. Journal of the American Chemical Society 1991;113(15) 5914-5915.

[157] Subczynski WK., Swartz HM. EPR Oxymetry in Biological and Model Samples. In: Eaton SS., Eaton GR., Berliner L. (eds.) Biomedical EPR-PartA, Vol.23, Chapter 10. New York: Springer; 2004. p229-282.

[158] Tikhonov AN., Subczynski WK. Application of Spin Labels to Membrane Bioenergetics. In: Eaton SS., Eaton GR., Berliner L. (eds.) Biomedical EPR-PartA, Vol.23, Chapter 8. New York: Springer; 2004. p147-194.

Applications

Chapter 12

Bioenergy

Daniel K. Y. Tan and Jeffrey S. Amthor

Additional information is available at the end of the chapter

1. Introduction

Bioenergy is the chemical energy contained in organic materials that can be converted into direct, useful energy sources via biological (including digestion of food), mechanical or thermochemical processes [1]. Over the past 150 years, fossil fuel combustion provided the energy for industrialisation and development of the modern economy. Around 1900, total energy consumption by humanity was about 20 EJ yr^{-1} [exajoules (10^{18} J); see Appendix for list of abbreviations and units], mainly supplied by wood [2]. By 2009, the world total primary energy supply was about 510 EJ yr^{-1}, or the equivalent of 12 150 million Mg of oil per year (Mtoe yr^{-1}), which was almost double that of the 6 111 Mtoe supply in 1973 (Figure 1) [3]. This 2009 value is equivalent to global energy consumption by humans of 16 TW (16×10^{12} W), and global energy demand is projected to increase to 23 TW by 2030 [4]. In 2011, 87% of total energy consumption was derived from fossil fuels, with only 8.5% from renewable energy sources [5]. Unfortunately, we are running out of fossil fuels, which originated from plant material produced in ancient times, and combustion of these fossil fuels leads to emissions of CO_2 and the consequent global warming. Current proven reserves of oil would last only 50-55 years, natural gas 60-65 years and coal 110-115 years at 2011 rates of consumption [5]. Although some experts claim that peak oil will occur in about 20 years, others argue that the world is already at peak oil production [6]. In either case, fossil fuels are created at a slower rate than they are now being consumed and cannot be considered as the world's main source of energy for more than one or two more generations. This review gives an overview of the amount of energy that can be harvested from the sun for contemporary biomass production, both for food and for bioenergy.

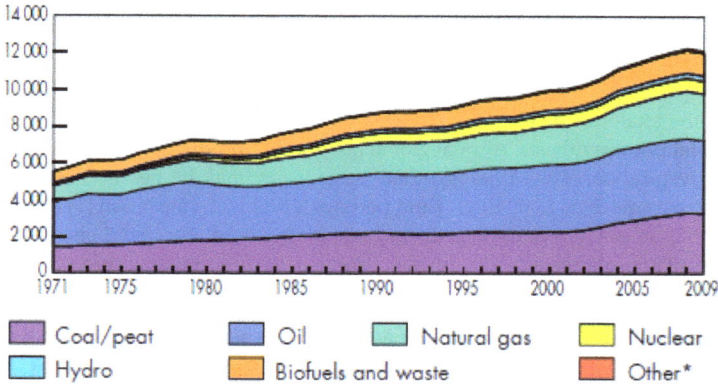

Figure 1. World total primary energy supply from 1971 to 2009 by fuel (Mtoe) [3] *Other includes geothermal, solar, wind, heat, etc.

2. Photosynthesis as a source of food and bioenergy

Oxygenic photosynthesis occurs in cyanobacteria, algae and land plants and is summarised by the equation:

$$CO_2 + H_2O + \text{light energy} \rightarrow [CH_2O] + O_2 \tag{1}$$

where [CH$_2$O] indicates a carbohydrate product of photosynthesis such as sucrose or starch.

Photosynthesis is the source of global food, feed, fibre and timber production as well as biomass-based bioenergy. Each of these products of photosynthesis are renewable. Starch and sucrose are the main products of photosynthesis and sucrose is the main form of carbon translocated from leaves to other organs in plants.

The total energy from sunlight reaching the earth's surface is about 101 000 TW[8] (nearly 3 000 000 EJ per year), about 6 000 times the 2011 annual global human primary energy consumption of 500 EJ [7]. However, solar energy is geographically diffuse and this makes the efficiency of conversion of sunlight important to capture this energy in useful forms. Due to the Carnot limit, the maximum theoretically possible conversion efficiency for sunlight into electricity is 93% [8]. Photovoltaic cells have efficiencies around 15 – 20% for converting sunlight into electricity, but are limited to a maximum conversion efficiency of ~30% due to the Shockley-Queisser limit [9]. Actual solar energy conversion by photosynthesis and subsequent plant growth (biomass production) is much lower at around 2 – 4% for productive plant communities [10].

3. Global primary production

Earth's land plants assimilate about 10 Pmol of atmospheric CO_2 each year, storing (at least temporarily) nearly 5 000 EJ yr^{-1} in sucrose, starch and other carbohydrates. For marine organisms the total may be 7-8 Pmol of CO_2 yr^{-1}. Those are values of *gross primary production* (GPP) or total photosynthesis. Annual global *net* primary production (NPP), defined as GPP minus the respiration of the photosynthetic organisms, may be nearly 5 Pmol C on land (equivalent to as much as 2 000 EJ yr^{-1}) and perhaps 4 Pmol C in the ocean [11]. NPP is critical to present life on earth because it is the organic matter (and associated energy) potentially available to all non-photosynthetic organisms for use in support of their growth, maintenance and reproduction. NPP also contains the bioenergy potentially available to society.

Land plants may have an advantage over aquatic plants as they are able to photosynthesise using leaves that can make use of the rapid diffusion of gases in air which is about 10 000 times faster than that in water [12-14]. Thus, cyanobacteria and algae in water may need to be well stirred to support rapid photosynthesis and growth [12]. Fortunately for oceanic photosynthesisers, surface waters are often vigorously mixed and where nutrients are available primary production can proceed rapidly. Since oceanic NPP is dominated by phytoplankton, most of the "plant" biomass there is photosynthetic as they do not have non-photosynthetic structures such as roots and woody stems. Those oceanic primary producers represent only about 0.2% of global (ocean + land) primary producer biomass due to rapid turnover time in the oceans (average 2 to 6 days) compared to much slower turnover on land (average of about 20 years) [11]. Large ocean area provides a significant *potential* for biomass production, though nutrients are often limiting and harvesting oceanic biomass is difficult and challenging. Because of relative ease of harvesting, and the longer life time of land plants, nearly all current bioenergy harvesting is from terrestrial plants.

4. Converting solar energy to biomass

To describe the component processes associated with the use of solar energy to produce biomass, Monteith's equation [15] can be used:

$$Y = 0.5 \times S_t \times \varepsilon_i \times \varepsilon_c \times \varepsilon_p \tag{2}$$

Where Y is biomass energy yield (J m^{-2} of ground); S_t (J m^{-2}) is the total incident solar radiation during the growing season; ε_i is light interception efficiency (fraction of incident radiation absorbed by a plant's photosynthetic apparatus, J J^{-1}); ε_c is photosynthetic conversion efficiency, including metabolic costs of growing new biomass from products of photosynthesis (J J^{-1} in resulting biomass); and ε_p is partitioning efficiency or harvest index (J J^{-1}).

Photosynthetically active radiation (PAR) is approximately confined to the 400-700 nm waveband [16] which contains about 50% of total solar energy reaching Earth's surface (S_t) [16].

Thus, about half of the incident solar energy is unavailable to higher-plant photosynthesis, which is accounted for in the coefficient 0.5 in the equation above (see Figure 3). In addition, the fraction of solar radiation absorbed by plants or $\varnothing_i$, depends on leaf area and orientation. A full canopy can potentially absorb about 93% of incident PAR with perhaps 92% of that absorption associated with chloroplasts [17]. Partitioning efficiency ($\varnothing_p$) or harvest index is the amount of total biomass energy partitioned into the harvested portion of the crop; for a biomass crop that may approach 100%, but for a seed crop can be as low as 30%. The amount of energy in a unit mass of plant material also varies, being about 17-18 MJ kg^{-1} for typical biomass, but as much as 35-40 MJ kg^{-1} for oilseeds [17, 18]. During the Green Revolution, the dwarfing of the crop-plant stem improved partitioning efficiency ($\varnothing_p$) [19] and selection of larger-leaved cultivars improved light interception efficiency ($\varnothing_i$), but there has been little apparent improvement in photosynthetic conversion efficiency ($\varnothing_c$).

5. Potential and actual photosynthetic conversion efficiency

The observed minimum quantum requirement of 9-10 mol photons per mol CO_2 assimiliated in C3 photosynthesis represents an absolute limit on biofuel production from sunlight, in spite of claims for biomass production (usually by algal systems) that would correspond to significantly smaller quantum requirements [20]. That range corresponds to C3 photosynthesis in the absence of photorespiration, which in the current atmosphere increases minimum quantum requirement to about 14 mol mol^{-1}. But due to light saturation, and other factors (below), biomass production, especially over an annual cycle, cannot approach limits set solely by minimum quantum requirements [17, 20].

The potential maximum efficiency of converting solar energy to biomass energy is estimated at about 4.5% for algae [12, 20], 4.1-5.3% for C3 land plants and 5.1-6.0% for C4 land plants at 20-30°C and present atmospheric [CO_2] (see Figure 2) [10, 17]. C4 plants can be more efficient than C3 plants because they are able to suppress photorespiration through a combination of biochemical and anatomical innovations that arose relatively recently in plant evolution. These innovations presumably were a response to declining global atmospheric [CO_2] during the past 100 million years.

Actual maximum conversion efficiency is generally lower than the calculated potential efficiency at around 3.2% for algae [12], 2.4% and 3.7% for C3 and C4 crops [10], respectively, across a full growing season (see Figure 3) due to insufficient capacity to utilise all radiation incident on a leaf, and photoprotective mechanisms that impair efficiency. The actual photosynthetic efficiency of mature C3 forest stands was also calculated to be between 2.2 to 3.5% [21]. Of course, plants are self-regenerating and self-maintaining whereas photovoltaic cells are not.

The low yields from biomass energy production are frustrating compared with photovoltaic cells that have efficiencies of up to 20%, and this is due to the following limitations in plants [10, 12, 17, 20, 22]:

1.   Two photosystems (Photosystems I and II in series);

2. Dependence on photons limited to the approximate waveband 400–700 nm;

3. Inherent inefficiences of enzymes and biochemical processes;

4. Light saturation under bright conditions and associated photoinhibition in Photosystem II;

5. Respiration, an absolutely essential process for life and growth [23], which consumes 30-60% of the energy contained in the products of photosynthesis; and

6. Plants are living organisms that spend about half of each day in the dark, when they need to use previously generated carbohydrate stores to keep themselves metabolically active and growing.

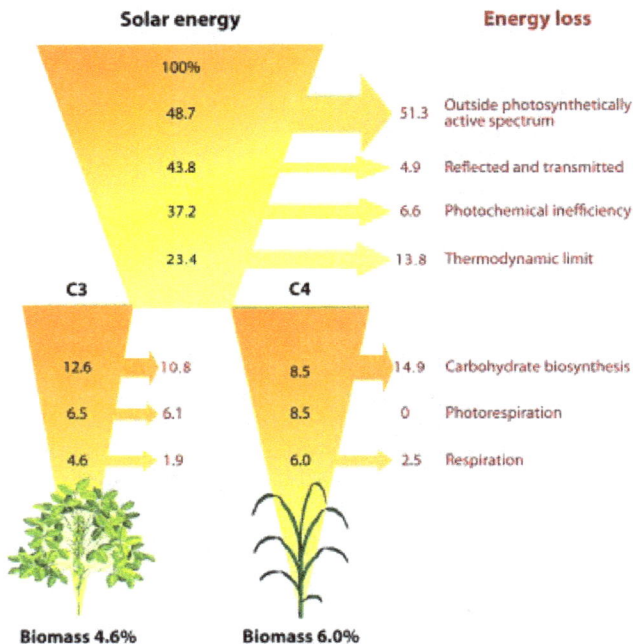

Figure 2. Minimum energy losses associated with biomass production. Wedges show the percentage of energy from solar radiation remaining (inside arrows) and percentage losses (at right) from an original 100% calculated for several stages of photosynthetic and biosynthetic energy transduction from sunlight incident on a plant community to new plant biomass [18]. This analysis indicates that a theoretical maximal photosynthetic energy conversion efficiency is 4.6% for C3 and 6.0% for C4 plants at 25-30 °C.

(a) Potential (maximal) efficiency

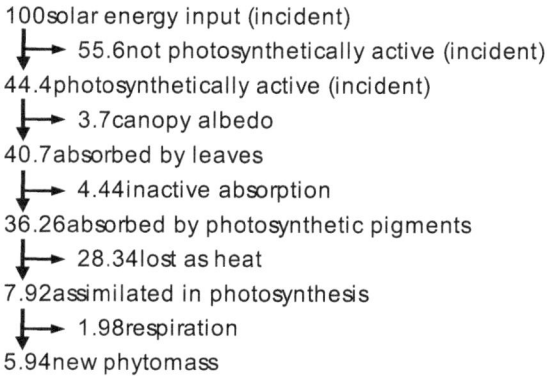

100solar energy input (incident)
→ 55.6not photosynthetically active (incident)
44.4photosynthetically active (incident)
→ 3.7canopy albedo
40.7absorbed by leaves
→ 4.44inactive absorption
36.26absorbed by photosynthetic pigments
→ 28.34lost as heat
7.92assimilated in photosynthesis
→ 1.98respiration
5.94new phytomass

(b) Actual (observed) efficiency

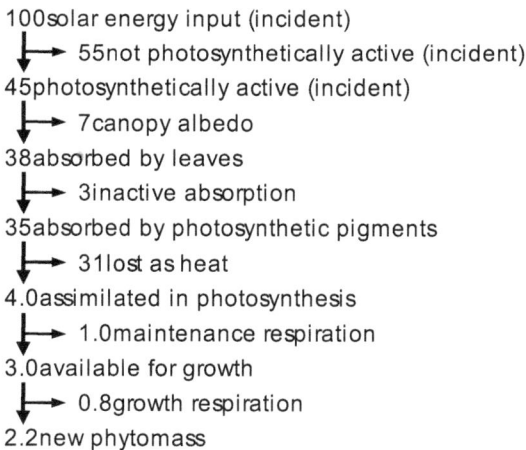

100solar energy input (incident)
→ 55not photosynthetically active (incident)
45photosynthetically active (incident)
→ 7canopy albedo
38absorbed by leaves
→ 3inactive absorption
35absorbed by photosynthetic pigments
→ 31lost as heat
4.0assimilated in photosynthesis
→ 1.0maintenance respiration
3.0available for growth
→ 0.8growth respiration
2.2new phytomass

Figure 3. estimates of (a) potential efficiency (theoretically maximal) and (b) actually efficiency for biomass production of a healthy crop [17]. These results are still generally applicable.

The current bioenergy enterprises are focussing on C4 crops such as sugarcane, maize, sweet sorghum, switchgrass and miscanthus presumably due to the higher energy conversion efficiency. However, this advantage of C4 over C3 will disappear as atmospheric $[CO_2]$ approaches 700 µL L^{-1}.

6. Improving energy conversion efficiency

Due to the apparently low actual energy conversion efficiency in whole-plant photosynthesis (i.e., 2-4%), much discussion has focused on improving photosynthesis to improve crop yield potential including [17, 18, 24, 25]:

1. Engineering C3 crops to use C4 photosynthesis. This would potentially suppress photo-respiration and increase net photosynthetic efficiency by as much as 50% and increase both water and nitrogen use efficiencies [26, 27]. There is an ongoing ambitious research program, led by the International Rice Research Institute (IRRI), to convert the normally C3 rice to a C4 system by transforming rice to express Kranz anatomy and the C4 metabolic enzymes [28].

2. Improving both rubisco's catalytic rate of carboxylation (k_{cat}) and specificity for CO_2 relative to O_2 (τ). This would improve the efficiency of rubisco as a catalyst of CO_2 assimilation [18, 24]. Unfortunately, τ and k_{cat} are inversely related across many rubiscos found in nature [29]. Another complication for engineering an improved rubisco is that it is composed of eight large chloroplast-encoded subunits and eight small nuclear-encoded subunits, and assembling modified subunits in chloroplasts remains a challenge [30].

3. Minimising, or truncating, the chlorophyll antenna size of chloroplast photosystems. This would potentially improve solar conversion efficiency by up to 3-fold in high light, which normally saturates photosynthesis [31]. Individual cells or chloroplasts would have a reduced probability of absorbing sunlight, allowing greater transmission to leaves lower in a canopy and a more uniform distribution of light across leaves within a canopy, hence reducing dissipation and loss of "excess" photons in non-photochemical quenching (NPQ).

4. Improving the recovery rate from the photoprotected state. This would potentially increase carbon uptake by crop canopies in the field [32]. The xanthophyll photoprotection system protects plants from damage from absorption of excess light (the reduction of photosynthesis by dissipation of photons by NPQ). High-yielding rice are reported to recover more quickly from photoinhibition than traditional varieties [33].

7. C3 photosynthesis

C3 crops include wheat, rice, cotton, barley, soybean, bean, chickpea, algae, palm and peanut. C3 photosynthesis of CO_2 forming fructose 6-P can be summarised by:

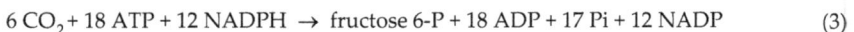

$$6\ CO_2 + 18\ ATP + 12\ NADPH \rightarrow \text{fructose 6-P} + 18\ ADP + 17\ Pi + 12\ NADP \qquad (3)$$

In principle, the ATP and NADPH required to assimilate one CO_2 molecule can be produced during absorption of eight photons of PAR. Because 8 mol photons (PAR) contain on average

1.75 MJ, and because about 0.47 MJ of energy per mol C is stored in carbohydrates, the potential efficiency of converting *absorbed* PAR to biomass approaches 27% for C3 photosynthesis (or about 13% for total solar radiation). That efficiency occurs only in low light, however; under a bright sun, C3 photosynthesis becomes light-saturated. In addition, the process of photorespiration, which is relatively rapid in C3 plants, especially at higher temperature, is a significant constraint on CO_2 assimilation in C3 plants. As much as one third of the C assimilated in C3 photosynthesis can be almost immediately lost to photorespiration with present atmospheric CO_2 concentration (higher CO_2 concentration not only stimulates photosynthesis, but inhibits photorespiration). In sum, at about 20°C, the efficiency of converting *absorbed* PAR into carbohydrate may be about 18% in C3 plants, accounting for photorespiration, but ignoring light saturation. Moreover, that efficiency does not account for plant respiration, and some respiration is essential for growth and maintenance processes.

8. C4 photosynthesis

C4 plants include maize, sugarcane, sorghum, millet, miscanthus and switchgrass. The C4 system involves the specialised metabolism and Kranz leaf anatomy to concentrate CO_2 in the bundle sheath cells. Normal C3 photosynthesis takes place in the bundle sheath cells in C4 plants, but because the CO_2 concentration there is quite high, photorespiration is greatly suppressed. The C4 cycle, which concentrates the CO_2 in bundle sheath cells, requires two ATP to assimilate a CO_2 in the mesophyll, release it in the bundle sheath and regenerate the CO_2 acceptor in the mesophyll. Some CO_2 leakage from the bundle sheath is inevitable, and this requires that the C4 cycle operates more quickly than the C3 cycle in C4 plants. Hence, C4 photosynthesis may require at least 2.2 ATP CO_2^{-1} more than C3 photosynthesis, based on a modest CO_2 leak rate of 10% from bundle sheath cells [17]. In spite of the extra energy cost of the C4 cycle, C4 photosynthesis responds better to bright sunlight and to higher temperatures than C3 photosynthesis because of suppressed photorespiration. At cooler temperatures (e.g., 10-15°C), however, C3 photosynthesis is superior because photorespiration operates slowly at low temperature. In addition, many C4 plants are sensitive to low temperature. The C4 plant miscanthus (*Miscanthus × giganteus*) is relatively tolerant of low temperature, and it may be a good source of germplasm for improving the low temperature tolerance of other C4 plants [34]. In terms of efficiency, C4 photosynthesis might retain as much as 16-17% of the energy in *absorbed* PAR in carbohydrate products, again before any required respiration and ignoring light-saturation. That efficiency is relatively insensitive to temperature, at least over the normal range experienced by typical C4 crops during daylight hours.

9. CAM photosynthesis

Commonly cultivated CAM (crassulacean acid metabolism) plants include agave (*Agave* spp.), Opuntia (*Opuntia* spp.), pineapple (*Ananas comosus*), *Aloe vera*, and vanilla (*Vanilla planifolia*). CAM plants are well adapted to arid and semi-arid habitats. They open their stomata at night and take up CO_2 in the dark to form malic acid, which is then metabolised to release CO_2 for photosynthesis during the following day, but with their stomata closed [35, 36]. By closing the stomata during the day, less water is lost, resulting in high water use efficiencies with a trade-off of lower growth rates. CAM imposes an additional metabolic cost of ~10% compared with the standard C3 pathway due to the transport of malic acid into the vacuole at night and conversion of C3 residue back to the level of storage carbohydrate during the daytime [37]. CAM plants have been suggested to have potential for food, fibre and biofuel production in dry marginal lands [38, 39].

10. World food energy demand

The energy contained in food consumed per person is only about 10 MJ day^{-1} (equivalent to 2 500 kcal per day, 10 000 Btu or 120 W) [40]. Hence, the food energy needed to feed the world's current seven billion persons is ~25 EJ yr^{-1}, which is only about 5% of the world's ~510 EJ of annual energy consumption, but more than 10% of global land NPP. The world's food production system consumes about 95 EJ yr^{-1} and hence, it takes about 4 units of fossil energy to produce 1 unit of food energy [41]. In the United States, the overall energy input/food output ratio is even larger, around 7 to 1 [42]. Most of the energy consumption (~80%) occurs after the farm gate, during transportation, processing and retail. Globally, one third of food, around 1.3 billion Mg, is discarded (including spoilage) each year, and a similar share of the total energy inputs are embedded in these losses [41].

Global population is projected to increase to 9-10 billion within 40-50 years [43]. In developing countries, food consumption per person is rising with increased consumption of animal protein with the livestock revolution [44]. Average annual meat consumption is projected to rise from 32 kg person^{-1} in 2011 to 52 kg person^{-1} by 2050 [45]. Grazing livestock already occupy a quarter of the world's land surface, and the production of livestock feed uses a third of arable cropland [44]. With future increases in global population and per capita food consumption, global food production will have to increase by as much as 70% to meet the increased demand in 2050, an annual growth rate in food supply of 1.1% yr^{-1} [46]. In principle, this means that by 2050 the energy consumption for global food production may increase by 162 EJ yr^{-1} from today's 67 EJ yr^{-1} assuming the energy conversion efficiency remains constant.

In ancient civilisations, most of the energy used for farming was provided by animals and the nutrients were derived from animal manure. During and after the Green Revolution, depend-ence on non-renewable fossil fuels resulted in a conversion of fossil energy into food energy, but in an inefficient way. Agriculture uses about 4% of the global fossil fuel energy of which 50% is used for the production of nitrogen fertiliser from natural gas and atmospheric N_2 using the Haber-Bosch process [45] with a stoichiometry of about 60 MJ kg^{-1} N [47]. The dependency of agriculture on fossil fuels has resulted in commodity (food) prices being closely linked with global energy prices [48]. Hence, food prices tend to fluctuate and trend (upwards) in parallel with energy prices. It is instructive to compare maize production in Mexico using human labour (with a hoe and sickle) returning 10.7 times as much energy in the harvested crop as used in production of that crop with a return of less than 4 times for mechanised maize production in the United States (Table 1). The U.S. crop was, however, more than nine times as productive.

Crop	Country	Tillage	Yield (Mg ha⁻¹)	Inputs (MJ ha⁻¹)	Output (MJ ha⁻¹)	Energy ratio
Groundnut	Thailand	Buffalo	1.28	8 048	20 892	2.60
Groundnut	USA	Mechanised	3.72	45 817	64 051	1.40
Maize	Mexico	Human	1.94	2 687	28 881	10.70
Maize	Mexico	Oxen	0.94	3 222	13 982	4.34
Maize	USA	Mechanised	8.66	33 961	130 396	3.84
Rice	Borneo	Human	2.02	4 327	30 626	7.08
Rice	Philippiines	Carabou	1.65	7 638	25 126	3.29
Rice	Japan	Mechanised	6.33	34 405	96 163	2.80
Rice	USA	Mechanised	7.37	49 542	110 995	2.24
Soybean	USA	Mechanised	2.67	12 609	40 197	3.19
Wheat	USA	Mechanised	2.67	17 740	35 354	2.13

Table 1. Energy use in grain and legume production [49].

In developing countries, populations tend to have a cereal-based diet and are effectively at a lower trophic level in the food chain, while populations in developed countries tend to consume more meat and operate at a higher trophic level. Production of livestock, on average, may require 4 kg of wheat for the production of 1 kg of meat [40, 50]. Therefore, in developed countries where 400 kg of cereal and 100 kg of meat are consumed per year, the total need for food and feed is 800 kg of cereal per person per year [40]. Overfishing of the ocean predators (e.g., killer whales, tuna, salmon) at high trophic levels has also led to the decline in ocean fisheries yield [51]. It is important that cereal crops supply 70% of the calories consumed by humans on the global scale with the remainder supplied by potatoes, beans and other crops, with marine animals now contributing only 2% of the human food supply [52]. To increase the energy efficiency of our primary food production system, we should focus on primary production in agriculture (e.g., cereals) and aquaculture (e.g., algae, phytoplankton) rather than secondary production (e.g., livestock, fish).

11. Biofuels

In addition to providing food and feed, plants are an important source of fuel. Indeed, biofuels are not a new concept. In 300 B.C., the Syrian city of Antioch had public street lighting fuelled by olive oil. More recently, the German inventor Rudolph Diesel demonstrated his engine that ran on peanut oil at the 1900 World Fair in Paris. In simple terms, the nearly 5 000 EJ contained in annual global NPP is about 10 times current global energy demand (~ 510 EJ) [53]. That NPP, however, includes vast amounts of biomass that cannot be physically or economically harvested (including national parks). In 2009, biomass, including agricultural and forest products and organic wastes and residues, accounted for nearly 10% of the world's total primary energy supply [3], with fraction less than 10% in developed countries, but as high as 20-30% in developing countries [1]. Replacing fossil fuels with renewable energy sources derived from sunlight, such as solar, hydro or biomass is very challenging as these energy sources have a lower energy density than fossil fuels and are generally more expensive [54]. In some developing countries, as much as 90% of total energy consumption is supplied by biomass [54]. Solid biomass such as firewood, charcoal and animal dung represent up to 99% of all biofuels [54].

Since the beginning of civilisation, humans have depended on biomass for cooking and heating, and many developing countries in Asia and Africa are still dependent on traditional sources of biomass. Liquid biofuels account for only 2% of total bioenergy, and they are mainly significant in the transportation sector. Transportation accounts for 28% of global energy consumption and 60% of global oil production, and liquid biofuels supplied only 1% of total transport fuel consumption in 2009 [3]. The automative industry currently uses relatively energy inefficient internal combustion engines to burn liquid fuels (e.g., gasoline and diesel). Electric car motors have a 7.5 times higher energy efficiency than internal combustion engines, but the lightness and compactness of liquid fuels still have a fifty-fold higher energy storage than the best available batteries [1]. Hydrogen fuel cells may replace electric motors in the future but this is still in the developmental phase. In the meantime, liquid biofuels are the transition renewable alternative to fossil fuels. Globally, liquid biofuels can generally be classified into three production sources; maize ethanol from the United States, sugarcane ethanol from Brazil and rapeseed biodiesel from the European Union. In 2010, Brazil and the United States produced 90% of the 86 billion L of global bioethanol and the European Union produced 53% of the 19 billion L of global biodiesel [55]. For the rest of this chapter, we use the term biofuels to refer to liquid biofuels. First generation biofuels refer to the traditional or conventional supply chains based on food crops, whereas second generation biofuels require more complex and expensive processes and are generally operating in pilot plants and not yet widely available on the market.

12. First generation biofuels

The first generation of biofuels is produced from starches, sugars and oils of agricultural food crops, including maize, sugarcane, rapeseed and soybean. Carbohydrates are fermented to bioethanol, which is mixed with gasoline as a transporation fuel. Bioethanol, produced mainly from sugarcane, replaced 40% of gasoline used in Brazil in 2008, with the introduction of flex-fuel vehicles allowing high-blending of bioethanol with petrol (all petrol blends in Brazil contain 25% bioethanol) [56]. In the United States, up to 40% of the maize crop was used for bioethanol production in 2011. If all the main cereal and sugar crops (wheat, rice, maize, sorghum, sugar cane, cassava and sugar beet) representing 42% of global cropland were to be hypothetically converted to ethanol, this would correspond to only 57% of total petrol use in 2003 [57], and would leave no cereals or sugar for human consumption, although the reduced sugar in the human diet would have health benefits. Oils/fats (i.e., a mixture of triglycerides, free fatty acids, and/or phospholipids) are converted to biodiesels, potentially competing with food and feed production from oilseed crops such as rapeseed (including canola) and soybean. Biodiesel, a supplement or replacement to traditional diesel, is also produced from animal fats (tallow).

13. Second generation and advanced biofuels

Due to food and energy security concerns, many countries are promoting bioenergy crops that can be grown on land not suited for food production, so that the two systems are complementary rather than competitive [58, 59]. Second generation biofuels refer to the range of feedstocks (e.g., dedicated energy crops such as miscanthus, switchgrass, jatropha, pongamia, agave, Indian mustard, sweet sorghum, algae, carbon waste), conversion technologies (e.g., fast pyrolysis and supercritical water), and refining technologies (e.g., thermo-chemical Fischer-Tropsch methods) used to convert biomass into useful fuels (Figure 4) [60]. There is a fine line between a first and second generation biofuel. For example, sugarcane is a first generation biofuel feedstock (sucrose) but co-generation for electricity using sugarcane residue (bagasse) as a fuel is also possible, and sugarcane residues may serve as future feedstocks in second generation ligno-cellulosic bioethanol production [61]. Ligno-cellulosic bioethanol is based on the conversion of lignocellulosic compounds, made up of chains of about 10 000 glucose and other small organic molecules, into sugars with sophisticated methods of acid or enzymatic hydrolysis. Those sugars can then be converted to fuel using tradiational methods. This means that non-food products such as cereal and wood residues can be converted to ethanol instead of remaining as a waste by-product. These lignocellulosic residues are mainly cell walls that make up 60-80% and 30-60% of the stems of woody and herbaceous plants, respectively, and about 15-30% in their leaves, and consists of around 40-55% cellulose, 20-50% hemicelluloses and 10-25% lignins [1]. There are a few examples of commercial ligno-cellulosic plants. For example, Swiss company Clariant opened a ligno-cellulosic plant in Germany in 2012 that can produce up to 1 000 Mg of cellulosic ethanol from 4 500 Mg of wheat straw [62]. Where lignin cannot be converted to small sugars easily through biochemical processes, it can be burnt for co-generation of bio-electricity.

Another potential bioethanol feedstock is agave (*Agave* spp.) which is adapted to semi-arid land unsuitable for food production [63, 64]. Agaves are well-suited for biofuel production as they can be grown in sandy soil with little or no irrigation and are less likely to be weedy. Agave have above-ground productivities similar to that of the most efficient C3 and C4 crops (25-38 Mg ha^{-1} yr^{-1} dry biomass), but with only 20% of the water required for cultivation [38].

Sisal (*Agave sisalana*) is mainly produced in east African countries of Kenya, and Tanzania, as well as in Brazil, China and Madagascar. The sisal leaf contains about 4% by weight of extractable hard fibre (vascular tissue), the remaining 96% being water and soluble sugars which is disposed of during the decortication process into rivers and the sea, causing pollution, eutrophication and water contamination [65]. Production of ethaonol and bioenergy from sisal juice from the sisal leaves and stems is under pilot testing at the Institute for Production Innovation at the Uninversity of Dar es Salaam and Aalborg University [65]. The first field experiment of blue agave (*Agave tequilana*) as a biofuel crop was planted in 2009 in the Burdekin

River Irrigation Area of Queensland, Australia [35]. Blue agave can acheive strong growth rates by potentially switching from CAM to C3 photosynthesis if there is sufficient water supply [66]. Approximately 0.6 Mha of arid land was used to grow sisal for coarse fibres (sisal) but this has fallen out of production or abandoned due to competition with synthetic fibre [63]. In theory, this crop area (0.6 Mha) alone could provide 6.1 billion L of ethanol if agave were re-established as a biofuel feedstock without causing indirect land use change [63].

In the meantime, new and novel feedstock conversion technologies are being developed such as fast pyrolysis and supercritical water treatment that can now convert nearly any biomass feedstock, such as wood residues, agricultural residues (e.g., wheat and maize stalks), woody plants, and C_4 grasses [e.g., switchgrass (*Panicum virgatum*), miscanthus and sweet sorghum] into a green biocrude that can be processed into jet fuel, biodiesel, or bioethanol [60]. Hydrogen (H_2) is designated as a third generation biofuel, when it is produced from biomass by algae or enzymes [1]. H_2 is a fuel whose combustion produces only water, although future technological breakthroughs are needed before H_2 can be produced economically.

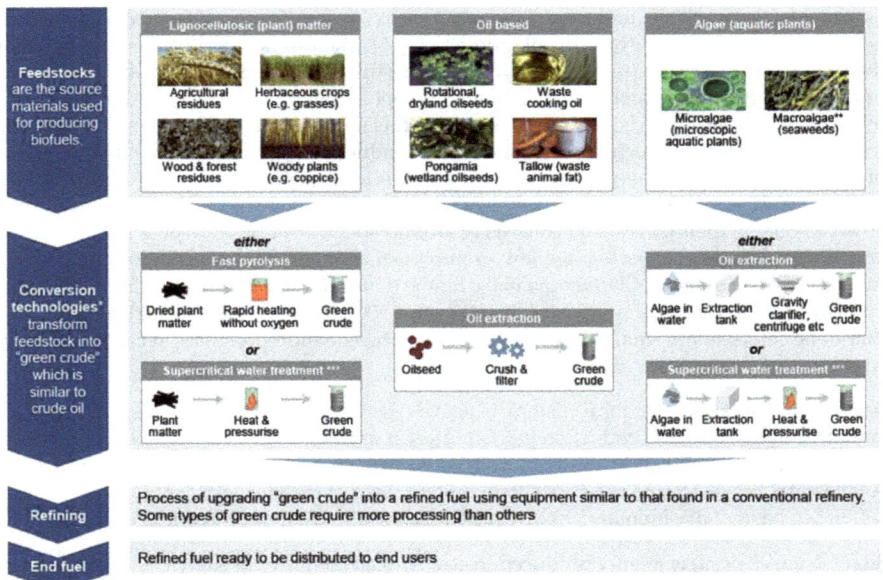

Source: L.E.K. research, interviews and analysis

Figure 4. The advanced biofuels value chain [60]. *Conversion technologies include fast pyrolysis and supercritical water treatment that transform feedstock into "green crude" which is similar to crude oil. ** Macroalgae are multicellular organisms (seaweeds) with low lipid content but are high in carbohydrates. *** Supercritical water treatment is a thermochemical process which involves subjecting the biomass to controlled temperature and pressure conditions in the presence of appropriate catalysts to produce a "green crude".

There is also a move to source oilseed from non-food dedicated energy crops grown on marginal land. These crops might include jatropha (*Jatropha curcas*), pongamia (*Millettia pinnata*), Indian mustard (*Brassica juncea*), and microalgae. The recent failure of jatropha as an energy crop in India and other developing countries due to a lack of bioenergy policy highlights the need for investment in research and policy development before starting on large-scale investments [67]. Pongamia is a tropical tree legume (Fabaceae family) and is a native of India and northern Australia. It has been used as a biofuel crop in India for some time, and is well-suited to marginal land as it is regarded as both a saline- and drought-tolerant species. The seeds contain about 40% extractable oil, predominantly in the form of triglycerides; is rich in C18:1 fatty acid (oleic acid); and has relatively low amounts of palmitic and stearic acid, making it useful for the manufacture of biodiesel [68]. In India, the de-oiled cake of pongamia (i.e., the leftover component of seeds following solvent extraction, and containing up to 30% protein) is used as a feed supplement for cattle, sheep, and poultry [69]. Opportunities exist for a sustainable pongamia agroforestry program to supply biodiesel in northern Australia, although substantial infrastructure investment in processing plants would be needed [59]. Indian mustard is another potential annual oilseed crop being developed in India and Australia. It is drought-tolerant (annual rainfall 300-400 mm) and many varieties can express greater osmotic adjustment than canola [70]. Indian mustard was up to 50% more productive than canola under dry conditions, but not under normal rainfall conditions in northwest New South Wales, Australia [71]. An Indian mustard breeding program for biodiesel production was commenced in 2006 at the University of Sydney's I.A. Watson Grains Research Centre at Narrabri, Australia. Indian mustard is now part of a four year rotation at the Watson Centre.

Microalgae can be cultivated in open raceway ponds or closed photobioreactors, harvested, extracted and then converted into a suitable biofuel such as biodiesel. Raceway ponds are shallow (no more than 30 cm deep) raceways and contents are cycled continuously around the pond circuit using a paddlewheel. Most commercial algal producers are currently using open raceway ponds as these require lower capital costs to set up but may result in increased evaporation and risks of contamination [72]. Photobioreactors are closed systems which offer better control over contamination and evaporation but have higher capital and operating costs than open raceway ponds [72]. The surface area/volume ratios of photobioreactors are also almost double that of the open pond, hence doubling the energy recovered as biomass and potential productivity [73]. Surface fouling due to competitors (e.g., other algae), grazers and pathogens (e.g., bacteria) are a major problem with photobioreactors and cleaning can be a major design and operational problem [72].

Despite the development of microalgae as a feedstock for biodiesel production, there are problems scaling up from laboratories to commercial production [74, 75]. Key limitations to algal production in raceways or photobioreactors are (1) the need for stirring, (2) provision of nutrients for optimal growth and (3) very large surface areas required to capture significant amounts of sunlight [12]. Other problems are pathogen attack, ageing of algal cultures, and lack of system optimisation [76]. The need to de-water and dry the algal biomass can consume up to 69% of the energy input of the process [77]. Despite their potential productivity per unit surface area, and containing up to 30% lipids as storage products, algal biodiesel is not yet economically competitive with petroleum diesel; algal diesel was recently priced at USD 2.76 kg^{-1} compared with petroleum-based diesel at USD 0.95 kg^{-1} [78].

14. Lifecycle analysis and energy balance

Life cycle analysis (LCA) is a tool to take into account the inputs and outputs of a food or biofuel crop production system, including the growing of the crop and its subsequent processing; the technique is also used to assess the energy efficiency and impact of food and biofuel crops on greenhouse gases [79]. Ecologists can relate an LCA to a foodweb or ecosystem model that traces the fluxes of energy through the system. Net energy value (NEV) is an efficiency term calculated as the difference between the usable energy produced from a crop and the amount of energy required for the production of that crop [79].

Three annual crop management systems, conventional (several tillage operations for weed control, seedbed preparation, seeding), conservation (reduced, minimum and no-till systems), and organic (intensive tillage for seeding, weed control) were compared in Canada and Spain [80-82]. Generally, energy inputs for the conservation system were 10% lower than for the conventional system (due to lower fuel and machinery use from reduced tillage) [80]. However, fertiliser and pesticide rates were often increased in response to increased soil water, resulting in a similar total energy use by conventional compared with conservation systems [80-82]. In contrast, there was a reduction in energy input in organic systems due to the use of organic fertilisers instead of synthetic pesticides and fertilisers [47]. In terms of energy output:input ratio, organic farming in Spain was 2.3 times more energetically efficient (5.36:1) than either the conventional or conservation systems (2.35:1 and 2.38:1, respectively) [80]. Inclusion of a leguminous forage crop (e.g., vetch, chickpea) into canola and cereal (e.g., wheat, barley) rotations increased the energy efficiency and output under all management systems [80, 83]. Legume-rhizobial associations are effective solar-energy-driven systems fixing atmospheric N_2 into ammonia with minimal CO_2 emissions compared to industrial nitrogen-fertiliser production. Legumes fix nitrogen and thus reduce synthetic N fertiliser use in farming systems; they also enhance the productivity of subsequent crops through breaks in the disease cycle [84]. Pulses contribute about 21 million Mg of fixed-N per year globally, accounting for one third of the toal biological N_2 fixation in agroecosystems [85].

The energy efficiency of biofuels can also be termed the fossil energy ratio (FER) expressed as the ratio of the amount of fuel energy produced to the amount of fossil fuel energy required for that production [79]. An FER < 1 indicates a net energy loss, whereas an FER > 1 represents a net energy gain. Life cycle assessments for biofuels have also shown that Brazilian sugarcane, agave, and switchgrass ethanol could achieve positive energy balances and substantial greenhouse gas offsets, while maize in the United States and China offers modest or no offsets [64]. The bioenergy created in sugarcane and agave ethanol, and in palm oil, is at least four times the amount required to produce it, while maize in the United States and China release almost as much energy when they are burnt as the energy that is consumed in growing and processing them (Figures 5 and 6) [54, 86]. Sugar crops usually produce more ethanol per ha with a better energy balance than starch crops because sugar crops produce higher sugar amounts per ha than starch crops; and sugar (sucrose) can be directly fermented, whereas starch polymers have to be hydrolysed before being fermented by yeast [1]. In general the energy gain and conversion of solar energy into biomass in the sub-tropics is substantially greater than any achievable in temperate zones [87], possibly due to the longer growing season and higher levels of solar energy over an annual cycle. For example, the FER of sugarcane in Brazil was 8.1-10 in 2009, compared with 1.4 for maize-ethanol in the United States and 2.0 for sugarbeet in Europe [61]. There is already evidence of a land-grab with countries (e.g., China and the Middle East) securing their own energy and food security by acquiring large areas of subtropical land in Africa and Asia [88]. Many countries may never be able to establish a position of energy or food independence or anywhere near approaching it [88]. For example, Sweden is importing Brazilian bioethanol as its main source of renewable transportation energy, due to the climatic constraints of growing biomass for liquid fuels within Sweden [88]. FER of microalgal-based (*Chlorella vulgaris*) biodiesel produced in raceways is 0.31, which is 2.5 times as energy intensive as conventional diesel (FER of 0.83] in the United States [89]. This current negative energy balance is unacceptable unless the production chain can be fully optimised with heating and electricity inputs decarbonised [89].

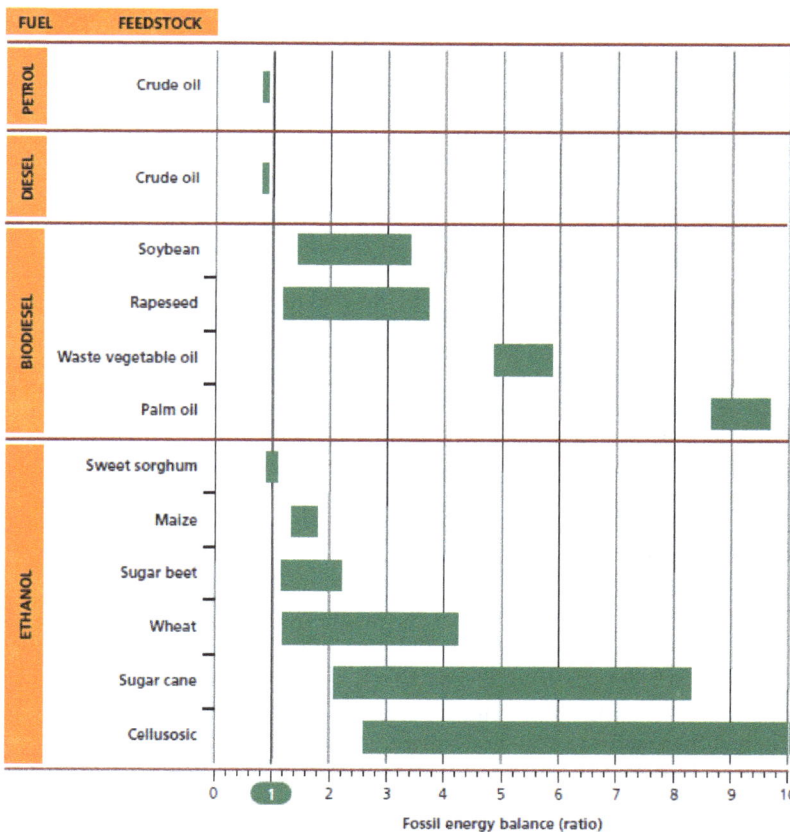

Figure 5. Estimated ranges of fossil energy ratio (FER) of selected fuel types [54, 86]. Note: The ratios for cellulosic biofuels are theoretical.

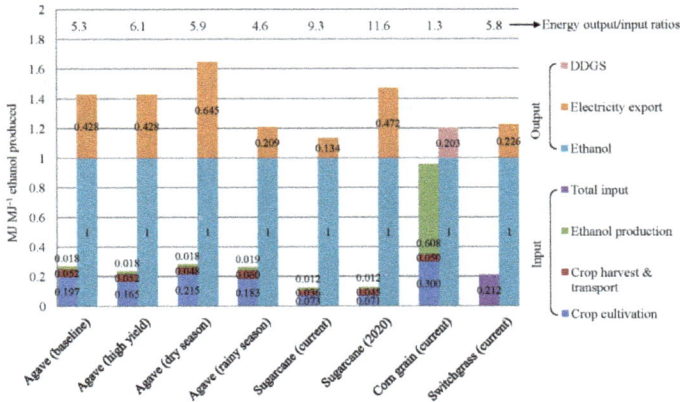

Figure 6. Energy input and output per MJ ethanol produced for various feedstocks including Mexican agave, Brazilian sugarcane, and United States maize (corn) and switchgrass [64]. Abbreviation: DDGS, Dried Distillers Grains with Solubles.

15. Carbon footprint of food and biofuels

The world's food production system is inefficient, globally taking four units of energy to produce one unit of food energy. Globally, agriculture also accounts for ~60% of nitrous oxide (N_2O) and 50% of methane (CH_4) emissions [90]. On-farm, N_2O emissions are mainly associated with the use of nitrogenous fertiliser, while CH_4 emissions are mainly from digestion in ruminant livestock (e.g., cattle and sheep). The carbon footprint is the total amount of greenhouse gases (GHG) associated with the production, processing and distribution of food and biofuel crops expressed in carbon dioxide equivalents (CO_2e). This can range from 0.15 to 0.20 kg CO_2e per kg of wheat in New South Wales, Australia [91], to 0.46 kg CO_2e per kg of wheat in Canada [83]. In Canada, legumes like chickpea, dry pea, and lentil may emit only 0.20 to 0.33 kg CO_2e per kg of grain, about half that of wheat, due to biological N fixation by the legumes. As with the energy balance, durum wheat emitted 20% lower CO_2e when preceded by an N-fixing crop, compared to when the crop was preceded by a cereal, highlighting the benefits of including a legume in a cereal rotation on GHG balance [83].

Since nearly 80% of the energy used in the food supply chain is in the postharvest phase [41], there is a fashionable trend for an eat-local movement to reduce `food miles`. Food miles refers to the distance a food commodity travels from the point of production to the point of consumption and the related energy and CO_2 emitted during transportation [92]. However, shipping food for long distances may sometimes require less energy and emit less CO_2. For example, even when shipping was accounted for, New Zealand dairy products imported into the UK used half the energy of their UK counterparts, and in the case of lamb, a quarter of the energy due to grass-fed conditions in New Zealand compared with the energy-intensive system used in the UK [93]. New Zealand was 10% more energy efficient for apples, and the energy costs of shipping of onions was less than the cost of storage in the UK, making New Zealand onions more energy efficient overall [93].

Over 30% of the food energy is lost through wastage in both developed and developing countries [41]. In developing countries, food is mainly lost due to pests, and spoilage due to the lack of cold storage and food-chain infrastructure. In developed countries, food safety issues have resulted in the over-reliance of `use by` dates resulting in good food being discarded in landfills (instead of being used, e.g., for animal feed or compost) [94]. This wastage can be reduced through improved education, better legislation and research in postharvest technology to reduce food wastage. We can invest in better diagnostics that monitor food spoilage such as temperature- and time-sensitive inks on food package that cause labels to change colour if the food has been exposed to the wrong temperature for too long [95]. Restaurants can stop serving super-sized portions.

GHG emissions during agricultural production of biofuel crops contribute 34-44% to the GHG balance of maize ethanol in the United States [96] and more than 80% in pure vegetable oils [1]. In theory, biofuel feedstocks remove CO_2 from the air and can potentially reduce greenhouse gas emissions. However, clearing of undisturbed native ecosystems such as rainforest and savanna or grassland for biofuel production also increases net GHG production due to change in land use [97]. For example, a hectare of maize grown for bioethanol can sequester 1.8 Mg ha^{-1} yr^{-1} of CO_2e, but each hectare of forest converted to maize field has up-front emissions of 600-1100 Mg CO_2e and each hectare of grassland converted to crop releases 75-300 Mg CO_2e equivalents. Hence, maize-based bioethanol production might double GHG emissions over 30 years and increase GHGs for 167 years [97]. Converting lowland tropical rainforest in Indonesia and Malaysia to oil palm biodiesel crops would result in a carbon debt of 610 Mg CO_2 ha^{-1}. This might take 85-90 years to repay, while sugarcane bioethanol produced on Brazilian Cerrado woodland-savanna might take 17 years to repay [98].

16. Water footprint of food and biofuel

Agriculture accounts for about 86% of global fresh water consumption [99]. Energy is also needed to pump water for irrigation. The water footprint (WF) of a product, such as food or biofuel, is the total volume of fresh water used for production and processing, through to eventual use of the the product [99]. In general, the WF of biofuels is up to 2-5 times larger than the WF of fossil fuels [100]. For example, water consumption of bioethanol processed from rainfed maize grain is 0.71 L km^{-1} travelled by a light duty vehicle compared with 0.24 L km^{-1} for fossil fuel-based gasoline [100]. Most of the water used in gasoline is for the oil refining process while most of the water used in bioethanol production is water used to grow the crop. The water footprint includes three components: green, blue, and gray WFs. Green WFs refer to rainwater transpired and blue WFs to surface and groundwater evaporated following their use in irrigation. Gray WF refers to water that becomes polluted during crop production and includes the amount of water necessary to reduce pollutants (through dilution) discharged so that water quality meets appropriate standards [101].

The global annual mean WF was 9 087 Gm^3 yr^{-1} (74% green, 11% blue, 15% gray) and agricultural production contributed 92% to that total during the period 1996-2005, with the remainder accounted for by food processing and end-user consumption [102]. The average consumer in the United States has a WF of 2 842 m^3 yr^{-1}, whereas the average consumer in China and India have WFs of 1 071 and 1 089 m^3 yr^{-1}, respectively. The differences are mainly due to differences in meat consumption. Globally, consumption of cereals gives the largest contribution to WF of the average consumer (27%), followed by meat (22%) and milk products (7%) [102]. Approximately one third of the total WF of agriculture is related to livestock products. The average WF per calorie (or MJ) for beef is 20 times larger than cereals and starchy roots, and the WF g^{-1} of protein for chicken meat, eggs, and milk is 1.5 times larger than for legumes [103]. The WF of any livestock product is larger than any WF of crop products of equivalent nutritional value (e.g., calories, protein, and fat) due to the unfavourable feed conversion efficiency for livestock products [103]. The weighted global average WF of sugarcane was 209 m^3 Mg^{-1}; 133 m^3 Mg^{-1} for sugarbeet and 1222 m^3 Mg^{-1} for maize [104] for bioethanol. Incentives to switch from overhead sprinkler irrigation systems to drip irrigation can potentially use 40% less water and lower energy requirement for maize farmers by reducing energy needed to pump water, and by reducing evaporation losses [95].

17. Improving the energy efficiency of food and biofuel systems

The following are some suggestions to improve the energy efficiency of food and biofuel systems:

1. (i) Commercial hybrids of wheat and rice and (ii) increased crop stress tolerance. Increasing productivity during the Green Revolution was largely through a combination of crop genetic improvement through the development of F_1 hybrid varieties of maize and semi-dwarf, disease-resistant varieties of wheat and rice; and increased use of fertiliser and irrigation [105]. Unfortunately, crop yield increases are now slowing (or have halted) and input costs such as fuel and fertiliser are increasing. Future increases in crop production will have to come mainly from increased yield per hectare, higher cropping intensity (number of crops sown in the same field per year) and to a lesser extent from cultivation of new land [46, 106]. Hybrids are the result of heterosis or the favourable combination of dominant genes by crossing two genetically different parents. In general, hybrids provide around 15% yield advantage over open-pollinated parents in maize, and around 10% over inbred parents in wheat and rice [107]. Although hybrid maize has been widely utilised, adoption of hybrids in wheat and rice (with the exception of China where *O. indica* hybrids cover around 60% of area) is low [108]. Hence, future adoption of hybrid wheat and rice may increase yield by around 10%. Currently, most of the major commercial genetically modified (GM) crops are based on simple insertion of a gene for protective traits such as insect toxin (e.g., *Bacillus thuringiensis, Bt*) and/or herbicide resistance (e.g., glyphosate tolerance). The next phase will be the development of abiotic stress tolerance genes such as DroughtGard® (drought tolerant) hybrid maize released by Monsanto in 2012. Not all crop improvement initiatives need to be GM based on transgenic biotechnology. For

example, there are a number of heat tolerance breeding programs in wheat, chickpea, and cotton using conventional breeding together with marker assisted selection [109-111].

2. Reduced yield gaps. Yield gaps are the difference between realised farm yield and the potential yield that can be achieved using available technologies and management. In many irrigated cereal systems, actual yield tends to plateau at or around 80% of yield potential while, in rainfed systems, average actual yields are no more than 50% of yield potential [112]. In many instances, however, even the 80% of yield potential is not achieved, presumably due to technical, knowledge, climatic and biophysical constraints [113]. Reducing the yield gaps can potentially boost yields per unit of input, and hence, energy conversion efficiency. International aid programs such as the Gates Foundation and the Consultative Group on International Agricultural Research (CGIAR) are working to help to close these yield gaps.

3. Conservation agriculture. Zero or reduced tillage and retaining crop residues can potentially reduce energy use and fuel use for farm machinery in agriculture by 66-75% [108], as well as conserve soil moisture and sequester carbon in soil organic matter. Conversion from conventional tillage to zero/reduced tillage can reduce on-farm GHG emssions by 110-130 kg CO_2e ha^{-1} per season, since soil disturbance caused by tillage increases soil organic carbon losses through decomposition (accelerated oxidation) and physical erosion [114]. Crop residues produced worldwide are estimated at 3 Pg, equivalent to more than 1 Pg carbon per year [115]. With less than 10% of the global crop land under conservation tillage, there is an opportunity for wider adoption of this practice to improve energy- and water-use efficiency [108], as well as reducing net GHG emissions.

4. Legume rotations and N-fixing cereals. Replacing synthetic N fertilisers with legumes and organic fertilisers (e.g., animal manures and green manure crops) can reduce the fossil fuel combusted during fertiliser synthesis as well as reduce N_2O emissions. Soil microorganisms such as rhizobium N_2-fixers (in legumes), arbuscular mycorrhizal fungi (AM fungi, which can improve plant P and Zn uptake) and P solubilising fungi and bacteria can form a symbiotic relationship with plant roots and act as biofertilisers and biopesticides [116, 117]. Many bacteria can produce natural plant hormones such as ethylene, cytokinins, auxins and gibberellins that can stimulate plant growth, increasing root branching, or shoot development [118]. Microorganisms can also affect gene expression and activate plant defence mechanisms through systemic acquired resistance (SAR) [119]. A challenge going forward is to boost biological N fixation to levels that can substitute for the synthetic fertilisers now used. Effective nitrogen-fixing wheat, rice, and/or maize would be boons to reducing energy input to cropping systems for food, feed, and bioenergy [120]. Past attempts to develop such symbioses are not encouraging.

5. Improving overall annual solar radiation use efficiency in annual crops. In locations supporting only one crop per year, it may be possible to extend the photosynthetic period via stay-green traits to maximise annual solar energy capture [121, 122]. In other areas, it may be possible to shorten the crop cycle to fit a second or third crop into the annual rotation (i.e. increase crop intensification) [123].

6. Perennial crops. Perennial crops have greater potential for a more sustainable production due to their longer growing season than annuals, utilising more sunlight during the year, and reduced farming operations [1]. The development of perennial wheat by the Land Institute in Kansas provides the opportunity to crop continuously without tillage and reduce soil erosion [124]. Most LCA also show that perennial C4 biofuel crops such as miscanthus and switchgrass and CAM biofuel crops like agave can provide positive energy balances, supplement renewable energy demands, and mitigate GHG emissions [63, 79]. An important question is whether perennial wheat can attain yields of present wheats across a wide range of environments.

7. A lower-trophic-level society. Overall solar energy conversion efficiency in agriculture would be increased if humans consumed more crops directly, rather than after their processing through livestock. Even a modest replacement of energy-intensive meats with less-energy-intensive grains, fruits and vegetables [95] would be significant at the global scale. In marine systems, gains in sustainability could come from harvesting lower trophic level species such as algae, phytoplankton, and filter feeder organisms such as bivalves. For example, over 1.5 Tg yr^{-1} of macroalgae (seaweed) is produced in China to be used as food for humans, feed for marine animals, and industrial raw materials [51]. A factor to be overcome is the global trend toward eating more energy-costly food as a component of economic development.

18. Hydrogen production and artificial photosynthesis

Hydrogen cells might be used to fuel future cars. Although H_2 contains three times the energy of petrol on a mass basis, 4.6 L of H_2, compressed at 70 MPa, are needed to substitute for 1 L of petrol. H_2 is also highly flammable and it is 50% more expensive to transport than natural gas [1]. There is interest in both biotic and abiotic sytems that mimic the biological production of H_2 gas via the breakdown of water (analogous to its electrolysis, $2H_2O \rightarrow 2H_2 + O_2$, which is carried out in photosystem II using solar energy). Certain algae contain the enzymes hydrogenase or nitrogenase (a key enzyme in nitrogen fixation) which can produce H_2 from CO or organic waste [125]. The combination of microalgae (harvesting radiation in the waveband 400-750 nm) and purple bacteria (using the waveband 400-1100 nm) allow a more complete utilisation of the solar energy spectrum. There are still many unresolved problems in producing H_2 using algae and bacteria, including how to combine microalgal and bacterial biological processes [125].

Artificial photosynthesis involves mimicking natural systems using molecular photocatalytic systems for light-driven water oxidation and H_2 production [126]. Artificial photosynthesis was only an academic activity until the development of the first practical artificial leaf by Nocera in 2011 [127]. The key to this breakthrough was the discovery of new, cheaper photocatalysts made from nickel and cobalt that are capable of splitting water into H_2 and O_2 efficiently. This artificial leaf was claimed to be potentially 10 times more efficient in photosynthesis than a natural leaf [127]. Commercialisation of artificial photosynthesis is yet to be proven.

19. Conclusions

Solar radiation is the ultimate source of renewable energy for human use, and bioenergy will continue to be a major vehicle for its use. Solar-energy conversion efficiency by even the most productive plant communities are less than 5%, however, while photovoltaic cells may approach 20%. 'Average' plant communities operate at considerably lower efficiencies, but there are opportunities to substantially increase the average efficiency in crop systems. Photosynthesis is now used extensively in agriculture to produce food, feed, fibre, and biofuels, but the current biofuels (bioethanol and biodiesel), mainly produced from first generation feedstocks (e.g., sucrose from sugarcane, carbohydrates from maize seeds, and lipids from rapeseed seeds), constitute only a small fraction (1%) of present transportation energy, and a much smaller fraction of total human energy supply. The future second generation biofuels will come from dedicated perennial energy crops (e.g., miscanthus, switchgrass, agave, pongamia), and in the near future, hydrogen gas may be produced from algae, bacteria, or artificial photosynthesis to fuel hydrogen-cell powered cars.

Abbreviations and units

Abbreviation	Term represented
Bt	*Bacillus thuringiensis*
Btu	British thermal unit (equivalent to 1.055 kJ)
CAM	Crassulacean acid metabolism
CO_2e	Carbon dioxide equivalents
ε_c	Photosynthetic conversion efficiency (J J^{-1} in resulting biomass)
ε_i	Light interception efficiency (fraction of incident radiation absorbed by a plants photosynthetic apparatus, J J^{-1})
EJ	Exajoules (10^{18} J)
ε_p	Partitioning efficiency or harvest index (J J^{-1}).
FER	Fossil energy ratio
GHG	Greenhouse gas
GM	Genetically modified
GPP	Gross primary production
kcal	Kilocalories
K_{cat}	Catalytic rate of carboxylation (reactions catalysed per second by each enzymatic site)
LCA	Life cycle analysis
Mtoe	Million Mg of oil equivalent

NEV	Net energy value
NPP	Net primary production
NPQ	Non-photochemical quenching
PAR	Photosynthetically active radiation (radiation in the 400-700 nm wave band)
Pg	Petagram (1×10^{15} g)
Pmol	Petamole (1×10^{15} mol)
SAR	Systemic acquired resistance
S_t	Total incident solar radiation across the growing season ($J\ m^{-2}$)
Tg	Teragram (1×10^{12} g)
TL	Trophic level
TW	Terawatt (1×10^{12} W)
WF	Water footprint
Y	Biomass energy yield ($J\ m^{-2}$ of ground)
τ	Specificity for CO_2 relative to O_2

Author details

Daniel K. Y. Tan* and Jeffrey S. Amthor

*Address all correspondence to: daniel.tan@sydney.edu.au

Faculty of Agriculture and Environment, The University of Sydney, Sydney, NSW, Australia

References

[1] Bessou, C, Ferchaud, F, Gabrielle, B, & Mary, B. Biofuels, greenhouse gases and climate change. A review. Agron Sustain Dev. (2011). Jan;, 31(1), 1-79.

[2] Grubler, A, & Nakicenovic, N. Decarbonizing the global energy system. Technological Forecasting and Social Change. (1996). , 53(1), 97-110.

[3] IEAKey World Energy Statistics (2011). In: OECD/IEA, editor. IEA Publications2011.

[4] IEOInternational Energy Outlook, US Department of Energy. Washington, DC: US Department of Energy; (2007).

[5] BP Statistical Review of World Energy 2012 [database on the Internet]2012 [cited 17 August 2012]. Available from: http://www.bp.com/assets/bp_internet/globalbp/glob-

albp_uk_english/reports_and_publications/statistical_energy_review_2011/STAG-ING/local_assets/pdf/statistical_review_of_world_energy_full_report_2012.pdf.

[6] Kerr, R. A. Peak oil production may already be here. Science. [News Item]. (2011). Mar;, 331(6024), 1510-1.

[7] Cho, A. Energy's Tricky Tradeoffs. Science. [News Item]. (2010). Aug;, 329(5993), 786-7.

[8] Henry, C. H. Limiting efficiencies of ideal single and multiple energy gap terrestrial solar cells. Journal of Applied Physics. (1980). , 51(8), 4494-500.

[9] Shockley, W, & Queisser, H. J. Detailed balance limit of efficiency of P-N junction solar cells. Journal of Applied Physics. [Article]. (1961). , 32(3), 510.

[10] Zhu, X-G, Long, S. P, & Ort, D. R. What is the maximum efficiency with which photosynthesis can convert solar energy into biomass? Current Opinion in Biotechnology. (2008). Apr;, 19(2), 153-9.

[11] Field, C. B, Behrenfeld, M. J, Randerson, J. T, & Falkowski, P. Primary production of the biosphere: integrating terrestrial and oceanic components. Science. (1998). July 10, 1998;, 281(5374), 237-40.

[12] Larkum, A.W.D. Limitations and prospects of natural photosynthesis for bioenergy production. Current Opinion in Biotechnology. (2010). Jun;, 21(3), 271-6.

[13] Raven, J. A. Exogenous inorganic carbon sources in plant photosynthesis. Biological Reviews of the Cambridge Philosophical Society. (1970). , 45(2), 167.

[14] Raven, J. A. The evolution of vascular land plants in relation to supracellular transport processes(1977).

[15] Monteith, J. L. Climate and efficiency of crop production in Britain. Philosophical Transactions of the Royal Society of London Series B-Biological Sciences. [Article]. (1977). , 281(980), 277-94.

[16] Mccree, K. J. The action spectrum, absorptance and quantum yield of photosynthesis in crop plants. Agricultural Meteorology. (1971). , 9, 191-216.

[17] Amthor, J. S. From sunlight to phytomass: on the potential efficiency of converting solar radiation to phyto-energy. New Phytol. (2010). , 188(4), 939-59.

[18] Zhu, X-G, Long, S. P, & Ort, D. R. Improving photosynthetic efficiency for greater yield. In: Merchant S, Briggs WR, Ort D, editors. Annual Review of Plant Biology, (2010). , 61,, 235-261.

[19] Evans, L. T. Crop evolution, adaptation and yield. Cambridge: Cambridge University Press; (1993). Available from: <Go to ISI>://CABI:19930763163.

[20] Walker, D. A. Biofuels, facts, fantasy, and feasibility. Journal of Applied Phycology. [article; Conference paper]. (2009). , 21(5), 509-17.

[21] Hellmers, H. An evaluation of the photosynthetic efficiency of forests. Quarterly Review of Biology. (1964). , 39(3), 249-57.

[22] Radmer, R, & Kok, B. Photosynthesis- limited yields, unlimited dreams. Bioscience. (1977). , 27(9), 599-605.

[23] Amthor, J. S. The McCree-de Wit-Penning de Vries-Thornley respiration paradigms: 30 years later. Annals of Botany. (2000). Jul;, 86(1), 1-20.

[24] Amthor, J. S. Improving photosynthesis and yield potential. In: Ranalli P, editor. Improvement of Crop Plants for Industrial End Uses. Dordrecht: Springer; (2007). , 27-58.

[25] Skillman, J. B, Griffin, K. L, Earll, S, & Kusama, M. Photosynthetic Productivity: Can Plants do Better? Thermodynamics- Systems in Equilibrium and Non-Equilibrium: Intech; (2011).

[26] Covshoff, S, Hibberd, J. M, & Integrating, C. photosynthesis into C-3 crops to increase yield potential. Current Opinion in Biotechnology. (2012). Apr;, 23(2), 209-14.

[27] Sage, R. F, & Zhu, X. G. Exploiting the engine of C-4 photosynthesis. Journal of Experimental Botany. [Editorial Material]. (2011). May;, 62(9), 2989-3000.

[28] Hibberd, J. M, Sheehy, J. E, Langdale, J. A, & Using, C. photosynthesis to increase the yield of rice- rationale and feasibility. Current Opinion in Plant Biology. (2008). Apr;, 11(2), 228-31.

[29] Spreitzer, R. J, & Salvucci, M. E. Rubisco: Structure, regulatory interactions, and possibilities for a better enzyme. Annual Review of Plant Biology. (2002). , 53, 449-75.

[30] Parry MAJAndralojc PJ, Mitchell RAC, Madgwick PJ, Keys AJ. Manipulation of Rubisco: the amount, activity, function and regulation. Journal of Experimental Botany. (2003). May;, 54(386), 1321-33.

[31] Melis, A. Solar energy conversion efficiencies in photosynthesis: Minimizing the chlorophyll antennae to maximize efficiency. Plant Sci. (2009). Oct;, 177(4), 272-80.

[32] Zhu, X. G, Ort, D. R, Whitmarsh, J, & Long, S. P. The slow reversibility of photosystem II thermal energy dissipation on transfer from high to low light may cause large losses in carbon gain by crop canopies: a theoretical analysis. Journal of Experimental Botany. (2004). May;, 55(400), 1167-75.

[33] Wang, Q, Zhang, Q. D, Zhu, X. G, Lu, C. M, Kuang, T. Y, & Li, C. Q. PSII photochendstry and xanthophyll cycle in two superhigh-yield rice hybrids, Liangyoupeijiu and Hua-an 3 during photoinhibition and subsequent restoration. Acta Botanica Sinica. (2002). Nov;, 44(11), 1297-302.

[34] Naidu, S. L, & Long, S. P. Potential mechanisms of low-temperature tolerance of C-4 photosynthesis in Miscanthus x giganteus: an in vivo analysis. Planta. (2004). Nov;, 220(1), 145-55.

[35] Holtum JAMChambers D, Morgan T, Tan DKY. Agave as a biofuel feedstock in Australia. Global Change Biology Bioenergy. (2011). Feb;, 3(1), 58-67.

[36] Borland, A. M. Zambrano VAB, Ceusters J, Shorrock K. The photosynthetic plasticity of crassulacean acid metabolism: an evolutionary innovation for sustainable productivity in a changing world. New Phytol. [Review]. (2011). , 191(3), 619-33.

[37] Winter, K. Smith JAC. Crassulacean acid metabolism: Current status and perspectives. Winter K, Smith JAC, editors(1996).

[38] Borland, A. M, Griffiths, H, & Hartwell, J. Smith JAC. Exploiting the potential of plants with crassulacean acid metabolism for bioenergy production on marginal lands. Journal of Experimental Botany. (2009). Jul;, 60(10), 2879-96.

[39] Nobel, P. S. Crop ecosystem responses to climatic change: Crassulacean acid metabolism crops. Climate Change and Global Crop Productivity. (2000). , 2000, 2000-315.

[40] Nonhebel, S. Global food supply and the impacts of increased use of biofuels. Energy. [Article]. (2012). Jan;, 37(1), 115-21.

[41] FAOEnergy-smart food for people and climate. Rome: Food and Agriculture Organisation of the United Nations(2011).

[42] Heller, M. C, & Keoleian, G. A. Life cycle-based sustainability indicators for assessment of the U.S. Food System.: University of Michigan(2000). Contract No.: Report (CSS00-04), 00-04.

[43] UN PD, World Population Prospects, The (2010). Revision. United Nations Population Division. New York2010.

[44] Pica-ciamarra, U, & Otte, J. The'Livestock Revolution': rhetoric and reality. Outlook Agric. [Article]. (2011). Mar;, 40(1), 7-19.

[45] Foresight, G. O-S. c. i. e. n. c. e. The Future of Food and Farming: Challenges and Choices for Global Sustainability. Final Project Report. London: The Government Office for Science; (2011).

[46] Bruinsma, J. The resources outlook: by how much do land, water and crop yields need to increase by 2050? Conforti P, editor: Food and Agriculture Organization of the United Nations (FAO); (2011).

[47] Hoeppner, J. W, Entz, M. H, Mcconkey, B. G, Zentner, R. P, & Nagy, C. N. Energy use and efficiency in two Canadian organic and conventional crop production systems. Renewable Agriculture and Food Systems. (2006). Mar;, 21(1), 60-7.

[48] Kim, G. R. Analysis of global food market and food-energy price links- based on systems dynamics approach. Scribd, South Korea: Hankuk Academy of Foreign Studies(2010).

[49] Pimentel, D, & Pimentel, M. H. Food, energy and society, 3rd Edn. Boca Raton: CRC Press; (2008).

[50] Nonhebel, S. Energy from agricultural residues and consequences for land requirements for food production. Agricultural Systems. (2007). May;, 94(2), 586-92.

[51] Duarte, C. M, Holmer, M, Olsen, Y, Soto, D, Marba, N, Guiu, J, et al. Will the oceans help feed humanity? Bioscience. (2009). Dec;, 59(11), 967-76.

[52] FAOSTAT statistical database [database on the Internet]FAO. (2012). cited 9 August 2012]. Available from: http://faostat.fao.org/.

[53] Kapur, J. C. Available energy resources and environmental imperatives. World Affairs, Issue (2004). , 10N1(V10N1)

[54] FAOThe state of food and agriculture. Biofuels: prospects, risks and opportunities, 978-9-25105-980-7Rome(2008). p., 138.

[55] REN21Renewables 2011 Global Status Report. Paris: Renewable Energy Policy Network of the 21st Century; (2011). p., 115.

[56] Goldemberg, J. The Brazilian biofuels industry. Biotechnol Biofuels. [Review]. (2008). May;1.

[57] Rajagopal, D, Sexton, S. E, Roland-holst, D, & Zilberman, D. Challenge of biofuel: filling the tank without emptying the stomach? Environmental Research Letters. (2007). Oct-Dec;2(4).

[58] Odeh IOA, Tan DKY. Expanding biofuel production in Australia: opportunities beyond the horizon. Farm Policy Journal. (2007). May Quarter 2007;, 4(2), 29-39.

[59] Odeh IOA, Tan DKY, Ancev T. Potential suitability and viability of selected biodiesel crops in Australian marginal agricultural lands under current and future climates. Bioenergy Research. [Article]. (2011). Sep;, 4(3), 165-79.

[60] Consulting, L. E. K. Advanced Biofuels Study- Strategic Directions for Australia. Sydney: L.E.K. Consulting(2011).

[61] Goldemberg, J, & Guardabassi, P. The potential for first-generation ethanol production from sugarcane. Biofuels Bioprod Biorefining. [Article]. (2010). Jan-Feb;, 4(1), 17-24.

[62] ClariantClariant launches biofuel of the future. Netzwerk Biotreibstoffe. (2012).

[63] Davis, S. C, Dohleman, F. G, & Long, S. P. The global potential for Agave as a biofuel feedstock. Global Change Biology Bioenergy. (2011). Feb;, 3(1), 68-78.

[64] Yan, X. Tan DKY, Inderwildi OR, Smith JAC, King DA. Life cycle energy and greenhouse gas analysis for agave-derived bioethanol. Energy & Environmental Science. (2011). Sep;, 4(9), 3110-21.

[65] Bisanda ETNEnock J. Review on sisal waste utilisation: Challenges and opportunities. Discovery and Innovation. (2003). Jun;15(1-2):17-27.

[66] Hartsock, T. L, & Nobel, P. S. Water converts a CAM plant to daytime CO_2 uptake. Nature. (1976). , 262(5569), 574-6.

[67] Kant, P, & Wu, S. The extraordinary collapse of Jatropha as a global biofuel. Environmental Science & Technology. (2011). Sep 1;, 45(17), 7114-5.

[68] Kazakoff, S. H, Gresshoff, P. M, & Scott, P. T. editors. *Pongamia pinnata*, a sustainable feedstock for biodiesel production: Energy Crops: Royal Society of Chemistry; (2010).

[69] Scott, P. T, Pregelj, L, Chen, N, Hadler, J. S, Djordjevic, M. A, & Gresshoff, P. M. *Pongamia pinnata*: An Untapped Resource for the Biofuels Industry of the Future. Bioenergy Research. (2008). Mar;, 1(1), 2-11.

[70] Niknam, S. R, Ma, Q, & Turner, D. W. Osmotic adjustment and seed yield of Brassica napus and B. juncea genotypes in a water-limited environment in south-western Australia. Australian Journal of Experimental Agriculture. (2003). , 43(9), 1127-35.

[71] Robertson, M. J, Holland, J. F, & Bambach, R. Response of canola and Indian mustard to sowing date in the grain belt of north-eastern Australia. Australian Journal of Experimental Agriculture. (2004). , 44(1), 43-52.

[72] Darzin, A, Pienkos, P, & Edye, L. Current status and potential for algal biofuels production. A report to IEA Bioenergy Task 39 (Report T39T2). Commercialising 1st and 2nd Generation Liquid Biofuels from Biomass. (2010).

[73] Gonzalez-fernandez, C, & Molinuevo-salces, B. Cruz Garcia-Gonzalez M. Open and enclosed photobioreactors comparison in terms of organic matter utilization, biomass chemical profile and photosynthetic efficiency. Ecological Engineering. (2010). Oct;, 36(10), 1497-501.

[74] Campbell, P. K, Beer, T, & Batten, D. Life cycle assessment of biodiesel production from microalgae in ponds. Bioresource Technology. (2011). Jan;, 102(1), 50-6.

[75] Razif, H, Davidson, M, Doyle, M, Gopiraj, R, Danquah, M, & Forde, G. Technoeconomic analysis of an integrated microalgae photobioreactor, biodiesel and biogas production facility. Biomass and Bioenergy. [article]. (2011). , 35(1), 741-7.

[76] Wijffels, R. H, & Barbosa, M. J. An outlook on microalgal biofuels. Science. (2010). Nov;, 329, 796-9.

[77] Sander, K, & Murthy, G. S. Life cycle analysis of algae biodiesel. International Journal of Life Cycle Assessment. (2010). Aug;, 15(7), 704-14.

[78] Li, P, Miao, X, Li, R, & Zhong, J. In situ biodiesel production from fast-growing and high oil content Chlorella pyrenoidosa in rice straw hydrolysate. Journal of biomedicine & biotechnology. (2011). , 2011, 141207.

[79] Davis, S. C, Anderson-teixeira, K. J, & Delucia, E. H. Life-cycle analysis and the ecol-
      ogy of biofuels. Trends Plant Sci. [Review]. (2009). Mar;, 14(3), 140-6.

[80] Moreno, M. M, Lacasta, C, Meco, R, & Moreno, C. Rainfed crop energy balance of
      different farming systems and crop rotations in a semi-arid environment: Results of a
      long-term trial. Soil Tillage Res. [Article]. (2011). Jul;, 114(1), 18-27.

[81] Economics and energy use efficiency of alternative cropping strategies for the Dark
      Brown soil zone of Saskatchewan, Saskatchewan Agriculture Development Fund Fi-
      nal Report: Project 20070029. [database on the Internet](2009). cited 17 August 2012].
      Available   from: http://www.agri.gov.sk.ca/apps/adf/adf_admin/reports/
      20070029.pdf.

[82] Zentner, R. P, Stumborg, M. A, & Campbell, C. A. Effect of crop rotations and fertili-
      zation on energy balance in typical production systems on the Canadian prairies. Ag-
      riculture Ecosystems & Environment. (1989). Mar;25(2-3):217-32.

[83] Gan, Y. T, Liang, C, Hamel, C, Cutforth, H, & Wang, H. Strategies for reducing the
      carbon footprint of field crops for semiarid areas. A review. Agron Sustain Dev. [Re-
      view]. (2011). Oct;, 31(4), 643-56.

[84] Kirkegaard, J, Christen, O, Krupinsky, J, & Layzell, D. Break crop benefits in temper-
      ate wheat production. Field Crops Research. (2008). Jun 3;, 107(3), 185-95.

[85] Herridge, D. F, Peoples, M. B, & Boddey, R. M. Global inputs of biological nitrogen
      fixation in agricultural systems. Plant and Soil. (2008). Oct;311(1-2):1-18.

[86] Rajagopal, D, & Zilberman, D. Review of environmental, economic and policy as-
      pects of biofuels. World Bank Policy Research Working Paper Washington DC:
      World Bank; (2007). (4341)

[87] Woods, J, Black, M, & Murphy, R. Future feedstocks for biofuel systems. In: Howarth
      RW, Bringezu S, editors. Biofuels: Environmental Consequences and Interactions
      with Changing Land Use. Gummersbach, Germany: SCOPE; (2009).

[88] Harvey, M, & Pilgrim, S. The new competition for land: Food, energy, and climate
      change. Food Policy. [Article]. (2011). Jan;36:SS51., 40.

[89] Shirvani, T, Yan, X, Inderwildi, O. R, Edwards, P. P, & King, D. A. Life cycle energy
      and greenhouse gas analysis for algae-derived biodiesel. Energy & Environmental
      Science. (2011). Oct;, 4(10), 3773-8.

[90] Smith, P, Martino, D, Cai, Z, Gwary, D, Janzen, H, Kumar, P, et al. Chapter 8 Agricul-
      ture. In: Metz B, Davidson OR, Bosch PR, Dave R, Meyer LA, editors. Climate
      Change 2007: Mitigation Contribution of Working Group III to the Fourth Assess-
      ment Report of the Intergovernmental Panel on Climate Change. Cambridge, UK:
      Cambridge University Press; (2007). , 497-540.

[91] Brock, P, Madden, P, Schwenke, G, & Herridge, D. Greenhouse gas emissions profile for 1 tonne of wheat produced in Central Zone (East) New South Wales: a life cycle assessment approach. Crop and Pasture Science. (2012). , 63(4), 319-29.

[92] Kissinger, M. International trade related food miles- The case of Canada. Food Policy. (2012). , 37(2), 171-8.

[93] Saunders, C, & Barber, A. Carbon footprints, life cycle analysis, food miles: global trade trends and market issues. Polit Sci. [Article]. (2008). Jun;, 60(1), 73-88.

[94] Godfray HCJBeddington JR, Crute IR, Haddad L, Lawrence D, Muir JF, et al. Food security: the challenge of feeding 9 billion people. Science. (2010). Feb 12;, 327(5967), 812-8.

[95] Webber, M. E. More food, less energy. Scientific American. (2012). Jan;, 306(1), 74-9.

[96] Farrell, A. E. Ethanol can contribute to energy and environmental goals (pg 506, 2006). Science. (2006). Jun 23;312(5781):1748-., 311

[97] Searchinger, T, Heimlich, R, Houghton, R. A, Dong, F, Elobeid, A, Fabiosa, J, et al. Use of US croplands for biofuels increases greenhouse gases through emissions from land-use change. Science. (2008). Feb 29;, 319(5867), 1238-40.

[98] Fargione, J, Hill, J, Tilman, D, Polasky, S, & Hawthorne, P. Land clearing and the biofuel carbon debt. Science. (2008). Feb 29;, 319(5867), 1235-8.

[99] Hoekstra, A. Y, & Chapagain, A. K. editors. Globalization of water, sharing the planet's freshwater resources. Oxford, UK: Blackwell Publishing Ltd; (2008).

[100] King, C. W, & Webber, M. E. Water intensity of transportation. Environmental Science & Technology. (2008). Nov 1;, 42(21), 7866-72.

[101] Hoekstra, A. Y, Chapagain, A. K, Aldaya, M. M, & Mekonnen, M. M. Water Footprint Manual- State of the Art 2009. Enschede, The Netherlands: Water Footprint Network; November, (2009).

[102] Hoekstra, A. Y, & Mekonnen, M. M. The water footprint of humanity. Proc Natl Acad Sci U S A. (2012). Feb 28;, 109(9), 3232-7.

[103] Mekonnen, M. M, & Hoekstra, A. Y. A global assessment of the water footprint of farm animal products. Ecosystems. (2012). Apr;, 15(3), 401-15.

[104] Gerbens-leenes, W, & Hoekstra, A. Y. The water footprint of sweeteners and bioethanol. Environ Int. [Review]. (2012). Apr;, 40, 202-11.

[105] Evenson, R. E, & Gollin, D. Assessing the impact of the Green Revolution, 1960 to 2000. Science. (2003). May 2;, 300(5620), 758-62.

[106] Tilman, D, Balzer, C, Hill, J, & Befort, B. L. Global food demand and the sustainable intensification of agriculture. Proc Natl Acad Sci U S A. (2011). Dec 13;, 108(50), 20260-4.

[107] Bueno, C. S, & Lafarge, T. Higher crop performance of rice hybrids than of elite in-breds in the tropics: 1. Hybrids accumulate more biomass during each phenological phase. Field Crops Research. (2009). Jun 26;112(2-3):229-37.

[108] Fischer, R. A, Byerlee, D, & Edmeades, G. O. Can technology deliver on the yield challenge to 2050?(2009).

[109] Cottee, N. S. Tan DKY, Bange MP, Cothren JT, Campbell LC. Multi-level determina-tion of heat tolerance in cotton (Gossypium hirsutum L.) under field conditions. Crop Science. (2010). Nov-Dec;, 50(6), 2553-64.

[110] Devasirvatham, V. Tan DKY, Gaur PM, Raju TN, Trethowan RM. High temperature tolerance in chickpea and its implications for crop improvement. Crop and Pasture Science. (2012). , 63, 419-28.

[111] Trethowan, R. M, Turner, M. A, & Chattha, T. M. Breeding Strategies to Adapt Crops to a Changing Climate. In: Lobell D, Burke M, editors. Climate Change and Food Se-curity: Adapting Agriculture to a Warmer World(2010). , 155-174.

[112] Lobell, D. B, Cassman, K. G, & Field, C. B. Crop yield gaps: their importance, magni-tudes, and causes. Annual Review of Environment and Resources(2009). , 179-204.

[113] Neumann, K, Verburg, P. H, Stehfest, E, & Mueller, C. The yield gap of global grain production: A spatial analysis. Agricultural Systems. (2010). Jun;, 103(5), 316-26.

[114] Lal, R. Soil carbon sequestration to mitigate climate change. Geoderma. (2004). Nov; 123(1-2):1-22.

[115] Lal, R. The role of residues management in sustainable agricultural systems. Journal of Sustainable Agriculture. 1995 (1995). , 5(4), 51-78.

[116] Dennett, A. L, Burgess, L. W, Mcgee, P. A, & Ryder, M. H. Arbuscular mycorrhizal associations in Solanum centrale (bush tomato), a perennial sub-shrub from the arid zone of Australia. Journal of Arid Environments. (2011). Aug;, 75(8), 688-94.

[117] Kennedy, I. R, Rose, M. T, Kecskes, M. L, Roughley, R. J, & Marsh, S. Phan Thi C, et al. Future perspectives for biofertilisers: an emerging industry needing a scientific approach. In: Kennedy IR, Choudhury ATMA, Kecskes ML, Rose MT, editors. ACIAR Proceedings Series(2008). , 131-137.

[118] Van Loon, L. C. Plant responses to plant growth-promoting rhizobacteria. European Journal of Plant Pathology. (2007). Nov;, 119(3), 243-54.

[119] Bokshi, A. I, Jobling, J, & Mcconchie, R. A single application of Milsana (R) followed by Bion (R) assists in the control of powdery mildew in cucumber and helps over-come yield losses. Journal of Horticultural Science & Biotechnology. (2008). Nov;, 83(6), 701-6.

[120] Boddey, R. M, & Dobereiner, J. Nitrogen fixation associated with grasses and cereals: Recent progress and perspectives for the future. Nutrient Cycling in Agroecosystems. (1995). , 42(1), 241-50.

[121] Barry, C. S. The stay-green revolution: Recent progress in deciphering the mechanisms of chlorophyll degradation in higher plants. Plant Sci. [Review]. (2009). Mar;, 176(3), 325-33.

[122] Jordan, D. R, Hunt, C. H, Cruickshank, A. W, Borrell, A. K, & Henzell, R. G. The relationship between the stay-green trait and grain yield in elite sorghum hybrids grown in a range of environments. Crop Science. (2012). May-Jun;, 52(3), 1153-61.

[123] Tscharntke, T, Clough, Y, Wanger, T. C, Jackson, L, Motzke, I, Perfecto, I, et al. Global food security, biodiversity conservation and the future of agricultural intensification. Biol Conserv. [Article]. (2012). Jul;, 151(1), 53-9.

[124] Scheinost, P. L, Lammer, D. L, Cai, X. W, Murray, T. D, & Jones, S. S. Perennial wheat: the development of a sustainable cropping system for the U.S. Pacific Northwest. American Journal of Alternative Agriculture. (2001). , 16(4), 147-51.

[125] Tekucheva, D. N, & Tsygankov, A. A. Combined biological hydrogen-producing systems: A review. Applied Biochemistry and Microbiology. (2012). Jul;, 48(4), 319-37.

[126] Andreiadis, E. S, Chavarot-kerlidou, M, Fontecave, M, & Artero, V. Artificial photosynthesis: from molecular catalysts for light-driven water splitting to photoelectrochemical cells (pg 946, 2011). Photochemistry and Photobiology. (2011). Nov-Dec; 87(6):1478-., 87

[127] Nocera, D. G. The artificial leaf. Accounts of Chemical Research. (2012). May;, 45(5), 767-76.

Chapter 13

Mass Production of Microalgae at Optimal Photosynthetic Rates

Johan U. Grobbelaar

Additional information is available at the end of the chapter

1. Introduction

Microalgae have been studied in the laboratory and in mass outdoor cultures for more than a century and our initial understanding of photosynthesis became unravelled in the laboratories of Otto Warburg. The breakthrough in his laboratories came when he started using *Chlorella* as model organism [1]. As Grobbelaar [2] pointed out, applied phycology and the mass production of microalgae, became a reality in the 1940's. Since then, microalgae have been grown for a variety of potential applications, such as the production of lipids for energy using flue-gasses, anti-microbial substances, cheap proteins for human nutrition and the production of various bio-chemicals. At present the focus is on bioenergy [3], however, their only real success has been in wastewater treatment. A major frustration for microalgal biotechnologists has been the realization of much lower yields than what is potentially possible from laboratory measurements. The inability to operate photo-bioreactors including raceway ponds, at maximal photosynthetic efficiencies, impacts directly on the economies of scale. Because of this many large-scale projects have not delivered what was predicted and many investors have lost their investments. Richmond's [4] observation that "Microalga culture however is yet very far from supplying any basic human needs.." is as true today as then and he concluded that "the major reason for this stems from the failure to develop production systems which utilize solar energy efficiently". A consequence of low yields is high production costs rendering this technology only suited for exclusive high-priced products.

With the above in mind one can pose the question "whether this technology is a mere dream for cheap mass production of biomass or whether it is only suited for high valued components?" A container is required for the growth of microalgae and to date a distinction is made between open pond systems and photo-bioreactors. Grobbelaar [5] argued that open pond systems where microalgae are grown at high densities are in fact also photo-bioreactors.

However, photo-bioreactors are generally considered to be systems in which the culture has no or minimal contact with the atmosphere and they can be of a variety of designs, such as tubes, plates, coils, bags, etc. [6]. Open ponds, on the other hand, have a large area that is in contact with the atmosphere, but they can be enclosed in e.g. plastic covered greenhouse tunnels. Richmond [4] stated that the major weaknesses of open ponds are the absence of temperature control and the long light-path of about 15 cm. The latter results in large culture volumes and consequently low cell mass densities. The question of temperature control is debatable since the temperature fluctuations will be less in large culture volumes, compared to short light-path cultures with small areal volumes.

Grobbelaar [5] analysed the factors governing microalgal growth in "open" and "closed" systems and concluded that the culture depth (optical-depth/light-path) is the single most important factor that determines microalgal growth and photo-bioreactor productivity. Here we analyse the various variables that could impact on microalgal photosynthesis, especially in large commercial scale production systems, with the aim to develop high yielding micro-algal production systems that could be scaled-up. Results generated in small high density laboratory cultures have little value when mega ton production plants are required and it is generally agreed that open raceway ponds would be the means of large commercial outdoor cultivation. For this reason, this paper will focus mainly on open raceway production systems for the intensive production of microalgal biomass.

2. Open raceway ponds

A number of microalgal species have successfully been grown in open raceway ponds, such as *Chlorella*, *Scenedesmus*, *Spirulina*, *Heamatococcus* and *Nannochloropsis*. Commercial outdoor cultivation is mostly restricted to warm tropical and sub-tropical areas, preferably with low rainfall and cloud cover. Detailed construction details are not readily available because of commercial sensitivity, however, basic information can be found for open raceway systems in e.g. [7-9], centre pivot circular ponds have been used in Japan and Taiwan [10], and sloping cascade ponds in the Czech Republic [11]. Raceway ponds are by far the industrial choice, followed by horizontally stacked tubular systems.

The basic components of open raceway ponds are an oval basin with a central island around which the cultures are stirred. The basins could be excavated or constructed above ground. An impermeable PVC liner is commonly used to seal the ponds. However, open raceway ponds have been constructed using concrete and cement, fibre-glass and even epoxy-coated concrete. Determining factors that dictate the materials used are costs and the requirements of the specific application. For example *Spirulina* and *Dunaliella* are grown in high saline growth media that corrodes concrete over time. Another important consideration is the potential toxicity of the materials used, either for the microalgae as such or the quality of the products produced. Since the basins need to be cleaned from time to time, the materials used should be rigid enough to withstand some abrasive actions and repairs should be simple.

Culture depth varies from 10 to 50 cm with 15 cm as the most common [7]. Culture depth (optical depth/light-path) is an important factor because it influences pond construction, biomass density, harvesting costs and pond operation. Culture depths of <15 cm becomes an engineering challenge, especially when pond areas are > 500 m². Any variances can influence the areal density and light penetration, as well as flow and mixing (see below). The deeper the ponds the more dilute the algal suspension becomes because of the larger culture volumes and consequently the higher the harvesting costs. Larger culture volumes imply handling and moving large quantities of liquid, which in itself becomes an engineering challenge.

Various devices have been proposed to circulate and mix the cultures, such as low shear force pumps, air-lifts and paddle wheels [12]. Paddle wheels have become the industrial standard and although various designs and concepts have been used the basic requirement is to circulate the culture and not to lift or aerate it [13]. Flow velocities of 15 to 35 cm s^{-1} are common and Borowitzka [9] calculated that a 2 kW motor is sufficient to produce speeds of 30 cm s^{-1} in a 1000 m² raceway pond.

In the design of open ponds, a deeper portion is used as a sump for harvesting and cleaning. Also high photosynthetic rates require the addition of CO_2 and depending upon the size of the ponds, carbonation should be applied through pH-controlled sparging at various points to maintain a pre-determined pH range [14].

3. Growing microalgae in open outdoor raceway ponds

Growing microalgae in large outdoor open raceway ponds is very different to growing them under controlled optimal conditions in the laboratory and according to Grobbelaar [5], the questions applied phycologists need to resolve are:

1. How to improve the capture of light energy, uptake of nutrients and CO_2?

2. The role of excreted metabolites and auto-inhibition?

3. The differences/advantages/disadvantages between "open" and "closed" photo-bioreactors, if any?

4. The requirements of the specific application and the resources at hand.

Growing microalgae at their optimum temperature, light intensity, nutrient levels and CO_2 will result in high yields, but the growth rates may be low. The aim for applied phycologists, therefore, is to improve the rates or to realize the highest yields in the shortest possible time [5]. When growing algae the numbers of variables that can be controlled are limited. These are:

1. Culture depth or optical cross section. In open ponds light is attenuated with depth (Fig. 1), while the light attenuation and distribution become more complicated in closed vertically placed and transparent tube reactors.

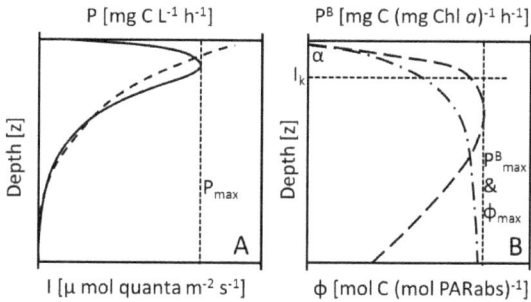

Figure 1. Productivity (solid line) and light attenuation (small dashed) depth profiles (A) in a microalgal culture showing photo-inhibition at the surface, maximum production (P_{max}) and a decrease in production with depth. The productivity profile is normalized with the chlorophyll a (Chl a) content (P^B) to give the Chl a specific productivity (long dashed line in B) and this gives the quantum efficiency (Φ) when normalized with the light intensity (dot dash line in B). Also shown in B is I_k the transition light intensity from light limited to light saturated growth and the photosynthetic efficiency (α) at light limited photosynthesis.

2. Mixing and the resultant turbulence. This differs markedly between open and closed systems, where much higher rates of turbulence can be achieved in closed tubular reactors, while laminar flow is often a problem in open raceway ponds.

3. Supply of nutrients and CO_2, and the prevention of deficit zones.

4. The biomass concentration or areal density. This determines the in-culture light climate, where a higher biomass will attenuate light energy more and *vice versa*. This also determines the light acclimated state of the microalgae.

5. The culture operation being either batch, semi-continuous or continuous.

6. Temperature within limits as well as the dissolved O_2 build-up in closed reactors.

4. Light

The production of mega-tons of biomass implicitly means that natural sunlight must be the source of light. Ironically only about 20 to 25 % of photosynthetically active radiation (PAR) from the sun saturates photosynthesis. Furthermore, the action spectrum of photosynthesis has a peak in the blue and red light regions [15], meaning that the green wavelengths have little photosynthetic value. In mass algal cultures light energy and its capture through photosynthesis is complex and it is governed by;

1. The intensity of the light.

2. The quality of the light and the selective absorption of specific wavelengths by the microalgal biomass.

3. The angle of the light impinging on the culture surface.

4. The condition of the culture surface, e.g. being smooth or rippled.

5. The optical density of the culture (assuming that only algal mass attenuates light energy).

6. The movement of individual cells through a light gradient caused by light attenuation in an optically dense culture (turbulence).

7. The light acclimated state of the culture.

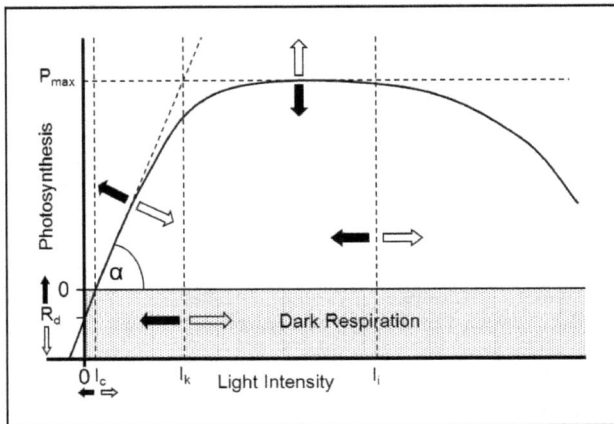

Figure 2. The photosynthetic irradiance (P/I) response of microalgae, where the open arrows indicate the high light acclimated response direction and the filled arrows the low light acclimated direction (figure after [15]). R_d is dark respiration P_{max} the maximum photosynthetic rate, I_c the maintenance light energy, α the maximum photosynthetic efficiency, I_k the transition light intensity from light limited to light saturated photosynthesis, and I_i the light intensity above which photosynthesis is photo-inhibited.

The photosynthetic versus irradiance response curve (P/I) also applies to microalgae and it has been used extensively to describe the response of algae to light energy. Three distinct regions are discernible; i.e. an initial light limited region at low light intensities where photosynthetic rates increase with increasing irradiance, a light saturated region where photosynthetic rates are independent on irradiance, and a region of photo-inhibition in which photosynthetic rates decreases with an increase in irradiance. In the light limited region the rate of photon absorption is correlated with the absorption of light energy and thus the rate of electron transport from water to CO_2, with the liberation of O_2. This initial slope or alpha (α) [16] when normalized with chlorophyll a or biomass is the maximum quantum efficiency of photosynthesis. The transition between light limited and light saturated photosynthesis could be gradual or abrupt [17], implying a non-linearity between absorbed light and photosynthetic rates. At light saturation photosynthetic rates reach a maximum (P_{max}, Fig. 1B and 2) irrespective of an increase in light intensity. The transition between light limited and light saturated photosynthesis is donated by I_k and it is defined as:

$$I_k = \frac{P_{max}}{\alpha} \tag{1}$$

At light saturation, the rate of photon absorption exceeds the rate of electron turnover in photosystem II (PS II), without damaging PSII. However, at even higher irradiancies light-induced depression of photosynthesis occurs commonly referred to as photo-inhibition. Over short term time scales, light induced photo-inactivation of PS II could be viewed as a survival strategy by reducing the number of redundant PS II units. If PS II reaction centres are not repaired through the continuous replacement of the D1 protein, then the damage would become permanent resulting in PS II inactivation [18]. Thus the light-dependent inactivation may be reversible or irreversible. An important finding is that P_{max} was unaffected when PSII reaction centres were reduced by almost 50 % [19]. The reason for this is that the acceptor side of photosynthesis is limited by the capacity of the Calvin-cycle, and the impact of photo-inhibition would only become apparent when the numbers of reaction centres are reduced to such a level that the capacity of the Calvin-cycle cannot be met.

Algae (and for that matter all plants) have developed several mechanisms to cope with changes in the quality and intensity of light. In essence the aim of a plant is to acclimate to the prevailing light climate by ensuring that the light reactions exceed the dark photosynthetic reactions. According to Sukenik et al. [20] the Rubisco levels remained relatively constant under varying light regimes, suggesting that the major regulation occurs in the light reactions, especially of PS II. This could be achieved through modulation of the light-harvesting capacity (various photosynthetic pigments) and/or changes in the number and sizes of PS II reaction centres (light harvesting units).

Photo-acclimation affects all components of the P/I curve (Fig. 2), where the open arrows show the direction of the response due to high light and the solid arrows the response due to low light acclimation [15]. The acclimation can take place in time-scales from seconds to hours, depending on the parameters measured [21]. It is important to realize that the microalgae present in a high density algal culture have acclimated to the average light intensity over the entire optical cross section. Furthermore, the microalgae will acclimate during the production cycle in batch cultures, being high light acclimated soon after inoculation of a new culture, to low-light acclimated at the end of the batch process when the cell density is high [22].

There is little that can be done with the above when the scales of mega-mass microalgal production systems are taken into consideration. It may be possible to maintain the biomass in a low light acclimated state by keeping the biomass concentration high. Such acclimated microalgae have high α, low P_{max} at saturating light intensities and low concentration of auxiliary photosynthetic pigments [22]. However, in small-scale production systems, it is possible to utilize the photo-acclimated state of microalgae to achieve very high areal yields. Grobbelaar and Kurano [23] tested a multi-layered flat plate photo-bioreactor, where the microalgae were acclimated to high light in the first layer facing the light source, followed by a layer where the microalgae became progressively more low light acclimated. In this reactor the properties of high and low light acclimated microalgae could simultaneously be exploited and productivities were almost 40 % higher compared to a single layer flat plate reactor of similar optical cross section.

5. Light/dark cycles

Grobbelaar [14] proposed that the key to high productivities lies in turbulence induced short L/D cycles. The three major effects of mixing are:

1. Prevention of the cells settling to the bottom of the pond. Should this happen, so-called "dead zones" occur where the accumulated of organic material leads to anaerobic decomposition and the release of unwanted substances and metabolites.

2. To prevent the formation of nutritional and gaseous gradients. Over and above the obvious lowering of oxygen super saturation, mixing would decrease the boundary layer around the cells [23-24]. This would increase the mass transfer rates between the cells and the culture medium for both nutrient uptake and exudation of metabolites. Grobbelaar [26] went on to show that a synergistic enhancement of productivity takes place between decreasing the boundary layer and L/D cycles, with increasing turbulence (see next point).

3. To move the cells through an optically dense gradient, with variations in the quantity and quality of light energy. In outdoor mass cultures the algae are subjected to the dynamic natural environment, with its variations in the quantity and quality of the light, both diurnally and seasonally, as well as through the mechanical means of mixing.

Kok [27] showed that photosynthesis is influenced by the intensity of light, L/D fluctuations and the ratio of dark time to light time. He used the term 'flashing light' to indicate the light time and it has often been confused with single turnover flashes. He concluded that the light time should be less than 4 ms in order to achieve 'full' efficiency and that the dark time should be at least ten times as long as the light time. For many decades, the enhancement of photosynthesis in intermittent light was interpreted as some residual light energy that was captured during the light time and then utilised in the dark until the next light flash is received. Today we know that this is not possible, because the time-scales for electron turnover in PSII and PSI range from femto- to milli-seconds [14].

Kok [27] also found that the pattern of intermittent light required, for high yields of *Chlorella* was dependent on turbulence manifested in a flowing culture. The increased production of biomass as a result of increased turbulence was eloquently demonstrated by Laws et al. [28] who placed aerofoil type of devices in the channels with flowing cultures. These caused vortices of 0.5–1 Hz at a flow rate of 30 cm/s, which resulted in photosynthetic conversion efficiencies increasing from an average of 3.7 to up to 10%.

Increasing photosynthetic rates and efficiencies in intermittent light has been shown by several authors, e.g. [21, 26] and Grobbelaar [2] found that the increase in specific production rates were consistently exponential with increasing L/D frequencies. The observed increases in productivities with increasing L/D frequencies are directly linked to electron turnover rates in the electron transport chains of photosynthesis. As the L/D frequencies approach the turnover rates, productivities and photosynthetic efficiencies increase until the L/D frequencies match the turnover rates, where production and photosynthetic efficiencies would be at their highest. These rates are in the millisecond range and in practical terms this should form part of the operational considerations when designing photo-bioreactors.

Grobbelaar [14] showed that the photosynthetic rates increased on average 2.1 times at equal L/D cycles when the frequencies were increased from 0.1 to 10 Hz and that the enhancement of photosynthetic rates only becomes significant at cycles > 0.1 Hz. In practical terms this should be achievable in SLP (short light path) reactors [14]. However, special devices would be needed in open raceway ponds, such as aerofoils [28], rippled floors and sides, and curvilinear end geometries.

6. Photo-inhibition

As discussed above, photo-inhibition is the light-induced depression of photosynthesis when the rate of photon absorption exceeds the rate of electron turnover in PS II. Over the short term, light-induced photo-inactivation of PS II is a survival strategy, whereby the number of redundant PS II units is reduced. Prolonged exposure to high light intensities eventually results in PS II reaction centres not being repaired, through the continuous replacement of the D1 protein. The damage then becomes permanent resulting in PS II inactivation [18]. Whether photo-inhibition occurs in dense mass algal cultures is open for debate, especially in turbulent tubular and plate photo-bioreactors. Congming and Vonshak [29] reported a midday maximum quantum efficiency (Φ_{max}) depression of dark adapted *Spirulina platensis* as a result of reaction centre inactivation, which they ascribed to photo-inhibition. Grobbelaar [30] measured a similar midday depression of Φ_{max} on open outdoor raceway *Spirulina platensis* cultures but concluded that it was as a result of light energy being lost as heat dissipation. This was particularly evident in low density cultures where more than 60 % of the reaction centres became "silent", meaning that they neither reduced Q_A, nor returned their excitation energy to the antenna.

Laminar flow is common in large open raceway ponds and Laws et al. [28] found that they had to space their foil arrays 1.2 m apart, because of the rapid dissipation of mixing in the raceway channels. Since raceway channels can be >100 m, it is reasonable to expect photo-inhibition to be a factor in the surface layers of the cultures. Grobbelaar et al. [31] modelled microalgal productivity in large outdoor raceway ponds. The model was calibrated against two years of data collected from five raceway ponds differing in surface area from 71 – 263 m². In a generalized form the model is written as:

$$PROD(mg\ (dw)m^{-2}h^{-1}) = PRD - RES - INB \qquad (2)$$

Where PROD = net production, PRD = productivity, RES = respiration, and INB = photo-inhibition.

PRD is calculated from inputs of the biomass concentration in the culture, culture temperature and light impinging on the surface of the culture. The equation for PRD is:

$$PRD = \left(A_1.X_1\left(A_2^T\right)\right).\left(\frac{I_z.I_s\left(A_3^T\right)}{I_z + I_s\left(A_3^T\right)}\right) \qquad (3)$$

where $A_1 - A_3$ are constants, X_1 is the biomass concentration in mg (dw) L^{-1}, I_z the irradiance in mol quanta m^{-2} h^{-1} at depth z in meters, I_s the light half saturation constant and T a temperature factor. Equation (3) has temperature/biomass and temperature/light energy terms. A_1 is the light utilization efficiency, A_2 the Q_{10} of photosynthesis and A_3 the Q_{10} for the light half saturation.

The component RES includes all losses due to respiration and exudation of organic compounds from the cells and is calculated from the following equation:

$$RES = X_1\left(\left(\frac{1.5^{T - 0.54}}{100}\right)\right) \qquad (4)$$

Photo-inhibition (INB) was included in the model and the equation was constructed such that it only took new production (PRD) into account, that it increases linearly with an increase in irradiance above a threshold irradiance of 1 mol quanta m^{-2} h^{-1}, and that temperature affected the overall rate. Photo-inhibition was calculated from:

$$INB = PRD\left(\left(\frac{2.5^T}{75}\right).I_z\right) \qquad (5)$$

Shown in Fig. 3 are the outcomes (predictions) for the day time interval of 12:00 to 13:00, in a 15 cm culture depth open raceway pond, with a daily irradiance of 60 mol quanta m^{-2} d^{-1}, minimum and maximum temperatures of 10 °C and 30 °C, 12 hour day-light length and culture biomass concentrations of 200 (Fig 3A), 400 (Fig. 3B), 600 (Fig. 3C), and 800 mg (dw) L^{-1} (Fig. 3D). The exponential decrease in irradiance with increasing depth is seen where all the light energy is absorbed at different depths, depending on the biomass concentration and the resultant attenuation of light energy. Grobbelaar et al. [31] defined the condition where all the light energy is absorbed over the depth profile (optical cross section) as the optimal areal density for maximum productivity (Fig. 2 B). Net productivity typically showed photo-inhibition at the surface, where the depth of P_{max} depended on the areal biomass density, ranging from 6 cm at an areal density of 30 g m^{-2} to 2 cm at an areal density of 120 g m^{-2}. Below P_{max} productivity decreased as the light energy was attenuated. The impact of photo-inhibition is seen at optical depths deeper than that at P_{max} when the plots of net production (PROD) are compared to the plots where photo-inhibition is excluded (PRD – RES, the dot dot dash lines in Fig. 3).

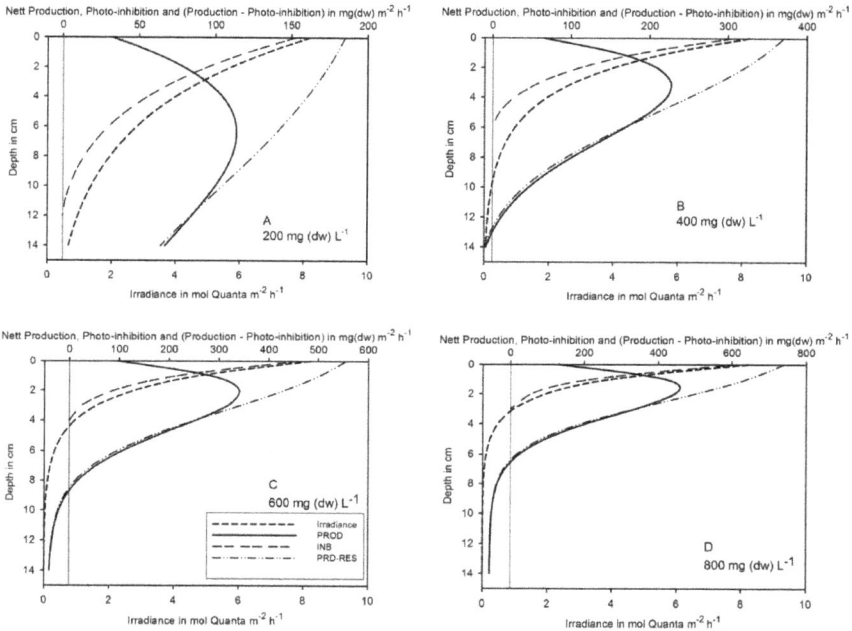

Figure 3. Modelled response of irradiance, PROD, INB and PRD-RES at four areal densities.

The significant impact that photo-inhibition has on the overall productivity is evident in all four examples (Figs. 3A – 3D), when the depth integrated rates are compared (Table 1). Losses due to photo-inhibition are clearly dependent on areal density and light attenuation where it is the lowest at the optimal areal density. At an areal density of 120 g m^{-2} losses due to photo-inhibition can be as high as 66 % (Table 1). Respiratory losses are also dependent on areal density and increases linearly with an increase in areal density. However, as the "dark" zone increases with an increase in areal density a larger percentage of the biomass is lost due to RES (Fig. 3 and Table 1). For example the total RES loss at a biomass concentration of 200 mg L^{-1} is 9.6 %, whereas at 800 mg L^{-1} it is 28 % (Table 1).

With such losses, it is surprising that very little research has gone into eliminating or limiting the losses due to photo-inhibition. Several authors, e.g. [25, 27 and 31], have already reported significant increases in volumetric and areal productivities where mixing was increased. As mentioned above, Laws et al. [28] attributed an increase of up to 10 % in photosynthetic efficiencies in their systems where foils created vortices with rotation rates of about 0.5–1 Hz. However, Grobbelaar et al. [22] showed minimal enhancement at light/dark cycles ≤1 Hz. This

Biomass in mg (dw) L⁻¹	200	400	600	800
PRD	2167	2719	2860	2966
RES	209	419	628	837
INB	613	665	752	848
PROD	1345	1635	1481	1280
PRD – RES	1958	2300	2232	2128
% loss due to INB	45	40	50	66

Table 1. Predicted depth Integrated results in mg (dw) m^{-2} h^{-1} in four cultures all, 15 cm deep, minimum and maximum temperatures of 10 °C and 30 °C, irradiance from 12:00 to 13:00 of 8.2 mol quanta m^{-2} h^{-1}, and at biomass concentrations of 200, 400, 600 and 800 mg (dw) L^{-1}. The abbreviations are as defined in equations 2 to 5.

was acknowledged by Laws and Berning [33] and they attributed the increased productivity in flumes with foils, rather to the re-suspension of settled cells and breaking down of nutrient gradients. As stated before, Grobbelaar [26] clearly showed how increased turbulence enhanced the exchange rates of nutrients and metabolites between the cells and the growth medium, and this, together with increased light/dark frequencies, synergistically increases productivity and photosynthetic efficiency.

It is thus suggested that mixing microalgal cells in dense cultures other than the obvious re-suspension of the cells are important for:

1. Moving the cells in a light energy attenuating medium resulting in various L/D cycles and patterns.

2. Altering the boundary layers on the cells and thus the uptake and release of substances.

3. Shortening the exposure time allowing above saturating light intensities to be utilized. It is suggested that as the L/D frequencies approach the electron turnover rate full sunlight could be captured through photosynthesis, if the areal density of the cultures are high enough.

4. Ensuring an "average" photo-acclimated state of the entire culture. Grobbelaar et al. [34] clearly demonstrated how the photo-acclimated state of an outdoor culture changes depending on the cell density and how this affects the culture productivity.

7. Conclusions

The success or failure of producing mega-quantities of microalgal biomass will depend on culture systems where high photosynthetic rates are maintained. Although conditions in closed photo-bioreactors (vertical tubular, helical tubular, vertical flat plate or horizontal flat panel) are generally conducive for high photosynthetic rates; fowling, hydrodynamic stress and scale-up limitations [35], essentially rule these kinds of systems out, for mega-scale production microalgal biomass.

It is clear that features of growth reactors that would have to be considered are:

1. the surface/volume (S/V) ratio where the aim should be to operate systems with the highest possible S/V ratio,

2. mixing and not only the devices used, e.g. paddle wheels, airlift or impellers, but critically the prevention of laminar flow,

3. mass transfer rates of nutrients and metabolites, where this is dependent on mixing,

4. the light exposure patterns, especially with the introduction of mixing devices to manifest L/D cycles of > 0.1 Hz and limiting of photo-inhibition losses,

5. nutrient supply and modes of dosing (see [36])

6. the scalability of the systems and potential of automation, and

7. the ease of operation.

Producing the quantities required for, e.g. bioenergy [3] production, would only become a reality when the above factors are systematically researched, analysed and optimized.

Author details

Johan U. Grobbelaar

Address all correspondence to: grobbeju@ufs.ac.za

Department of Plant Sciences, University of the Free State, Bloemfontein, South Africa

References

[1] Nickelsen K. Otto Warburg's first approach to photosynthesis. Photosynth. Res. 2007; 92 109-120.

[2] Grobbelaar JU. Microalgal biomass production: challenges and realities. Photosynth. Res. 2010; 106 135-144.

[3] Tan DKY., Amthor JS. Bioenergy. In: Dubinsky, Z. (ed.) *This publication*

[4] Richmond A. Efficient utilization of high irradiance for production of photoautotrophic cell mass: a survey. J. Appl. Phycol. 1996; 8 381-387.

[5] Grobbelaar JU. Factors governing algal growth in photobioreactors: the "open" versus the "closed" debate. J. Appl. Phycol. 2009a; 21 489-492.

[6] Tredici MR. (2004) Mass production of microalgae: Photobioreactors. In: Richmond, A. (ed.) Handbook of microalgal culture. Blackwell Science; 2004. 178-214.

[7] Dodd JC. (1986) Elements of pond design and construction. In: Richmond A (ed.) Handbook of microalgal mass culture. CRC Press, Boca Raton; 1986. 265-284.

[8] Belay A. (1997) Mass culture of *Spirulina* outdoors: The Earthrise Farms experience. In: Vonshak A. (ed.), *Spirulina platensis* (*Arthospira*): Physiology, Cell-biology and Biochemistry. Taylor & Francis, London; 1997. 131-158.

[9] Borowitzka MA. Culturing microalgae in outdoor ponds. In: Andersen RA. (ed.), Algal culturing techniques. Elsevier, Amsterdam; 2005. 205-218.

[10] Kawaguchi K. Microalgae production systems in Asia. In: Shelef G., Soeder CJ (eds.), Algae biomass production and use. Elsevier/North Holland Biomedical Press, Amsterdam; 1980. 25-33.

[11] Setlík I., Veladimir S., Malek I. Dual purpose open circulation units for large scale culture of algae in temperate zones. I. Basic design considerations and scheme of pilot plant. Algolog. Studies (Trebon) 1970; 1 111-164.

[12] Soeder CJ. Types of algal ponds. In: Grobbelaar JU., Soeder CJ., Toerien DF. (eds.) Wastewater for aquaculture. UFS Publication Series C, Bloemfontein; 1981. 131-135.

[13] Grobbelaar JU. From laboratory to commercial scale: a case study of a *Spirulina* (*Arthrospira*) facility in Musina, South Africa. J. Appl. Phycol. 2009b; 21 523-527.

[14] Grobbelaar JU. Upper limits of photosynthetic productivity and problems of scaling. J. Appl. Phycol. 2009c; 21 519-522.

[15] Grobbelaar JU. Photosynthetic Response and Acclimation of Microalgae to Light Fluctuations. In: Subba Rao DV. (ed.), Algal Cultures Analogues of Blooms and Applications. Science Publishers Enfield (NH), USA, Plymouth,UK; 2006. 671-683.

[16] Jassby AD., Platt T. Mathematical formulation of the relationship between photosynthesis and light for phytoplankton. Limnol. Oceanogr. 1976; 21 540-547.

[17] Leverenz, JW. Chlorophyll content and the light response curve of shade-adapted conifer needles. Physiol. Plant 1987; 71 20-29.

[18] Prasil O., N. Adir N., I. Ohad I. Dynamics of photosystem II: mechanisms of photoinhibition and recovery processes. In: Barber J. (ed.), Topics in photosynthesis: The photosystems: Structure, Function and Molecular Biology. Elsevier, Amsterdam; 1992. 11 295-348.

[19] Behrenfeld MJ., Prasil O., Kolber ZS., Babin M., P.G. Falkowski PG. Compensatory changes in Photosystem II electron turnover rates protect photosynthesis from photoinhibition. Photosynthesis Res. 1998; 58 259-268.

[20] Sukenik A., Bennett J., Falkowski PG. Light-saturated photosynthesis-limitation by electron transport or carbon fixation. Biochim. Biophys. Acta 1987; 891 205–215.

[21] Cullen JJ., Lewis MR. The kinetics of algal photoadaptation in the context of vertical mixing. J. Plankton Res. 1988;10 1039-1063.

[22] Grobbelaar JU., L. Nedbal L., Tichy V. Influence of high frequency light/dark fluctuations on photosynthetic characteristics of microalgae photo-acclimated to different light intensities and implications for mass algal cultivation. J. Appl. Phycol. 1996; 8(4-5) 335-343.

[23] Grobbelaar JU., Kurano N. A novel photobioreactor for achieving extreme high yields. J. Appl. Phycol. 2003; 15 121-126.

[24] Grobbelaar JU. Do light/dark cycles of medium frequency enhance phytoplankton productivity? J. App. Phycol. 1989; 1 333-340.

[25] Grobbelaar JU. 1991. The influence of light/dark cycles in mixed algal cultures on their productivity. Bioresource Technol. 1991; 38 189-194.

[26] Grobbelaar JU. Turbulence in mass algal cultures and the role of light/dark fluctuations. J. Appl. Phycol. 1994; 6 331–335.

[27] Kok B. Experiments on photosynthesis by *Chlorella* in flashing light. In: Burlew JS. (ed.) Algal culture: from laboratory to pilot plant. Vol. 600, Carnegie Institution of Washington Publ., Washington; 1953. 63–75.

[28] Laws EA., Terry KL., Wickman J., Chalup MS. A simple algal production system designed to utilize the flashing light effect. Biotechnol. Bioeng. 1983; 25 2319–2335.

[29] Congming L., Vonshak A. Photoinhibition in outdoor *Spirulina platensis* cultures assessed by polyphasic chlorophyll fluorescence transients. J. Appl. Phycol. 1999; 11 355–359.

[30] Grobbelaar JU. Photosynthetic characteristics of *Spirulina platensis* grown in commercial-scale open outdoor raceway ponds: what do the organisms tell us? J. Appl. Phycol. 2007; 19 591-598.

[31] Grobbelaar JU., Soeder CJ., Stengel E. Modeling algal productivity in large outdoor cultures and waste treatment systems. Biomass 1990; 21 297-314.

[32] Richmond A., Vonshak, A. *Spirulina* culture in Israel. Arch. Hydrobiol. Beih. Ergebn. Limnol. 1978; 11 274–280.

[33] Laws EA., Berning JL. A study of the energetic and economics of microalgal mass culture with the marine chlorophyte *Tetraselmis suecica*: implications for use of power plant stack gasses. Biotechnol. Bioeng. 1991; 37 936–947.

[34] Grobbelaar JU., Nedbal L., Tichy V., Setlik I. Variations in some photosynthetic char-
      acteristics of microalgae cultured in outdoor thin-layered sloping reactors. J. Appl.
      Phycol. 1995; 7 243-260.

[35] Kumar K., Dasgupta CN., Nayak B., Lindblad P., Das D. (2011) Development of suit-
      able photobioreactors for CO_2 sequestration addressing global warming using green
      algae and cyanobacteria. Bioresource Technol. 2011; 102 4945–4953.

[36] Grobbelaar JU. (2004) Algal Nutrition. In: Richmond A. (ed.), Handbook on microal-
      gal culture. Blackwell Sci.; 2004. 97-115.

Chapter 14

Biosynthesis of Lipids and Hydrocarbons in Algae

Masato Baba and Yoshihiro Shiraiwa

Additional information is available at the end of the chapter

1. Introduction

Lipids function as important storage compounds to maintain cellular activities. Lipids store high reducing power and energy since those biosynthetic processes require high amount of reducing cofactors and ATP. Storage lipids do not cause any chemical effect on cellular activity such as osmolarity, pH and ion strength because of its hydrophobicity. Membrane lipids such as phospholipids, carotenoids, and cholesterols play a housekeeping role. In addition, some of lipids function as protein modifiers or signaling molecules.

Recently, plant oils are gathering keen interest as a source of renewable energy according to rapid increase in social demands for establishing a low-carbon-society. However, oil production for biofuels and biorefinery using higher plants and crops is strongly worried for competing with food production and to increase those market prices. Therein, algae came into play a new oil-producing organism since algae do not compete with food production. According to their high productivity per unit area, prokaryotic photoautotrophs such as cyanobacteria and eukaryotic algae such as protists are expected to become a promising feedstock in future (Gong & Jiang, 2011).

Although numerous kinds of lipids exist in nature, main carbon chain of the molecules is almost derived from limited numbers of precursor molecules such as fatty acids and isoprenoids. Interestingly, some parts of the synthetic pathways of lipids are quite different among animals, higher plants, cyanobacteria and some eukaryotic microalgae. Although there is quite few information on lipid biosynthesis and metabolism in algae, it is noteworthy that most of biosynthetic pathways of hydrocarbons such as fatty acid- and isoprene-derived hydrocarbon have been well characterized in microalgae.

In this chapter, we will introduce recent progresses on lipid and hydrocarbon biosynthetic pathways in microalgae: First, unique features of algal lipid synthetic pathways mostly hypothesized by advanced DNA sequencing technique although those are not well proved experimentally yet. Second, two factors for hydrocarbon biosynthesis in microalgae characterized recently by a combination of expressed sequence tags (EST) analysis and novel enzyme characterization. Those are: (1) decarbonylase to produce fatty acid-derived hydrocarbons in cyanobacteria and (2) isoprene-derived hydrocarbon biosynthetic pathway in a representative oil-producing colonial microalga, *Botryococcus braunii*. Third, the mechanism for carbon flow and energy balance in lipid and hydrocarbon biosynthetic pathways. Finally, we will describe a future perspective of algal lipid biosynthetic pathways and its application to biofuel production.

2. Unique features of algal lipid biosynthetic pathways

Although there is no agreed definition and classification of "lipids" (The AOCS Lipid Library, http://lipidlibrary.aocs.org/), here we define a term "lipid" as follows: 1) it is biological component of and derived from organisms; 2) it is basically very soluble in organic solvents but not in water; 3) it contains hydrocarbon group in its structure. We adopt biosynthetic classification to categorize lipids such as fatty acid, isoprene or others of unique lipids as shown in Table 1, instead of a conventional lipid classification such as simple lipid, derived lipid, complex lipid, and so on. Here we used the term "lipids" for compounds composed of only carbon, hydrogen, and oxygen.

Table 1 indicates the list of various lipids and their functions. Although numberless lipids exist in nature, main carbon chain of the molecules is mostly derived from fatty acids, isoprenes and their homologous compounds via some synthetic pathways. Recent progress in genome/transcriptome sequencing technology and its computational analysis on similarity of those base sequences among organisms, namely *in silico* analysis, enable us rapid prediction of metabolic pathway even in uncharacterized organism. Owing to the modern *in silico* analyses, some unique features of algal lipid biosynthesis are starting to be enlightened. Interestingly, some parts of the synthetic pathways are quite different among animals, higher plants, cyanobacteria and eukaryotic algae.

Name	Structure	Function
Fatty acid	C_{2n}-/straight-carbon chainwith carboxyl group	Membrane component;Bioactivity
Polyketide	Various carbon chain with polyketone group	Antibiotic; Bioactivity
Glyceride	Ester of fatty acid & glycerol	Common storage lipid
Terpenoid	C_{5n}./branched-carbon chain;isoprene derivate	Bioactivity
Steroid	Tri-terpenoid derivate	Common hormone
Carotenoid	Tetra-terpenoid derivate;conjugated double bond; absorbent	Pigment

Table 1. Various lipids and their functions

2.1. Fatty acid biosynthesis

Acetyl-coenzyme A (CoA) is a universal carbon donor for fatty acid biosynthesis. Acetyl-CoA is supplied via multiple paths from various origins and then subsequently metabolized into malonyl-acyl carrier protein (ACP) by sequential reactions. One molecule of ATP (1ATP) is used for the carboxylation of acetyl unit to produce one malonyl unit. In general, fatty acid biosynthesis utilizes acetyl-CoA and malonyl-ACP as starting substrates and acetyl unit donors. Primarily, butyryl(C_4)-ACP is synthesized from acetyl(C_2)-CoA and malonyl(C_3)-ACP via sequential reactions of condensation, decarboxylation, and reduction of non-malonyl-ACP derived keto unit. Two molecules of NADPH (2NADPH) is used for the reduction of keto group. Accordingly, 1ATP and 2NADPH are consumed to elongate chain of fatty acid molecule by adding C_2-saturated carbon unit in fatty acid biosynthesis. Acyl-ACP is elongated up to acyl($C_{16 \text{ or } 18}$)-ACP. Molecules with C_{2n}-carbon chain are widely distributed among various organisms and those of C_{2n-1}-carbon chain are synthesized from C_{2n}-compounds by carbon-loss (Řezanka & Sigler, 2009). In bacteria, *iso-* and *anteiso-*fatty acids (branched-chain fatty acids) are synthesized from amino-acid-derived precursors with branch (Kaneda, 1991). ACP is subsequently removed from acyl($C_{16 \text{ or } 18}$)-ACP to form fatty acid($C_{16 \text{ or } 18}$). The fatty acids synthesized on the plastid envelopes are excreted into cytosol by accompanying probably with the process of binding of CoA (Joyard et al., 2010). Oppositely, fatty acid is necessary to be activated by the binding of ACP for passing through the cell membrane in the cyanobacteria *Synechocystis* species.

In any step, a synthesized carbon chain can be metabolized into various products including glycerolipids, triacylglycerides (TG), phospholipids and glycolipids (Joyard et al., 2010). Fatty acids synthesized excessively are stored as TG in most eukaryotes. Usually prokaryotes do not accumulate TG although *Actinomycetes* and a few other bacteria exceptionally synthesize TG (Alvarez & Steinbüchel, 2002).

Same fatty acids as metabolites are widely observed in various organisms but their biosynthetic pathways are different depending on classification. There are four known groups of enzyme(s) for fatty acid biosynthesis; type-I fatty acid synthase (FAS), type-II FAS, particular elongases, and enzymes for catalyzing the reversal of ß-oxidation. Typically, animals and fungi possess type-I FAS which is a large multi-functional enzyme with multiple functional domains (Chan & Vogel, 2010; Joyard et al., 2010). Bacteria, plastids and mitochondria have type-II FAS which is composed of four subunit proteins such as ß-ketoacyl-ACP synthase (KAS), ß-ketoacyl-ACP reductase, ß-hydroxyacyl-ACP dehydratase and enoyl-ACP reductase (Chan & Vogel, 2010; Joyard et al., 2010; Hiltunen et al., 2010). A trypanosomatid *Leismania major* possesses three elongases instead of above mentioned FASs for fatty acid biosynthesis (Lee et al., 2006). In a microalga *Euglena gracilis* fatty acids are synthesized *de novo* via the reversal pathway of ß-oxidation under anaerobic conditions (Hoffmeister et al., 2005; Inui et al., 1984). As such, wide variation seems to exist among organisms but detailed information on fatty acid biosynthesis is not well understood in algae.

One of model organism *Chalmydomonas reinhardtii* (Chlorophyta) possesses type-II FAS gene which is homologous to that of land plants (Riekhof et al., 2005). Land plants have both plastidal- and mitochondrial-FASs which exhibit different substrate specificity (Yasuno et al.,

2004). Mitochondrial ACP-type enzymes are well characterized especially in the yeast *Saccharomyces cerevisiae* (Hiltunen et al., 2010). In the green alga *C. reinhardtii*, FAS is thought to be localized both in the plastid and the mitochondrion individually since respective FAS gene is represented as single gene in each organelle (Riekhof et al., 2005). Expressed sequence tags (ESTs) of Chlorophyta *Botryococcus braunii* Bot-88-2 (race A) contained a partial sequence of type-I FAS (accession number: FX056119) which is partially similar to animal FAS (Baba et al., 2012a). However, further information of complete genome sequence is necessary to confirm it. In *E. gracilis* which has triple-layer envelope, there are five fatty acid synthetic pathways located in four subcellular compartments; the type-I FAS system in the cytosol, two type-II FAS systems in the plastid, fatty acid synthesis associated partly with wax ester fermentation in the microsomes (see 3.1.4. *Wax ester*) and a malonyl-ACP independent process located in the mitochondria. *E. gracilis* synthesizes wax esters when cells were grown under either heterotrophic or anaerobic conditions (Hoffmeister et al., 2005). Wax ester fermentation under anaerobic conditions produces ATP in net profit while general malonyl-ACP-dependent fatty acid synthesis consumes ATP (Inui et al., 1982). Fatty acids supplied for wax ester production are *de novo* synthesized by acetyl-CoA condensing reaction which is similar to the reversal pathway of ß-oxidation, where *trans*-2-enoyl-CoA reductase contributes instead of acyl-CoA dehydrogenase (mitochondrial ß-oxidation enzyme) or acyl-CoA oxidase (peroxisomal ß-oxidation enzyme) (Hoffmeister et al., 2005; Inui et al., 1984). The *trans*-2-enoyl-CoA reductase reaction is a key step in the reversal pathway of ß-oxidation. The direction of the pathway is O_2-dependently determined to proceed irreversibly under aerobic conditions since the key enzyme pyruvate:NADP$^+$ oxidoreductase is sensitive to O_2 (Tucci et al., 2010). The reversal pathway may be essential to maintain the redox balance in the mitochondria under anaerobic conditions (Tucci et al., 2010). Such hypothesis is supported by another report which demonstrated that ß-oxidation can be reversed in genetically engineered *Escherichia coli* (Dellomonaco et al., 2011). In mammalian mitochondria, the reversal pathway functions for only elongation process due to substrate specificity of mammal enoyl-CoA reductase and therefore the pathway contributes to short-chain fatty acid elongation process (Inui et al., 1984).

2.2. Fatty acid elongation

C_{18}-Fatty acid can be further elongated via the fatty acid elongation pathway. Fatty acid elongation process is very similar to that of the fatty acid synthesis although acyl-CoA and malonyl-CoA are used as a substrate. In the process, 1ATP and 2NADPH are required for C_2-unit elongation of saturated carbon chain since CoA-activation is not essential as suggested by another study (Hlousek-radojcic et al., 1998). Fatty acid elongation reaction site was shown to be located in the endoplasmic reticulum (Kunst & Samuels, 2009). In contrast to FAS system, all known elongation systems are basically compatible and functions simultaneously. Fatty acid elongase constitutes an enzyme complex of four subunits which is similar to type-II FAS; namely, ß-ketoacyl-CoA synthase (KCS), ß-ketoacyl- CoA reductase, ß-hydroxyacyl-CoA dehydratase and enoyl-CoA reductase. There are two different KCSs; namely "elongation of very long-chain fatty acid" (ELOVL or merely ELO)-type elongase which contributes to sphingolipid biosynthesis and "fatty acid elongation" (FAE)-type elongase which contributes to plant seed TG or wax biosynthesis (Venegas-Calerón et al., 2010). Typically, animals and fungi possess ELO-type while land plants possess FAE-type. In some cases, ELO and FAE subunits are inaccurately referred as mere "elongase" since heterologous expression of single gene for KCS often results in successful elongation of acyl-CoA by the help of the other subunit of the host (typically, yeast and land plant *Arabidopsis thaliana*).

Poly unsaturated very long-chain fatty acid (PUVLCFA, PULCA, VLC-PUFA, etc.) is one of elongated fatty acids (e.g. Arachidonic acid, Eicosapentaenoic acid, and Docosahexaenoic acid). PUVLCFA is commonly observed in algae such as Euglenophytes, diatoms (*Phaeodactylum tricornutum, Thalassiosira pseudonana*) and haptophytes (*Emiliania huxleyi, Isochrysis galbana, Pavlova salina*) which possess desaturase/elongases to produce EPA and DHA from C_{18} fatty acid and derivatives (Venegas-Calerón et al., 2010). This elongation is shown to be catalyzed by ELO-type fatty acid elongase. On the other hand, contribution and physiological function of FAE-type elongase has not been proved in microalgae yet. Macroalgae produce short and long-chain aldehydes (Moore, 2006). These metabolites are suggested to be produced from PUVLCFA although the detailed synthetic pathway is almost unknown. Long-chain aldehyde producing reaction was reported to be catalyzed by lipooxygenase which functions to oxygenize and cleave PUVLCFA at a specific position to form short-chain metabolites (Moore, 2006).

2.3. Polyketide biosynthesis

Polyketide includes various complex compounds such as antibiotics (e.g. erythromycin, tetracycline, lovastatin) (Staunton & Weissman, 2001). Polyketide biosynthesis is similar to C_4 and longer fatty acid synthesis except successive reduction of *keto*-group and utilizes various ACP/CoA compounds as its substrate. Mycolic acids, extremely large fatty acids (ca. C_{90}) in the cell envelope of mycobacteria (Verschoor et al., 2012), are synthesized by condensation of two VLCFA acyl-CoA molecules via polyketide biosynthesis (Portevin et al., 2004).

There are three types of polyketide syntheses (PKSs): type-I PKS which is a large multi-functional enzyme in the consequence of multiple functional domains, type-II PKS which is composed of monofunctional proteins to form complex and type-III PKS which resembles chalcone synthase catalyzing the committed step in flavonoid biosynthesis in higher plants and some bryophytes (Shen, 2003). Type-I PKSs are further classified into two, namely iterative and non-iterative (modular) types. Bacteria possess type-I to III of PKSs. Fungi and animal typically possess type-I iterative PKS which is closely related each other (Jenke-Kodama & Dittmann, 2009). Interestingly, there is evolutionary connection between PKSs and FASs (Jenke-Kodama & Dittmann, 2009; John et al., 2008; Sasso et al., 2011).

In the genomes of chlorophyta (*C. reinhardtii, Chlorella variabilis*, two *Ostreococcus* species, etc.), heterokontophyta (*Aureococcus anophagefferens*), and haptophyceae (*Emiliania huxleyi* and *Chrysochromulina polylepis*), "non-iterative" type-I PKS is coded but not in land plant, rhodophyta (*Cyanidioschyzon merolae* and *Galdieria sulphuraria*), Stramenopiles (*Thalassiosira pseudonana, Phaeodactylum tricornutum*, two *Phytophthora* species), and Euglenozoa (*Trypanosoma brucei, Trypanosoma cruzi* and *Leishmania major*) (Sasso et al., 2011). Polyketide synthesis plays a role in biosynthesis of cyanobacterial toxin, microcystin (Jenke-Kodama & Dittmann, 2009). Dinoflagellate toxins are also polyketides although its synthetic pathway remains unknown because of technical restrictions (Sasso et al., 2011). In addition, biosynthetic processes of polyketides in macroalgae are one of important targets to be urgently elucidated.

2.4. Terpenoid biosynthesis

Terpenoid which is composed of branched C_{5n} carbon unit are synthesized by condensation of C_5 isoprene units (as isopentenyl diphosphate (IPP) and its isomer dimethylallyl diphosphate (DMAPP) *in vivo*) (Bouvier et al., 2005). In general, isoprene unit is supplied via either or both mevalonic acid (MVA) pathway or/and methylerythritol phosphate (MEP) pathway (or non-mevalonic acid pathway). MVA pathway is located in the cytosol of Archaea and eukaryotes or in the peroxisome (Lohr et al., 2012). In MVA pathway, three molecules of acetyl-CoA are condensed into DMAPP through MVA and then IPP by sequential reactions, using 3ATP and 2NADPH. The MEP pathway is known to be located in the cytosol of bacteria including cyanobacteria and in the stroma of plastids in plants and eukaryotic algae (Joyard et al., 2009). In MEP pathway, pyruvate and glyceraldehyde-3-phosphate (GAP) react to form IPP or DMAPP via MEP using energy of three high-energy phosphate bonds on either ATP or CTP and reducing power from at least 1NADPH and four reducing coenzymes (not completely identified yet) (Hunter, 2007). One molecule of CO_2 is released in MVA and MEP pathway respectively. In land plants, isoprene units can be exchanged through chloroplast envelope by that MVA and MEP pathways complement each other. However, any protein for isoprene unit exchange has not been isolated so far (Bouvier et al., 2005; Joyard et al., 2009; Lohr et al., 2012).

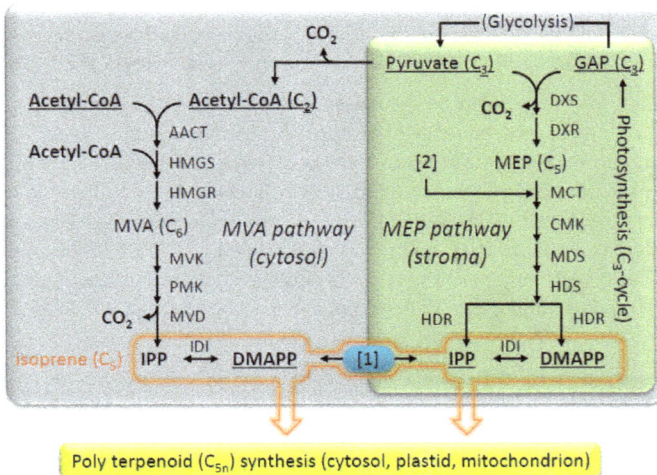

Figure 1. Pathway for terpenoid biosynthesis

Primary and terminal molecules are underlined respectively. Substrates multiply used are shown in bold. [1]: putative Isoprene transporter. [2]: a predicted junction from the pentose phosphate pathway to the MEP pathway in cyanobacteria. AACT, acetoacetyl-CoA thiolase; CMK, 4-(cytidine 5'-diphospho)-2-C- methylerythritol kinase; DXR, 1-deoxy-D-xylulose 5-phosphate reductoisomerase; DXS, 1-deoxy-D-xylulose 5-phosphate synthase; HDR, 4-hydroxy-3-methylbut-2-en-1-yl diphosphate reductase; HDS, 4-hydroxy-3-methylbut-2-en-1-yl diphosphate synthase; HMGS, 3-hydroxy-3-methylglutaryl-CoA synthase; HMGR, 3-hydroxy-3-methylglutaryl-CoA reductase; IDI, isopentenyl diphosphate:dimethylallyl diphosphate isomerase; MCT, 2-C-methyl-D-erythritol 4-phosphate cytidylyltransferase; MDS, 2-C-methyl-D-erythritol 2,4-cyclodiphosphate synthase; MVD, mevalonate-5-diphosphate decarboxylase; MVK, mevalonate kinase; PMK, 5-phosphomevalonate kinase.

In silico analysis suggested that algae have enzymes for terpenoid biosynthesis which resembles with those in land plants (Lohr et al., 2012; Sasso et al., 2011). On the other hand, MVA pathway genes are often lost in some algae although such algal unique system remains to be understood in future work (Lohr et al., 2012; Sasso et al., 2011). EST analysis on race A and B of *B. braunii* revealed that MEP pathway genes are actively expressed but not MVA pathway genes. Some secondary symbiotic algae possess a mosaic MVA pathway which involves enzymes originated from both primary and secondary hosts (Lohr et al., 2012). MEP pathway connected with the pentose phosphate pathway was observed in a species of cyanobacteria although detailed mechanism is still not clear (Poliquin et al., 2004).

IPP and DMAPP are metabolically conjugated by condensation and dephosphorylation to produce polyterpenoid. No ATP or reducing power is required when isoprene units get into condensation reaction by head-to-tail conjunction (e.g. farnesyl pyrophosphate formation while 1NADPH is required in case of tail-to-tail condensation (e.g. squalene formation). Polyterpenoid is individually or cooperatively synthesized either in the cytosol, plastid or mitochondrion (Bouvier et al., 2005; Joyard et al., 2009; Lohr et al., 2012). Each terpenoid condensation enzyme has particular specific to isoprene molecules such as mono-/sesqui-/di-/tri-/tetra-terpene, respectively. Unlikely land plant, the biosynthesis of isoprene in green macroalgae proceeds via MEP pathway in the plastid (Lohr et al., 2012) and it functions to produce special natural products such as bioactive halogenated poly terpenoid (Moore, 2006). Vanadium bromoperoxidase is an abundant enzyme to produce brominated products in all classes of marine macroalgae and vanadium iodoperoxidase is also identified and characterized (Moore, 2006). However, vanadium chloroperoxidase is not yet identified despite the abundance of chlorinated compounds in algae. These haloperoxidases catalyze both halogenation and cyclization to produce various unique halogenated cyclic terpenoid in macrolagae, but such unique isoprene condensing enzyme is not yet identified in microalgae (Sasso et al., 2011).

3. Hydrocarbon biosynthesis in algae

Table 2 shows a list of lipids and hydrocarbons which can be candidates for renewable energy sources. These compounds are metabolites derived from the elemental lipids shown in Table

Name	Structure	Function
Odd-chain fattyhydrocarbon	Hydrocarbonfrom fatty acid (C_{2n-1})	Unknown
Wax ester	Ester of fatty acid & fatty alcohol	Cuticle component
Alkenones	*trans*-unsaturated-/straight-carbon chainwith ketone group	Storage lipid?
Heterocyst glycolipid	Alcohol-/ketone-glycoside	Cell wall component
Even-chain fatty hydrocarbon	Hydrocarbon from fatty acid (C2n)	Unknown
Olefinichydrocarbon	Hydrocarbon from fatty acidwith multiple double bonds	Unknown
Terpenoid hydrocarbon	Hydrocarbon from terpenoid	Unknown

Table 2. Lipids and hydrocarbons for renewable energy source

1. Their pool sizes of metabolites in cells and production capability largely varies among species and even strains of a certain species. The most extreme example can be seen in a colonial oil-producing green alga *B. braunii*: a certain strain dominantly produces odd-chain fatty hydrocarbons while another produces terpenoid derived hydrocarbons and those strains are classified as race A, B and L.

Microalgal species/strains nominated as oil-producer are simply classified into three groups by their main products: namely, hydrocarbons, TG/free fatty acids and the other lipids. For example, bacteria (Schirmer et al., 2010), a unicellular green alga *Pseudochoricystis ellipsoidea* (Satoh et al., 2010), a colonial green alga *B. braunii* race A (Yoshida et al., 2012) accumulate fatty hydrocarbons. According to a recent review (Yoshida et al., 2012), *B. braunii* race B and L and a heterotrophic Labyrinthulea *Aurantiochytrium sp.* produce terpenoid-derived hydrocarbons. Green algae *C. reinhardtii*, *Chlorella vulgaris*, *Chlorella protothecoides*, diatoms, *Nannochloropsis* spp. usually produce and accumulate TG and free fatty acids. *E. gracilis* Z produces wax ester and accumulate it in the cell (Inui et al., 1982). Haptophytes, but only five species, produce long-chain ketones, called as "alkenones" (Laws et al., 2001; Eltgroth et al., 2005; Toney et al., 2012). Red and green macroalgal species, *Gracilaria salicornia* and *Ulva lactuca*, respectively, contain only limited amounts of lipids including arachidonic acid and Docosapentaenoic acid ranging 1~2% of dry weight (Tabarsa et al., 2012). Recent big news was that genetic transformation by the particle-gun bombardment method was successfully established in alliphatic hydrocarbon producing photosynthetic eukaryote *Pseudochoricystis ellipsoidea* (Trebouxiophyceae) (Imamura et al., 2012). This achievement may open a new trail toward the genetic manipulation of metabolism for algal biofuel production.

3.1. Fatty hydrocarbons and the other fatty acid derivates

Alliphatic carbon-chain is a ubiquitous structure which exists in the molecules produced via fatty acid biosynthesis in organisms. In this part we introduce some fatty acid derivatives and their molecular properties and biosynthetic pathways.

3.1.1. Odd-chain fatty hydrocarbon

Bacteria, microalgae and land plants produce odd-chain hydrocarbons (Řezanka & Sigler, 2009; Tornabene, 1981). Plant wax constitutes of odd-chain hydrocarbons without any branching, namely fatty hydrocarbons (Jetter & Kunst, 2008). This type of hydrocarbons is suggested to be produced via the decarbonylation pathway (Jetter & Kunst, 2008; Schirmer et al., 2010). First, acyl-CoA is reduced to form fatty aldehyde using 1NADPH as a reductant cofactor (Schirmer et al., 2010; Willis et al., 2011). In pea, the decarbonylation reaction is catalyzed by a membrane-bound enzyme, fatty acyl-CoA reductase (Cheesbrough & kolattukudy, 1984; Vioque & Kolattukudy, 1997) which is also present in the race A of *B. braunii* (Wang & Kolattukudy, 1995).

A cyanobacterium *Synechococcus elongatus* PCC7942 has fatty acyl-ACP reductase which prefers acyl-ACP, not acyl-CoA, as substrate (Schirmer et al., 2010). Bacterial gene for the aldehyde-forming fatty acyl-CoA reductase was identified, but not eukaryotic gene yet. Fatty aldehyde is decarbonylated to form odd-chain fatty hydrocarbons with a release of carbon monoxide. Such aldehyde decarbonylase activity was successfully determined in land plants (Jetter & Kunst, 2008), a colonial green alga *B. braunii*, and bacteria (Schirmer et al., 2010), but its gene was identified only in bacteria at present (Schirmer et al., 2010). Using microsomal preparations of *B. braunii*, alkane was proved to be synthesized from fatty acid and aldehyde only under anaerobic conditions (Dennis & Kolattukudy, 1991). The aldehyde decarbonylase is a cobalt porphyrin enzyme which was suggested to locate in the microsomes (Dennis & Kolattukudy, 1992). However, it is not known yet how intracellular hydrocarbons are transferred to extracellular space in race A of *B. braunii* (Casadevall et al., 1985; Largeau et al., 1980; Templier et al., 1992). In addition, an enzyme for the synthesis of odd-chain fatty hydrocarbons is unidentified yet since the product of the enzyme was different from native hydrocarbons of *B. braunii* by lacking terminal double bond seen in the natural product. Although decarbonylation reaction does not require any reductant, the reaction of *in vitro* decarbonylation from octadecanal to heptadecane was observed only in the presence of ferredoxin, ferredoxin reductase and NADPH. This result suggest essential requirement of reductant for aldehyde decarbonylase to exhibit activity in the cyanobacterium *Nostoc punctiforme PCC73102* (Schirmer et al., 2010).

3.1.2. Wax ester

Wax esters consist of fatty acids (acyl-CoAs *in vivo*) and fatty alcohols and are one of components of plant cuticles or seed oils (Jetter & Kunst, 2008; Kunst & Samuels, 2009). *E. gracilis*
produces wax esters under either heterotrophic or anaerobic conditions (Inui et al., 1982). Wax
esters are produced by condensation of fatty acids and primary alcohols which are synthesized
from acyl-CoAs (see also *3.1.5. Even-chain fatty hydrocarbon*). Wax ester synthase/acyl-CoA:di-
acylglycerol acyltransferase (WSD1) is a condensation enzyme identified in *A. thaliana* (Li et
al., 2008). In *E. gracilis*, both NADH-requiring alcohol-forming fatty acyl-CoA reductase
(EgFAR) and wax synthase (EgWS) are already identified and the sequences of those genes
showed similarity with those of jojoba (land plant) (Teerawanichpan & Qiu, 2010). EgWS
utilizes a broad range of fatty acyl-CoAs and fatty alcohols as substrates with the preference
towards myristic acid and palmitoleyl alcohol (Teerawanichpan & Qiu, 2010). Those substrates
are suggested to be produced via various fatty acid synthetic pathways in this alga (see 2.1.
Fatty acid biosynthesis).

3.1.3. Alkenones

At least five species of haptophyceae (*Chrysotile lamellose, Emiliania. huxleyi, Gephyrocapsa
oceanica, Isochrysis galbana, Pseudoisochrysis* sp.) were reported to accumulate highly alkenes,
alkenoates (PUVLCFA-methyl/ethyl esters) and alkenones (PUVLC ketones) (Eltgroth et al.,
2005; Laws et al., 2001; Toney et al., 2012). In this section, we call those compounds "alkenones"
for our convenience. "Alkenones" are discriminated from the other lipids by 2 to 4 *trans*-carbon
double bonds stocked under low temperature conditions, and by its remarkable length (about
C_{38}). "Alkenones" are suggested to be synthesized near the chloroplast, and then stored in the
intracellular lipid body (Eltgroth et al., 2005). Mechanisms for alkenone biosynthesis and its
desaturation are not known yet. Alkenones may act as a storage lipid since behavior of
alkenones was shown to be similar to TG in the other algae functioning storage lipids (Eltgroth
et al., 2005).

3.1.4. Heterocyst glycolipid

The heterocyst of cyanobacterium *Anabaena* sp. PCC 7120 is surrounded by cell wall involving
unique glycolipids as a component (heterocyst glycolipid: HGL) (Bauersachs et al., 2009; Awai
& Wolk, 2007). HGL may be a good source of biofuel since such sugar-conjugated molecule
(aglycon) is changed to C_{26-28} fatty polyhydric alcohol or ketone by removing sugar residues.
HGL is known to be biosynthesized from fatty polyhydric alcohol via an unknown pathway
which may be composed of fatty acid and/or polyketide synthetic enzymes (Awai & Wolk,
2007; Bauersachs et al., 2009) as its mechanisms including transportation of HGL during
heterocyst development has been well studied (Bauersachs et al., 2009; Nicolaisen et al., 2009).
Fatty polyhydric alcohol is metabolized to glycoside by the catalysis of glycosyltransferase
(HGL formation protein: HglT) (Awai & Wolk, 2007).

3.1.5. Even-chain fatty hydrocarbon

The bacterium *Vibrio furnissi* M1 was suggested to synthesize even-chain fatty hydrocarbons by the reduction of fatty acids through primary fatty alcohols (Park, 2005). Primary fatty alcohols are commonly synthesized from acyl-CoA by either one-step or two-step reductions. The one-step reduction is catalyzed by "alcohol-forming" fatty acyl-CoA reductase of which gene was already identified in bacteria (Willis et al., 2011) and various eukaryotes such as land plants and *Euglena* (Teerawanichpan & Qiu, 2010; Vioque & Kolattukudy, 1997). The two-step reduction reactions are catalyzed by two enzymes: namely, "aldehyde-forming" fatty acyl-CoA reductase and then fatty aldehyde reductase of which gene is identified only in bacteria (Wahlen et al., 2009). In the two-step reductive reactions, 2NADPH are necessary to reduce acyl-CoA to primary fatty alcohol. However, mechanism for consequential reactions to form even-chain fatty hydrocarbon by reducing fatty alcohol has not been identified yet (Park, 2005) but elucidation of the mechanism and knowledge on its distribution among species are very important for the progress of biofuel production.

3.1.6. Olefinic hydrocarbon

Olefinic hydrocarbons contain many unsaturated bonds in the molecule. *Ole* (olefin) *ABCD* is a gene set harbored in bacteria and catalyzes the production of an olefinic hydrocarbon molecule by head-to-head condensation reaction (Beller et al., 2010; Sukovich et al., 2010a; Sukovich et al., 2010b). The head-to-head condensation reaction is summarized as follows: 1) a carboxyl group (R_1-COOH) of fatty acid X is reduced to a carbonyl group (R_1-CHO) ; 2) the carbonyl (or thioester (R_1-CO-S-CoA with carbon number x)) group of X reacts with an alpha-carbon of another fatty acid Y (R_2- H_2-COOH with carbon number y) to combine by forming a hydroxyl group (-OH) and releasing one carbon molecule; 3) the hydroxyl group changes into a carbon double bond (R_1-C=- R_2); 4) the reduction of the carbon double bond results to produce hydrocarbon molecule with carbon number of [x + y – 1] (Albro & Dittmer, 1969). So, acyl-CoA can react with an alpha-carbon of fatty acid instead of carbonyl unit of the aldehyde (Sukovich et al., 2010a). As the gene set *ole*ABCD was recently identified, further kinetic and phylogenic studies can be performed.

3.2. Hydrocarbon biosynthesis from isoprene: A novel terpenoid hydrocarbon biosynthetic pathway in a colonial green alga *Botryococcus braunii* (race B)

A colonial green alga *B. braunii* produces hydrocarbon up to 75% of its dry weight (Yoshida et al., 2012). *B. braunii* is generally classified into three races by its products: race A produces odd-chain fatty hydrocarbons (alkadiene, alkatriene); race B (triterpene) and L (tetraterpene) produces hydrocarbons from isoprene, namely terpenoid hydrocarbons (Yoshida et al., 2012). *B. braunii* cells are botryoidally-aggregated by a network of covalently and/or non-covalently conjugated hydrocarbon molecules to build colony structure: namely, as structural hydrocarbons (Metzger et al., 1993; Metzger et al., 2007; Metzger et al., 2008). Extracellular space is filled with liquid hydrocarbons being discriminated from the structural hydrocarbons (Weiss et al., 2012). Hydrocarbon biosynthesis in the race A is partially understood but no recent progress was reported (see 3.1.1. *Odd-chain fatty hydrocarbon*). Properties of the race B hydrocarbons are mostly well-known although the race L is still enigmatic.

The race B hydrocarbons are methylsqualene and botryococcene which are specifically produced by *B. braunii*. The amount of the race B hydrocarbons accumulated in the colony did not decrease in the dark, suggesting that extracellular hydrocarbons are a physiologically inactive storage compounds (Sakamoto et al., 2012). C_5-isoprene unit for synthesizing hydrocarbon molecules is supplied via MEP pathway in *B. braunii* (Sato et al., 2003) (see 2.4. *Terpenoid biosynthesis*). The enzyme 1-deoxy-D-xylulose-5-phosphate synthase catalyzes the first step of the MEP pathway and its three distinct isoforms are well-characterized (Matsushima et al., 2012). ESTs of most genes in the MEP pathway, but never the MVA pathway, are already obtained by transcriptome analysis although those are not completely cloned and characterized yet (Ioki et al., 2012a; Ioki et al., 2012b). Botryococcene is known to be synthesized via a similar pathway to squalene biosynthesis besides of cleavage manner of cyclopropane base in a precursor, presqualene pyrophosphate (PSPP) (Metzger & Largeau, 2005; Banerjee et al., 2002). Squalene synthase, which is widely observed in eukaryotes, catalyzes two step reactions: namely, (1) condensation of two farnesyldiphosphates (FPP, triterpene) to form PSPP and (2) dephospholylation, cyclopropane cleavage, carbon bond reformation and NADPH-dependent reduction of PSPP to produce squalene (Okada, 2012; Jennings et al., 1991). These two reactions are catalyzed at the domain 3 and 4 for 1st reaction and the domain 5 for 2nd reaction of six domains sequentially (Gu et al., 1998). *Botoryococcus* squalene synthase (BSS) was isolated by methods of homology screening of cDNA library based on its putative homologous sequence obtained by the degenerate PCR method to already known squalene synthase (Okada et al., 2000). BSS expressed in *E. coli* produced only squalene but not botryococcene (Okada et al., 2000). Squalene synthase like-1 (SSL-1) is a protein which is homologous to BSS possessing quite different amino acid sequence at the domain 5 (Niehaus et al., 2011). The purified SSL-1 did not function to produce neither botryococcene nor squalene *in vitro*. Instead, SSL-1 stimulates botryococcene production when it was added to *B. braunii* cell extracts. Furthermore, over expression of SSL-1 in FPP accumulating yeast resulted in the accumulation of presqualene alcohol (dephospholylated PSPP) in the cells. These results suggested that SSL-1 functions as PSPP synthase but subsequent reactions for squalene synthesis are catalyzed by other enzymes (Niehaus et al., 2011). SSL-2 and SSL-3 are proteins which are also homologous to BSS (Niehaus et al., 2011). SSL-2 catalyzes two NADPH-dependent reactions for the production of squalene from PSPP and bisfarnesyl ether with a little squalene from FPP. SSL-3 catalyzes a NADPH-dependent reaction to form botryococcene from PSPP. These results suggest that the pathway for terpenoid hydrocarbon biosynthesis in *B. braunii* is quite unique although cooperation of BSS, SSL-1, SSL-2, and SSL-3 is still unclear (Niehaus et al., 2011). After the synthesis, both squalene and botryococcene are subsequently methylated but the number of methylation is variable. The name of botryococcene was originally designated to methylated botryococcene but now it is used for both compounds. The methyl group is transferred from S-adenosyl methionine by triterpene methyltransferases (TMTs) although completely methylated (tetra-methylated) squalene and botryococcene is not yet produced *in vitro* (Niehaus et al., 2012).

4. Carbon flow and energy balance in lipid and hydrocarbon biosynthetic pathways

In Table 3 and Fig. 1, lipid and hydrocarbon biosynthetic pathways are summarized. All hydrocarbons are produced from precursors (namely acyl-ACP or IPP/DMAPP) which are produced from three primary metabolites; acetyl-CoA, pyruvate and GAP. GAP should be the primary metabolite during photosynthesis and transported into the cytosol. Then acetyl-CoA and pyruvate are sequentially produced from GAP in the glycolysis. On the other hand, acetyl-CoA is primarily produced by the degradation of various lipids via β-, α-, and ω- oxidation (Graham & Eastmond, 2002). Any pathway for hydrocarbon production includes decarboxylation of carbon chain supplied as substrate and consumption of ATP and reducing power (see the MVA/MEP pathway and glycolysis in Fig. 1). GAP production mostly depends on carbon fixation rate by the photosynthetic C_3 cycle and the process seems to be the most effective limiting factor for hydrocarbon production. Gene expression level for fatty acid synthesis is relatively higher in race A (fatty hydrocarbon) than race B (terpenoid hydrocarbon) in *B. braunii* while the expression of isoprene synthetic genes showed opposite trend (Ioki et al., 2012c). Transcriptional regulation network of fatty acid metabolism is well studied in *E. coli* (Fujita et al., 2007) but not yet in the other organisms. Carbon allocation into lipids is known to be affected by environmental change via metabolic regulation; e.g. wax ester fermentation *in E. gracilis* under anaerobic conditions (Tucci et al., 2010), TG accumulation in algae under nutrient deficiency such as nitrogen (Hu et al., 2008; Miao & Wu, 2006) and under cold-stress (Li et al., 2011; Renaud et al., 1995). On the other hand, such environmental changes do not affect carbon allocation into hydrocarbon in *B. braunii* (Baba et al., 2012b; Metzger & Largeau, 2005; Sakamoto et al., 2012).

Supply of inorganic and organic carbon sources, nutrient deficiency and low-temperature are empirically known to be stimulating factors for lipid biosynthesis. Enrichment of CO_2 as inorganic carbon source stimulated lipid biosynthesis and cell growth by accelerating photosynthetic carbon fixation in microalgae (Kumar et al., 2010). Neutral lipid production and accumulation was strongly accelerated in the presence of exogenous organic carbon source by accompanying with abolishing chlorophylls in a unicellular green alga *Chlorella protothecoides* (Miao & Wu, 2006). Nitrogen is the most effective factor for changing carbon/nitrogen metabolism and stimulates neutral lipid accumulation under N-deficient conditions (Hu et al., 2008). However, nitrogen deficiency diminishes whole cellular productivity and therefore the metabolic regulation to stimulating lipid biosynthesis does not always result in the increase in gross productivity of lipids. Lipid accumulation under cold (Li et al., 2011; Renaud et al., 1995) or other stress conditions (e.g., nutrient limitation at the stationary growth phase) also induce change in metabolisms. Either cold or high-salinity stress (Lu et al., 2009) stimulates lipid desaturation catalyzed by various lipid desaturases. Desaturation degree influences properties of lipids such as melting point, reactivity, odor, degradability and so on. Lipid biosynthesis by microalgae was shown to be affected by changing wavelength, namely stimulation of lipid production by red light, via the modulation of nitrogen and carbon metabolism in the cells (Miyachi et al., 1978).

Name	Reactions	Products	Notes
Carbon chain biosynthesis			
Fatty acid biosynthesis (C2)	acyl(n)-ACP + acetyl-CoA (+ CO_2) + ATP + 2NADPH + $2H^+ \rightarrow$ acyl(n +2)-ACP (+ CO_2) + H_2O + CoA + ADP + Pi + $2NADP^+$	Fatty acid(C4~18)	Reference pathway
Fatty acid biosynthesis (C2)	acyl(n)-CoA + acetyl-CoA + 2NADPH + $2H^+ \rightarrow$ acyl(n+2)-CoA + H_2O + CoA + $2NADP^+$	Fatty acid(C4~16)	*E. gracilis* mitochondria
Fatty acid biosynthesis (C2)	acyl(n)-CoA + acetyl-CoA + 2NADPH + $2H^+ \rightarrow$ acyl(n+2)-CoA + H_2O + CoA + $2NADP^+$	Fatty acid(C4~18)	Engineered *Escherichia coli*
MVApathway (C5)	3acetyl-CoA + H_2O + 3ATP + 2NADPH + $2H^+ \rightarrow$ IPP + 3CoA + CO_2 + 3ADP +Pi + $2NADP^+$; IPP $\rightleftharpoons$ DMAPP	Isoprene	Reference pathway
MEP pathway (C5)	pyruvate + GAP + ATP + CTP + NADPH + $4e^-$ + $5H^+ \rightarrow$ IPP (DMAPP) + CO_2 + $2H_2O$ + ADP + CMP + PPi + $NADP^+$	Isoprene	Incomplete about redox
Carbon chain elongation			
Fatty acid elongation (C2)	acyl(n)-CoA + acetyl-CoA (+ CO_2) + ATP + 2NADPH + $2H^+ \rightarrow$ acyl(n +2)-CoA (+ CO_2) + H_2O + CoA + ADP + Pi + $2NADP^+$	Fatty acid(C20~28)	Reference pathway
Fatty acid elongation (C2)	acyl(n)-CoA + acetyl-CoA + 2NADPH + $2H^+ \rightarrow$ acyl(n+2)-CoA + H_2O + CoA + $2NADP^+$	Fatty acid(C6~16)	Human mitochondria
Head-to-tail isoprene condensation (C5)	(n)isoprene $\rightarrow$ Poly terpenoid-PP + (n-1)PPi (n≥2)	Terpenoid	Reference reaction
Head-to-head isoprene condensation (C5)	2 isoprene + NADPH $\rightarrow$ Poly terpenoid + 2PPi + $NADP^+$ + H^+	Hydro-carbon	Reference reaction
Fatty acid (ACP/CoA) reduction			
Fatty aldehyde formation	acyl-CoA + NADPH + H^+ $\rightarrow$ fatty aldehyde + CoA + $NADP^+$	Fatty-aldehyde	Reference reaction
Fatty aldehyde formation	acyl-ACP + NADPH + H^+ $\rightarrow$ fatty aldehyde + ACP + $NADP^+$	Fatty-aldehyde	Cyanobacteria
Alcohol formation	acyl-CoA + 2NADPH + 2H+ $\rightarrow$ fatty alcohol + CoA + $2NADP^+$	Fatty-alcohol	Reference reaction
Alcohol formation	fatty aldehyde + NADPH + H^+ $\rightarrow$ fatty alcohol + $NADP^+$	Fatty-alcohol	Reference reaction
Odd-chain fatty hydrocarbon biosynthesis			
Aldehyde decarbonylation	fatty aldehyde(n) $\rightarrow$ fatty hydrocarbon(n-1) + CO	Hydro-carbon	Reference reaction
Even-chain fatty hydrocarbon biosynthesis			
Alcohol reduction	fatty alcohol (n) $\rightarrow$ fatty hydrocarbon (n)	Hydro-carbon	Incomplete about redox
Olefinic hydrocarbon			
Head-to-head acyl-CoA condensation	2acyl-CoA(n) $\rightarrow$ alkadiene (2n-1) + CO_2 + H_2O + 2CoA	Hydro-carbon	Incompletely understood
Wax ester			
Fatty acid/alcohole sterification	fatty acyl-CoA(X) + fatty alcohol(Y) $\rightarrow$ wax ester (X+Y) + CoA	Wax ester	Reference reaction

Table 3. Carbon flow, consumption of ATP and reducing power in lipid and hydrocarbon biosynthetic pathways

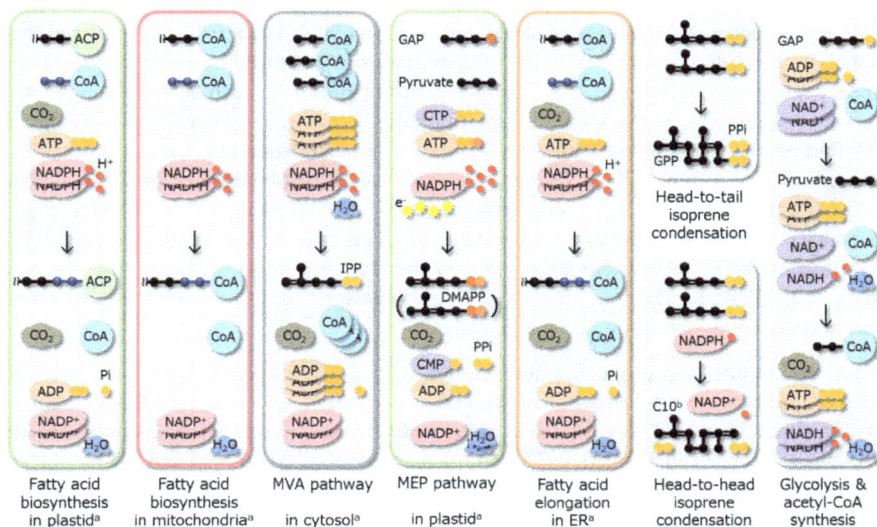

| Fatty acid biosynthesis in plastid[a] | Fatty acid biosynthesis in mitochondria[a] | MVA pathway in cytosol[a] | MEP pathway in plastid[a] | Fatty acid elongation in ER[a] | Head-to-head isoprene condensation | Glycolysis & acetyl-CoA synthesis |

[a]Representative localization of pathways in algae and plants.
[b]C10, mono-terpene is indicated as a model product which is typically synthesized via GPP *in vivo*.

Figure 2. Carbon flow, consumption of ATP and reducing power in lipid biosynthetic pathways. Black- or blue-chained spheres indicate C-C chain. Orange- or Yellow-colored circles indicate Pi in various compounds including IPP, DMAPP, ATP and so on. Difference in color of box-frames indicates difference in localization of pathways. Fatty acid biosynthesis in engineered *E. coli* and fatty acid (~C_4) elongation in human mitochondria are not shown here since those reactions are same as those in mitochondria of *E. gracilis*; namely, reversal pathway of β-oxidation. Detailed information is written in the text. GPP, geranyl diphosphate.

5. Future perspective of algal lipid biosynthetic pathways

Recent *in silico* analysis suggested the presence of some unique lipid-metabolic pathways in algae although those are not characterized yet (e.g., see Table3 and 3. *Hydrocarbon biosynthesis in algae*). Therefore, research on algal lipid biosynthetic pathways should be unavoidably worthy task to increase industrial algal-oil production. Study on lipid metabolism is also beneficial for useful and biologically active organic material production to achieve the invention of manufactural lipid synthesis. One of such good examples is the manipulation of β-oxidation to proceed for reversal direction in *E. coli* (Dellomonaco et al., 2011), the construction of recursive "+1" pathway by genetic engineering in contrast to "+2" fatty acid and "+5" isoprene pathways (Marcheschi et al., 2012). Increase in lipid production by eukaryotic algae strongly can be achieved by metabolic manipulation by controlling strict redox status and subcellular compartment of metabolisms.

Genetic engineering in eukaryotic algae is important technology to be established although it still is quite challenging (Gong et al., 2011; Radakovits et al., 2010). It is highly expected that algal oil is efficiently produced with high purity since it is produced by enzymatic reactions in homogenous productive cells. So, characteristics of products, such as chain length and number of double bond in the molecule, can be modified by genetic engineering (Gong & Jiang, 2011; Radakovits et al., 2010). Further, facilitation of lipid extraction (e.g. lipid auto-secretion from cells to the medium) (Cho & Cronan, 1995; Liu et al., 2010; Michinaka et al., 2003; Nojima et al., 1999) and cell precipitation control (Kawano et al., 2011)) are important to be improved since such processes consume vast energy at industrial process of production. It is noteworthy that direct extraction of oil from *B. braunii* was already achieved to reduce energy and cost (Frenz et al., 1989; Kanda et al., in press). In cyanobacteria, trial to generate hydrocarbon tolerant species was just started (Liu et al., 2012).

Finding of limiting step in whole photosynthetic CO_2 fixation process is also important to increase lipid productivity. Algae have evolved by developing ability to facilitate the utilization of ambient level of CO_2 by the action of innate CO_2 concentrating mechanisms (Giordano et al., 2005; Raven, 2010). Exogenous CO_2 supplementation recovers cells from CO_2-limitation when cells are exposed such conditions within few hours. However, the photosynthetic activity quickly changes to optimize their ability to exposed conditions since algal cells possess ability to adapt/acclimate to environmental change. The maximal carbon fixation rate and high-CO_2 tolerance are highly depend on microalgal species/strain and therefore CO_2-enrichment is not so beneficial for the improvement of cost and energy performance of microalgal production (Baba & Shiraiwa, 2012). Further investigation is necessary to produce newly-engineered algal cells which exhibit high and efficient CO_2-utilization and -fixation ability with enhanced photosynthesis and lipid productivity.

Acknowledgements

This work was financially supported, in part, by a Grant-in-Aid for Scientific Research from the Core Research of Evolutional Science & Technology (CREST) program from the Japan Science and Technology Agency (JST) (to YS).

Author details

Masato Baba[1,2] and Yoshihiro Shiraiwa[1,2]

1 Faculty of Life and Environmental Sciences, University of Tsukuba, Tsukuba, Ibaraki, Japan

2 CREST, JST, Japan

References

[1] Albro, P. W, & Dittmer, J. C. (1969). The Biochemistry of Long-chain Nonisoprenoid Hydrocarbons. III. The Metabolic Relationship of Long-chain Fatty Acids and Hydrocarbons and Other Aspects of Hydrocarbon Metabolism in Sarcina lutea. Biochemistry,May 1969), 0006-2960, 8(5), 1913-1918.

[2] Alvarez, H. M, & Steinbüchel, A. (2002). Triacylglycerols in Prokaryotic Microorganisms. Applied microbiology and biotechnology, December 2002), 0175-7598, 60(4), 367-376.

[3] Awai, K, & Wolk, C. P. (2007). Identification of the Glycosyl Transferase Required for Synthesis of the Principal Glycolipid Characteristic of Heterocysts of Anabaena Sp. Strain PCC 7120. FEMS Microbiology Letters, November 2007), 0378-1097, 226, 98-102.

[4] Baba, M, & Shiraiwa, Y. (2012). High-CO_2 Response Mechanisms in Microalgae, In: Advances in Photosynthesis- Fundamental Aspects-. Edited by Najafpour, M., In-Tech, 978-9-53307-928-8Rijeka, Croatia.

[5] Baba, M, Ioki, M, Nakajima, N, Shiraiwa, Y, & Watanabe, M. M. (2012a). Transcriptome Analysis of an Oil-rich Race A Strain of Botryococcus Braunii (BOT-88-2) by de novo Assembly of Pyrosequencing cDNA Reads. Bioresource technology, April 2012), 0960-8524, 109, 282-286.

[6] Baba, M, Kikuta, F, Suzuki, I, Watanabe, M. M, & Shiraiwa, Y. (2012b). Wavelength Specificity of Growth, Photosynthesis, and Hydrocarbon Production in the Oil-producing Green Alga Botryococcus braunii. Bioresource technology, May 2012), 0960-8524, 109, 266-270.

[7] Banerjee, A, Sharma, R, Chisti, Y, & Banerjee, U. C. (2002). Botryococcus braunii: A Renewable Source of Hydrocarbons and Other Chemicals. Critical reviews in biotechnology, 0738-8551, 22(3), 245-279.

[8] Bauersachs, T, Compaoré, J, Hopmans, E. C, Stal, L. J, Schouten, S, & Sinninghe, J. S. (2009). Phytochemistry Distribution of Heterocyst Glycolipids in Cyanobacteria. Phytochemistry, September 2009), 0031-9422, 70(17-18), 2034-2039.

[9] Beller, H. R, Goh, E, & Keasling, J. D. (2010). Genes Involved in Long-Chain Alkene Biosynthesis in Micrococcus luteus. Applied and environmental microbiology, February 2010), 0099-2240, 76(4), 1212-1223.

[10] Bouvier, F, Rahier, A, & Camara, B. (2005). Biogenesis, Molecular Regulation and Function of Plant Isoprenoids. Progress in lipid research, November 2005), 0163-7827, 44(6), 357-429.

[11] Casadevall, E, Dif, D, Largeau, C, Gudin, C, Chaumont, D, & Desanti, O. (1985). Studies on Batch and Continuous Cultures of Botryococcus braunii: Hydrocarbon

Production in Relation to Physiological State, Cell Ultrastructure, and Phosphate Nutrition. Biotechnology and bioengineering, March 1985), 0006-3592, 27(3), 286-295.

[12] Chan, D. I, & Vogel, H. J. (2010). Current Understanding of Fatty Acid Biosynthesis and the Acyl Carrier Protein. The Biochemical journal, August 2010), 0264-6021, 430(1), 1-19.

[13] Cheesbrough, M, & Kolattukudy, P. E. (1984). Particulate Preparation from Pisum sativum. Proceedings of the national academy of sciences of the United States of America, November 1984), 0027-8424, 81, 6613-6617.

[14] Cho, H, & Cronan, J. E. Jr. ((1995). Defective Export of a Periplasmic Enzyme Disrupts Regulation of Fatty Acid Synthesis. The Journal of biological chemistry, March 1995), 0021-9258, 270(9), 4216-4219.

[15] Dellomonaco, C, Clomburg, J. M, Miller, E. N, & Gonzalez, R. (2011). Engineered Reversal of the β-oxidation Cycle for the Synthesis of Fuels and Chemicals. Nature, August 2011), 0028-0836, 476(7360), 355-359.

[16] Dennis, M. W, & Kolattukudy, P. E. (1991). Alkane Biosynthesis by Decarbonylation of Aldehyde Catalyzed by a Microsomal Preparation from Botryococcus braunii. Archives of biochemistry and biophysics, June 1991), 0003-9861, 287(2), 268-275.

[17] Dennis, M, & Kolattukudy, P. E. Enzyme Converts a Fatty Aldehyde to a Hydrocarbon and CO. Proceedings of the national academy of sciences of the United States of America, June 1992), 0027-8424, 89(12), 5306-5310.

[18] Eltgroth, M. L, Watwood, R. L, & Gordon, V. (2005). Production and Cellular Localization of Neutral Long-Chain Lipids in the Haptophyte Algae Isochrysis galbana and Emiliania huxleyi. Journal of phycology, 0022-3646, 41, 1000-1009.

[19] Frenz, J, Largeau, C, Casadevall, E, Kollerup, F, & Daugulis, A. J. (1989). Hydrocarbon Recovery and Biocompatibility of Solvents for Extraction from Cultures of Botryococcus braunii. Biotechnology and bioengineering, September 1989), 0006-3592, 34(6), 755-762.

[20] Fujita, Y, Matsuoka, H, & Hirooka, K. (2007). Regulation of Fatty Acid Metabolism in Bacteria. Molecular microbiology, November 2007), 0095-0382X, 66(4), 829-839.

[21] Giordano, M, Beardall, J, & Raven, J. A. (2005). CO_2 Concentrating Mechanisms in Algae: Mechanisms, Environmental Modulation, and Evolution. Annual review of plant biology, June 2005), 1543-5008ISSN: 1543-5008., 56, 99-131.

[22] Gong, Y, & Jiang, M. (2011). Biodiesel Production with Microalgae as Feedstock: From Strains to Biodiesel. Biotechnology letters, July 2011), 0141-5492, 33(7), 1269-1284.

[23] Gong, Y, Hu, H, Gao, Y, Xu, X, & Gao, H. (2011). Microalgae as Platforms for Production of Recombinant Proteins and Valuable Compounds: Progress and Prospects.

Journal of industrial microbiology & biotechnology, December 2011), 1367-5435, 38(12), 1879-1890.

[24] Graham, I. A, & Eastmond, P. J. (2002). Pathways of Straight and Branched Chain Fatty Acid Catabolism in Higher Plants. Progress in lipid research, March 2002), 0163-7827, 41(2), 156-181.

[25] Gu, P, & Ishii, Y. Spencer, T. a & Shechter, I. ((1998). Function-structure Studies and Identification of Three Enzyme Domains Involved in the Catalytic Activity in Rat Hepatic Squalene Synthase. The Journal of biological chemistry, May 1998), 0021-9258, 273(20), 12515-12525.

[26] Hiltunen, J. K, Chen, Z, Haapalainen, A. M, Wierenga, R. K, & Kastaniotis, A. J. (2010). Mitochondrial Fatty Acid Synthesis- An Adopted Set of Enzymes Making a Pathway of Major Importance for the Cellular Metabolism. Progress in lipid research, January 2010), 0163-7827, 49(1), 27-45.

[27] Hlousek-radojcic, A, Evenson, K. J, Jaworski, J. G, & Post-beittenmiller, D. (1998). Fatty Acid Elongation Is Independent of Acyl-Coenzyme A Synthetase Activities in Leek and Brassica napus. Phytochemistry, 0031-9422, 116, 251-258.

[28] Hoffmeister, M, Piotrowski, M, Nowitzki, U, & Martin, W. (2005). Mitochondrial trans-2-enoyl-CoA Reductase of Wax Ester Fermentation from Euglena gracilis Defines a New Family of Enzymes Involved in Lipid Synthesis. The Journal of biological chemistry, February 2005), 0021-9258, 280(6), 4329-4338.

[29] Hu, Q, Sommerfeld, M, Jarvis, E, Ghirardi, M, Posewitz, M, Seibert, M, & Darzins, A. (2008). Microalgal Triacylglycerols as Feedstocks for Biofuel Production: Perspectives and Advances. The Plant journal, May 2008), 0960-7412, 54(4), 621-639.

[30] Hunter, W. N. (2007). The Non-mevalonate Pathway of Isoprenoid Precursor Biosynthesis. The Journal of biological chemistry, July 2007), 0021-9258, 282(30), 21573-21577.

[31] Imamura, S, Hagiwara, D, Suzuki, F, Kurano, N, & Harayama, S. (2012). Genetic Transformation of Pseudochoricystis ellipsoidea, an Aliphatic Hydrocarbon-producing Green Alga. Journal of general and applied microbiology, 0022-1260, 58(1), 1-10.

[32] Inui, H, Miyatake, K, Nakano, Y, & Kitaoka, S. (1982). Wax Ester Fermentation in Euglena gracilis. FEBS Letters, December 1982), 0014-5793, 150(1), 89-93.

[33] Inui, H, Miyatake, K, Nakano, Y, & Kitaoka, S. (1984). Fatty Acid Synthesis in Mitochondria of Euglena gracilis. European journal of biochemistry/FEBS, July 1984), 0014-2956, 142(1), 121-126.

[34] Ioki, M, Baba, M, Nakajima, N, Shiraiwa, Y, & Watanabe, M. M. (2012a). Transcriptome Analysis of an Oil-rich Race B Strain of Botryococcus braunii (BOT-22) by de novo Assembly of Pyrosequencing cDNA Reads. Bioresource technology, April 2012), 0960-8524, 109, 292-296.

[35] Ioki, M, Baba, M, Nakajima, N, Shiraiwa, Y, & Watanabe. M. M. (2012b). Transcrip-
      tome Analysis of an Oil-rich Race B Strain of Botryococcus braunii (BOT-70) by de
      novo Assembly of 5'-end Sequences of Full-length cDNA Clones. Bioresource tech-
      nology, April 2012), 0960-8524, 109, 277-281.

[36] Ioki, M, Baba, M, Bidadi, H, Suzuki, I, Shiraiwa, Y, Watanabe, M. M, & Nakajima, N.
      (2012c). Modes of Hydrocarbon Oil Biosynthesis Revealed by Comparative Gene Ex-
      pression Analysis for Race A and Race B Strains of Botryococcus braunii. Bioresource
      technology, April 2012), 0960-8524, 109, 271-276.

[37] Jenke-kodama, H, & Dittmann, E. (2009). Evolution of Metabolic Diversity: Insights
      from Microbial Polyketide Synthases. Phytochemistry, July 2009), 0031-9422,
      70(15-16), 1858-1866.

[38] Jennings, S. M, Tsay, Y. H, Fisch, T. M, & Robinson, G. W. (1991). Molecular Cloning
      and Characterization of the Yeast Gene for Squalene Synthetase. Proceedings of the
      National Academy of Sciences of the United States of America, July 1991), 0027-8424,
      88(14), 6038-6042.

[39] Jetter, R, & Kunst, L. (2008). Plant Surface Lipid Biosynthetic Pathways and Their
      Utility for Metabolic Engineering of Waxes and Hydrocarbon Biofuels. The Plant
      journal: for cell and molecular biology, May 2008), 0960-7412, 54(4), 670-683.

[40] John, U, Beszteri, B, & Derelle, E. Van de Peer, Y.; Read, B.; Moreau, H. & Cembella,
      A. ((2008). Novel Insights into Evolution of Protistan Polyketide Synthases Through
      Phylogenomic Analysis. Protist, January 2008), 1434-4610, 159(1), 21-30.

[41] Joyard, J, Ferro, M, Masselon, C, Seigneurin-berny, D, Salvi, D, Garin, J, & Rolland,
      N. (2009). Chloroplast Proteomics and the Compartmentation of Plastidial Isopre-
      noid Biosynthetic Pathways. Molecular Plant, November 2009), 1674-2052, 2(6),
      1154-1180.

[42] Joyard, J, Ferro, M, Masselon, C, Seigneurin-berny, D, Salvi, D, Garin, J, & Rolland,
      N. (2010). Chloroplast Proteomics Highlights the Subcellular Compartmentation of
      Lipid Metabolism. Progress in lipid research, April 2010), 0163-7827, 49(2), 128-158.

[43] Kanda, H, Li, P, Yoshimura, T, & Okada, S. in press) Wet Extraction of Hydrocarbons
      from Botryococcus braunii by Dimethyl Ether as Compared with Dry Extraction by
      Hexane. Fuel,0016-2361

[44] Kaneda, T. (1991). Iso- and Anteiso-fatty Acids in Bacteria: Biosynthesis, Function,
      and Taxonomic Significance. Microbiological reviews, June 1991), 0146-0749, 55(2),
      288-302.

[45] Kawano, Y, Saotome, T, Ochiai, Y, Katayama, M, Narikawa, R, & Ikeuchi, M. (2011).
      Cellulose Accumulation and a Cellulose Synthase Gene Are Responsible for Cell Ag-
      gregation in the Cyanobacterium Thermosynechococcus vulcanus RKN. Plant & cell
      physiology, June 2011), 0032-0781, 52(6), 957-966.

[46] Kumar, A, Ergas, S, Yuan, X, Sahu, A, Zhang, Q, Dewulf, J, Malcata, F. X, & Van Langenhove, H. fixation and biofuel production via microalgae: recent developments and future directions. Trends in Biotechnology, July), 0167-7799, 28(7), 371-380.

[47] Kunst, L, & Samuels, L. (2009). Plant Cuticles Shine: Advances in Wax Biosynthesis and Export. Current opinion in plant biology, October 2009), 1369-5266, 12, 721-727.

[48] Largeau, C, Casadevall, E, & Berkaloff, B. (1980). The Biosynthesis of Long-chain Hydrocarbons in the Green Alga Botryococcus braunii. Phytochemistry, 0031-9422, 19(6), 1081-1085.

[49] Laws, E. A, Popp, B. N, Bidigare, R. R, Riebesell, U, Burkhardt, S, & Wakeham, S. G. (2001). Controls on the Molecular Distribution and Carbon Isotopic Composition of Alkenones in Certain Haptophyte Algae. Geochemistry geophysics geosystems, January 2001), Paper 0025-3227, 2(2000GC000057)

[50] Lee, S. H, Stephens, J. L, Paul, K. S, & Englund, P. T. (2006). Fatty Acid Synthesis by Elongases in Trypanosomes. Cell, August 2006), 0092-8674, 126(4), 691-699.

[51] Li, F, Wu, X, Lam, P, Bird, D, Zheng, H, Samuels, L, Jetter, R, & Kunst, L. (2008). Identification of the Wax Ester Synthase/acyl-coenzyme A: Diacylglycerol Acyltransferase WSD1 Required for Stem Wax Ester Biosynthesis in Arabidopsis. Plant physiology, September 2008), 0032-0889, 148(1), 97-107.

[52] Li, X, Hu, H, & Zhang, Y. (2011). Growth and Lipid Accumulation Properties of a Freshwater Microalga Scenedesmus Sp. Under Different Cultivation Temperature. Bioresource technology, February 2011), 0960-8524, 102(3), 3098-3102.

[53] Liu, X, Brune, D, Vermaas, W, & Curtiss, R. III. ((2010). Production and Secretion of Fatty Acids in Genetically Engineered Cyanobacteria. Proceedings of the National Academy of Sciences of the United States of America, July 2010), 0027-8424, 6803(19)

[54] Liu, J, Chen, L, Wang, J, Qiao, J, & Zhang, W. (2012). Proteomic Analysis Reveals Resistance Mechanism Against Biofuel Hexane in Synechocystis Sp. PCC 6803. Biotechnology for biofuels, January 2012), 1754-6834, 5(1), 68.

[55] Lohr, M, Schwender, J, & Polle, J. E. W. (2012). Isoprenoid Biosynthesis in Eukaryotic Phototrophs: A Spotlight on Algae. Plant science : an international journal of experimental plant biology, April 2012) 0168-9452, 185-186, 9-22.

[56] Lu, Y. D, Chi, X, Yang, Q, Li, Z, Liu, S, Gan, Q, & Qin, S. (2009). Molecular cloning and stress-dependent expression of a gene encoding Delta(12)-fatty acid desaturase in the Antarctic microalga Chlorella vulgaris NJ-7. Extremophiles: life under extreme conditions, November), 1431-0651, 13(6), 875-884.

[57] Marcheschi, R. J, Li, H, Zhang, K, Noey, E. L, Kim, S, Chaubey, A, Houk, K. N, & Liao, J. C. (2012). A Synthetic Recursive "+1" Pathway for Carbon Chain Elongation. ACS chemical biology, April 2012), 1554-8929, 7(4), 689-697.

[58] Matsushima, D, Jenke-kodama, H, Sato, Y, Fukunaga, Y, Sumimoto, K, Kuzuyama, T, Matsunaga, S, & Okada, S. (2012). The Single Cellular Green Microalga Botryococcus braunii, Race B Possesses Three Distinct 1-deoxy-D-xylulose 5-phosphate Synthases. Plant science: an international journal of experimental plant biology, April 2012), 0168-9452, 185-186, 309-320.

[59] Metzger, P, & Largeau, C. (2005). Botryococcus braunii: a Rich Source for Hydrocarbons and Related Ether Lipids. Applied Microbiology, December 2004), 0003-6919, 66, 486-496.

[60] Metzger, P, Pouet, Y, Bischoff, R, & Casadeval, E. (1993). An Aliphatic Polyaldehyde from Botryococcus braunii (A Race). Phytochemistry, 0031-9422, 32(4), 875-883.

[61] Metzger, P, Rager, M, & Largeau, N. C. ((2007). Polyacetals Based on Polymethylsqualene Diols, Precursors of Algaenan in Botryococcus braunii Race B. Organic geochemistry, April 2007), 0146-6380, 38(4), 566-581.

[62] Metzger, P, Rager, M, Fosse, N, & Braunicetals, C. Acetals from Condensation of Macrocyclic Aldehydes and Terpene Diols in Botryococcus braunii. Phytochemistry, September 2008), 0031-9422, 69(12), 2380-2386.

[63] Miao, X, & Wu, Q. (2006). Biodiesel Production from Heterotrophic Microalgal Oil. Bioresource technology, April 2006), 0960-8524, 97(6), 841-846.

[64] Michinaka, Y, Shimauchi, T, Aki, T, Nakajima, T, Kawamoto, S, Shigeta, S, Suzuki, O, & Ono, K. (2003). Extracellular Secretion of Free Fatty Acids by Disruption of a Fatty acyl-CoA Synthetase Gene in Saccharomyces cerevisiae. Journal of bioscience and bioengineering, January 2003), 1347-4421, 95(5), 435-440.

[65] Miyachi, S, Miyachi, S, & Kamiya, A. (1978). Wavelength effects on photosynthetic carbon metabolism in Chlorella. Plant and Cell Physiology, 0032-0781, 19(2), 277-288.

[66] Moore, B. S. (2006). Biosynthesis of Marine Natural Products: Macroorganisms (Part B). Natural product reports, August 2006), 0265-0568, 23(4), 615-629.

[67] Nicolaisen, K, Hahn, A, & Schleiff, E. (2009). The Cell Wall in Heterocyst Formation by Anabaena Sp. PCC 7120. Journal of basic microbiology, February 2009), 0023-3111X., 49, 5-24.

[68] Niehaus, T. D, Okada, S, Devarenne, T. P, Watt, D. S, Sviripa, V, & Chappell, J. (2011). Identification of Unique Mechanisms for Triterpene Biosynthesis in Botryococcus braunii. Proceedings of the national academy of sciences of the United States of America, July 2011), 0027-8424, 108(30), 12260-12265.

[69] Niehaus, T. D, Kinison, S, Okada, S, & Yeo, Y. Bell, S. a; Cui, P.; Devarenne, T. P. & Chappell, J. ((2012). Functional Identification of Triterpene Methyltransferases from Botryococcus braunii Race B. The Journal of biological chemistry, March 2012), 0021-9258, 287(11), 8163-8173.

[70] Nojima, Y, Kibayashi, A, Matsuzaki, H, Hatano, T, & Fukui, S. (1999). Isolation and Characterization of Triacylglycerol-secreting Mutant Strain from Yeast, Saccharomyces cerevisiae. The Journal of general and applied microbiology, February 1999), 0022-1260, 45(1), 1-6.

[71] Okada, S, Devarenne, T. P, & Chappell, J. (2000). Molecular Characterization of Squalene Synthase from the Green Microalga Botryococcus braunii, Race B. Archives of biochemistry and biophysics, January 2000), 0003-9861, 373(2), 307-317.

[72] Okada, S. (2012). Chapter 7 Elucidation of hydrocarbon synthetic pathway in Botryococcus braunii, In: Technology of Microalgal Energy Production and its Business Prospect. Edited by Takeyama, H. CMC, 978-4-78130-657-5Osaka, Japan.

[73] Park, M. (2005). New Pathway for Long-Chain n-Alkane Synthesis via 1-Alcohol in Vibrio furnissii M1. Journal of bacteriology, February 2005), 0021-9193, 187(4), 1426-1429.

[74] Poliquin, K, Ershov, Y. V, & Cunningham, F. X. Jr; Woreta, T. T.; Gantt, R. R. & Gantt, E. & ((2004). Inactivation of sll1556 in Synechocystis Strain PCC 6803 Impairs Isoprenoid Biosynthesis from Pentose Phosphate Cycle Substrates In Vitro. Journal of bacteriology, July 2004), 0021-9193, 186(14), 4685-4693.

[75] Portevin, D. De Sousa-D'Auria, C.; Houssin, C.; Grimaldi, C.; Chami, M.; Daffé, M. & Guilhot, C. ((2004). A Polyketide Synthase Catalyzes the Last Condensation Step of Mycolic Acid Biosynthesis in Mycobacteria and Related Organisms. Proceedings of the national academy of sciences of the United States of America, January 2004), 0027-8424, 101(1), 314-319.

[76] Radakovits, R, Jinkerson, R. E, Darzins, A, & Posewitz, M. C. (2010). Genetic Engineering of Algae for Enhanced Biofuel Production. Eukaryotic cell, April 2010), 1535-9786, 9(4), 486-501.

[77] Raven, J. A. (2010). Inorganic Carbon Acquisition by Eukaryotic Algae: Four Current Questions. Photosynthesis research, June 2010), 1573-5079ISSN: 0166-8595., 106(1-2), 123-134.

[78] Renaud, S. M, Zhou, H. C, Parry, D. L, Thinh, L, & Woo, K. C. (1995). Effect of Temperature on the Growth, Total Lipid Content and Fatty Acid Composition of Recently Isolated Tropical Microalgae Isochrysis Sp., Nitzschia Closterium, Nitzschia Paleacea, and Commercial Species Isochrysis Sp. (Clone T.ISO). Journal of Applied Phycology, December 1995), 0921-8971, 7(6), 595-602.

[79] Rezanka, T, & Sigler, K. (2009). Progress in Lipid Research Odd-numbered Very-long-chain Fatty Acids from the Microbial, Animal and Plant Kingdoms. Progress in lipid research, March 2009), 0163-7827, 48(3-4), 206-238.

[80] Riekhof, W. R, Sears, B. B, & Benning, C. (2005). Annotation of Genes Involved in Glycerolipid Biosynthesis in Chlamydomonas reinhardtii: Discovery of the Betaine Lipid Synthase BTA1. Eukaryotic cell, February 2005), 1535-9786, 4(2), 242-252.

[81] Sakamoto, K, Baba, M, Suzuki, I, Watanabe, M. M, & Shiraiwa, Y. (2012). Optimization of Light for Growth, Photosynthesis, and Hydrocarbon Production by the Colonial Microalga Botryococcus braunii BOT-22. Bioresource technology, April 2012), 0960-8524, 110, 474-479.

[82] Sasso, S, Pohnert, G, Lohr, M, Mittag, M, & Hertweck, C. (2011). Microalgae in the Post-genomic Era: A Blooming Reservoir for New Natural Products. FEMS microbiology reviews, (September 2011), 0168-6445, 1-25.

[83] Sato, Y, Ito, Y, Okada, S, Murakami, M, & Abe, H. (2003). Biosynthesis of the Triterpenoids, Botryococcenes and Tetramethylsqualene in the B Race of Botryococcus braunii via the Non-mevalonate Pathway. Tetrahedron letters, September 2003), 0040-4039, 44(37), 7035-7037.

[84] Satoh, A, Kato, M, Yamato, K. T, Ikegami, Y, Sekiguchi, H, Kurano, N, & Miyachi, S. (2010). Characterization of the Lipid Accumulation in a New Microalgal Species,. Pseudochoricystis ellipsoidea (Trebouxiophyceae). Journal of the Japan institute of energy, September 2010), 0916-8753, 89(9), 909-913.

[85] Schirmer, A. Rude, M. a; Li, X.; Popova, E. & Del Cardayre, S. B. ((2010). Microbial Biosynthesis of Alkanes. Science, July 2010), 0036-8075, 329(599), 559-562.

[86] Shen, B. (2003). Polyketide Biosynthesis Beyond the Type I, II and III Polyketide Synthase Paradigms. Current Opinion in Chemical Biology, April 2003), 1367-5931, 7(2), 285-295.

[87] Staunton, J, & Weissman, K. J. (2001). Polyketide Biosynthesis: a Millennium Review. Natural product reports, June 2001), 0265-0568, 18(4), 380-416.

[88] Sukovich, D. J, Seffernick, J. L, Richman, J. E, Hunt, K. A, Gralnick, J. A, & Wackett, L. P. (2010a). Structure, Function, and Insights into the Biosynthesis of a Head-to-head Hydrocarbon in Shewanella oneidensis Strain MR-1. Applied and environmental microbiology, June 2010), 0099-2240, 76(12), 3842-3849.

[89] Sukovich, D. J, Seffernick, J. L, Richman, J. E, Gralnick, J. A, & Wackett, L. P. (2010b). Widespread Head-to-Head Hydrocarbon Biosynthesis in Bacteria and Role of OleA. Applied and environmental microbiology, June 2010), 0099-2240, 76(12), 3850-3862.

[90] Tabarsa, M, Rezaei, M, Ramezanpour, Z, & Waaland, J. R. (2012). Chemical Compositions of the Marine Algae Gracilaria salicornia (Rhodophyta) and Ulva lactuca (Chlorophyta) as a Potential Food Source. Journal of the science of food and agriculture, September 2012), 0022-5142, 92(12), 2500-2506.

[91] Teerawanichpan, P, & Qiu, X. (2010). Fatty acyl-CoA Reductase and Wax Synthase from Euglena gracilis in the Biosynthesis of Medium-chain Wax Esters. Lipids, March 2010), 0024-4201, 45(3), 263-273.

[92] Templier, J, Diesendorf, C, Largeau, C, & Casadevall, E. (1992). Metabolism of n-alka-dienes in the A race of Botryococcus braunii. Phytochemistry, 0031-9422, 31(1), 113-120.

[93] Toney, J. L, Theroux, S, Andersen, R. a, Coleman, A, Amaral-zettler, L, & Huang, Y. (2012). Culturing of the First 37:4 Predominant Lacustrine Haptophyte: Geochemical, Biochemical, and Genetic Implications. Geochimica et cosmochimica acta, February 2012), 0016-7037, 78, 51-64.

[94] Tornabene, T. G. (1981). Formation of Hydrocarbons by Bacteria and Algae. Basic life sciences, January 1981), 0036-8075, 18, 421-438.

[95] Tucci, S, Vacula, R, Krajcovic, J, Proksch, P, & Martin, W. (2010). Variability of Wax Ester Fermentation in Natural and Bleached Euglena Gracilis Strains in Response to Oxygen and the Elongase Inhibitor Flufenacet. The Journal of eukaryotic microbiolo-gy, December 2010), 1066-5234, 57(1), 63-69.

[96] Venegas-calerón, M, Sayanova, O, & Napier, J. A. (2010). An Alternative to Fish Oils: Metabolic Engineering of Oil-seed Crops to Produce Omega-3 Long Chain Polyunsa-turated Fatty Acids. Progress in lipid research, April 2010), 0163-7827, 49(2), 108-119.

[97] VerschoorJ. a; Baird, M. S. & Grooten, J. ((2012). Towards Understanding the Func-tional Diversity of Cell Wall Mycolic Acids of Mycobacterium tuberculosis. Progress in lipid research, October 2012), 0163-7827, 51(4), 325-339.

[98] Vioque, J, & Kolattukudy, P. E. (1997). Resolution and Purification of an Aldehyde-generating and an Alcohol-generating Fatty acyl-CoA Reductase from Pea Leaves (Pisum sativum L.). Archives of biochemistry and biophysics, April 1997), 0003-9861, 340(1), 64-72.

[99] Wahlen, B. D, Oswald, W. S, Seefeldt, L. C, & Barney, B. M. (2009). Purification, Char-acterization, and Potential Bacterial Wax Production Role of an NADPH-dependent Fatty Aldehyde Reductase from Marinobacter aquaeolei VT8. Applied and environ-mental microbiology, May 2009), 0099-2240, 75(9), 2758-2764.

[100] Wang, X, & Kolattukudy, P. E. (1995). Solubilization and Purification of Aldehyde-generating Fatty acyl-CoA Reductase from Green Alga Botryococcus braunii. FEBS letters, August 1995), 0014-5793, 370(1-2), 15-18.

[101] Weiss, T. L, Roth, R, Goodson, C, Vitha, S, Black, I, Azadi, P, Rusch, J, Holzenburg, A, Devarenne, T. P, & Goodenough, U. (2012). Colony Organization in the Green Al-ga Botryococcus braunii (Race B) Is Specified by a Complex Extracellular Matrix. Eu-karyotic cell, August 2012), 1535-9786, 11(12), 1424-1440.

[102] Willis, R. M, Wahlen, B. D, Seefeldt, L. C, & Barney, B. M. (2011). Characterization of a Fatty acyl-CoA Reductase from Marinobacter aquaeolei VT8: a Bacterial Enzyme

Catalyzing the Reduction of Fatty acyl-CoA to Fatty Alcohol. Biochemistry, December 2011), 0006-2960, 50(48), 10550-10558.

[103] Yasuno, R, Von Wettstein-knowles, P, & Wada, H. (2004). Identification and Molecular Characterization of the Beta-ketoacyl-[acyl Carrier Protein] Synthase Component of the Arabidopsis Mitochondrial Fatty Acid Synthase. The Journal of biological chemistry, February 2004), 0021-9258, 279(9), 8242-8251.

[104] Yoshida, M, Tanabe, Y, Yonezawa, N, & Watanabe, M. M. (2012). Energy Innovation Potential of Oleaginous Microalgae. Biofuels, November 2012), 1759-7269, 3(6), 761-781.

Permissions

The contributors of this book come from diverse backgrounds, making this book a truly international effort. This book will bring forth new frontiers with its revolutionizing research information and detailed analysis of the nascent developments around the world.

We would like to thank Prof. Emeritus Zvy Dubinsky, for lending his expertise to make the book truly unique. He has played a crucial role in the development of this book. Without his invaluable contribution this book wouldn't have been possible. He has made vital efforts to compile up to date information on the varied aspects of this subject to make this book a valuable addition to the collection of many professionals and students.

This book was conceptualized with the vision of imparting up-to-date information and advanced data in this field. To ensure the same, a matchless editorial board was set up. Every individual on the board went through rigorous rounds of assessment to prove their worth. After which they invested a large part of their time researching and compiling the most relevant data for our readers. Conferences and sessions were held from time to time between the editorial board and the contributing authors to present the data in the most comprehensible form. The editorial team has worked tirelessly to provide valuable and valid information to help people across the globe.

Every chapter published in this book has been scrutinized by our experts. Their significance has been extensively debated. The topics covered herein carry significant findings which will fuel the growth of the discipline. They may even be implemented as practical applications or may be referred to as a beginning point for another development. Chapters in this book were first published by InTech; hereby published with permission under the Creative Commons Attribution License or equivalent.

The editorial board has been involved in producing this book since its inception. They have spent rigorous hours researching and exploring the diverse topics which have resulted in the successful publishing of this book. They have passed on their knowledge of decades through this book. To expedite this challenging task, the publisher supported the team at every step. A small team of assistant editors was also appointed to further simplify the editing procedure and attain best results for the readers.

Our editorial team has been hand-picked from every corner of the world. Their multi-ethnicity adds dynamic inputs to the discussions which result in innovative outcomes. These outcomes are then further discussed with the researchers and contributors who give their valuable feedback and opinion regarding the same. The feedback is then collaborated with the researches and they are edited in a comprehensive manner to aid the understanding of the subject.

Apart from the editorial board, the designing team has also invested a significant amount of their time in understanding the subject and creating the most relevant covers. They scrutinized every image to scout for the most suitable representation of the subject and create an appropriate cover for the book.

The publishing team has been involved in this book since its early stages. They were actively engaged in every process, be it collecting the data, connecting with the contributors or procuring relevant information. The team has been an ardent support to the editorial, designing and production team. Their endless efforts to recruit the best for this project, has resulted in the accomplishment of this book. They are a veteran in the field of academics and their pool of knowledge is as vast as their experience in printing. Their expertise and guidance has proved useful at every step. Their uncompromising quality standards have made this book an exceptional effort. Their encouragement from time to time has been an inspiration for everyone.

The publisher and the editorial board hope that this book will prove to be a valuable piece of knowledge for researchers, students, practitioners and scholars across the globe.

List of Contributors

Snježana Jurić, Lea Vojta and Hrvoje Fulgosi
Division of Molecular Biology, Ruđer Bošković Institute, Zagreb, Croatia

Arthur M. Nonomura and Andrew A. Benson
Scripps Institution of Oceanography, University of California San Diego, La Jolla, California, USA

Masami Kobayashi, Shinya Akutsu, Daiki Fujinuma, Hayato Furukawa, Hirohisa Komatsu and Yuichi Hotota
Institute of Materials Science, University of Tsukuba, Tsukuba, Japan

Yuki Kato and Yoshinori Kuroiwa
Institute of Industrial Science, University of Tokyo, Tokyo, Japan

Tadashi Watanabe
Research Center for Math and Science Education, Organization for Advanced Education, Tokyo University of Science, Tokyo, Japan

Mayumi Ohnishi-Kameyama and Hiroshi Ono
National Food Research Institute, Tsukuba, Japan

Satoshi Ohkubo and Hideaki Miyashita
Graduate School of Human and Environmental Studies, Kyoto University, Kyoto, Japan

David Iluz
The Mina and Everard Goodman Faculty of Life Sciences, Bar-Ilan University, Ramat-Gan, Israel
Dept. of Environmental Science and Agriculture, Beit Berl College, Kfar-Saba, Israel

Zvy Dubinsky
The Mina and Everard Goodman Faculty of Life Sciences, Bar-Ilan University, Ramat-Gan, Israel

Alejandra Matiz, Paulo Tamaso Mioto, Adriana Yepes Mayorga, Luciano Freschi and Helenice Mercier
Botany Department, University of São Paulo, São Paulo, Brazil

Leandro Galon and Clevison L. Giacobbo
Federal University of the Southern Border, Brazil

Germani Concenço
Embrapa Western Region Agriculture, Brazil

Evander A. Ferreira
Federal University of Jequitinhonha e Mucuri Valleys, ANDA, Brazil

Ignacio Aspiazu
State University of Montes Claros, Brazil

Alexandre F. da Silva
Embrapa Maize and Sorghum, Brazil

André Andres
Embrapa Temperate Climate, Brazil

Gemma Kulk, Pablo de Vries, Ronald J. W. Visser and Anita G. J. Buma
Department of Ocean Ecosystems, Energy and Sustainability Research Institute Groningen, University of Groningen, Groningen, The Netherlands

Willem H. van de Poll
Department of Biological Oceanography, Royal Netherlands Institute for Sea Research, Den Burg, The Netherlands

Dilek Unal Ozakca
Bilecik Seyh Edebali University, Faculty of Science and Art, Department of Molecular Biology and Genetic, Bilecik, Turkey

Vivek Pandey
Plant Ecology & Environmental Science, CSIR-National Botanical Research Institute, Lucknow, India

Vivek Dikshit
Biochemistry Division, Jain Biotech Lab, Jalgaon, Maharashtra, India

Radhey Shyam
Plant Ecology & Environmental Science, CSIR-National Botanical Research Institute, Lucknow, India
Vikas Khand, Gomtinagar, Lucknow, India

Yulia Pinchasov-Grinblat
The Mina and Everard Goodman Faculty of Life Sciences, Bar-Ilan University, Ramat-Gan, Israel

František Šeršeň and Katarína Kráľová
Institute of Chemistry, Faculty of Natural Sciences, Comenius University, Bratislava, Slovakia

Daniel K. Y. Tan and Jeffrey S. Amthor
Faculty of Agriculture and Environment, The University of Sydney, Sydney, NSW, Australia

Johan U. Grobbelaar
Department of Plant Sciences, University of the Free State, Bloemfontein, South Africa

Masato Baba and Yoshihiro Shiraiwa
Faculty of Life and Environmental Sciences, University of Tsukuba, Tsukuba, Ibaraki, Japan
CREST, JST, Japan